5th International Conference on Mechanical, Automotive and Aerospace Engineering (ICMAAE 2021)

IOP Conference Series: Materials Science and Engineering
Volume 1244

Kuala Lumpur, Malaysia
21 - 23 June 2021

ISBN: 978-1-7138-5810-2
ISSN: 1757-8981

Printed from e-media with permission by:

Curran Associates, Inc.
57 Morehouse Lane
Red Hook, NY 12571

Some format issues inherent in the e-media version may also appear in this print version.

Institute of Physics
Dirac House, Temple Back
Bristol BS1 6BE UK

Phone: 44 1 17 929 7481
Fax: 44 1 17 920 0979

techtracking@iop.org

Additional copies of this publication are available from:

Curran Associates, Inc.
57 Morehouse Lane
Red Hook, NY 12571 USA
Phone: 845-758-0400
Fax: 845-758-2633
Email: curran@proceedings.com
Web: www.proceedings.com

TABLE OF CONTENTS

IOP Conf. Series: Materials Science and Engineering 1244 (2022) 011001 doi:10.1088/1757-899X/1244/1/011001

5th International Conference on Mechanical, Automotive and Aerospace Engineering 2021

(ICMAAE'21)

22 – 23 June 2021

Faculty of Engineering
International Islamic University Malaysia

IOP Conf. Series: Materials Science and Engineering 1244 (2022) 011001 doi:10.1088/1757-899X/1244/1/011001

Preface

The 5th International Conference on Mechanical, Automotive and Aerospace Engineering (ICMAAE) was initially scheduled to take place as a face-to-face conference in 11-13 August 2020 in Kuala Lumpur, Malaysia. However, due to the covid19 pandemic, the conference was finally conducted virtually a year later in 22 – 23 June 2021. Nevertheless, the conference has attracted fellow researchers, academicians, and industry professionals from 13 countries, including Bahrain, Bangladesh, Canada, India, Indonesia, Malaysia, Morocco, Nigeria, Pakistan, Saudi Arabia, Tunisia, UAE, UK and USA, to share their experiences and ideas while renewing old friendships and making new acquaintances.

The topics of the conference cover a wide range of fields in mechanical, automotive, and aerospace engineering. The conference received 134 papers where 96 papers were accepted for presentation. Among these papers, 25 papers are selected for publication in IOP Conference Series: Materials Science and Engineering. The other papers are submitted for publication in four indexed journals including CFD Letters, FME Transactions, IIUM Engineering Journal and Journal of Advanced Research in Fluid Mechanics and Thermal Sciences. The success of ICMAAE 2021 depends completely on the effort, talent, and energy of researchers in the field of Mechanical, Automotive and Aerospace engineering who have submitted papers on a variety of topics. We are indeed glad at the favourable response received from the scientific community around the world.

ICMAAE 2021 has been recognized by Board of Engineer Malaysia by approving 13 CPD points for attending the two-days conference session. We would like to extend our thanks to the members of the organizing committee for their hard work in organising this excellent event. We would like to take this opportunity to express our sincere gratitude and appreciation to all the reviewers who have helped in maintaining the high standard of the conference. Our thanks go to all the sponsors and all participants in making this conference a success. In particular, we would like to gratefully acknowledge the sponsorship and support by Malaysian Timber Council.

We hope that you'll have a chance to enjoy the city of Kuala Lumpur in the next edition of ICMAAE.

Prof Dr Meftah Hrairi (Chairman, ICMAAE'21)
Assoc. Prof Dr. Erwin Sulaeman (Chairman, ICMAAE'21 Technical Committee)
Assoc. Prof Dr. Moumen Idres (Chairman, ICMAAE'21 Publication)
Assoc. Prof Dr. Fadly J Darsivan (Head, Department of Mechanical and Aerospace Engineering, IIUM)

IOP Conf. Series: Materials Science and Engineering 1244 (2022) 011001 doi:10.1088/1757-899X/1244/1/011001

ICMAAE 2021 Conference Organizing Committee

Conference Manager
Prof. Dr. Meftah Hrairi (Chairman)
Prof. Dr. Waqar Asrar
Prof. Ir. Dr. Masjuki Hassan

Secretary
Dr. Amelda Dianne Andan

Technical Committee
Assoc. Prof. Dr. Erwin Sulaeman (Chairman)
Prof. Dr. Ahmad Faris Ismail
Prof. Dr. Waqar Asrar
Prof. Emeritus ir. Dr. Masjuki Hassan
Prof. Dr. Meftah Hrairi
Prof. Dr. Sher Afghan Khan
Prof. Dr. Waleed Faris
Prof. Dr. M. Ataur Rahman
Assoc. Prof. Dr. Sany I Ihsan
Assoc. Prof. Dr. Fadly J Darsivan
Assoc. Prof. Dr. Mohd Sultan Ibrahim
Assoc. Prof. Dr. Salmiah Ahmad
Assoc. Prof. Dr. Moumen Idris
Assoc. Prof. Dr. JS Mohammed Ali
Assoc. Prof. Dr. Mohammad Okasha
Dr. Sanisah Saharin
Dr. Hanan Mokhtar
Dr. Nabilah Ramli
Dr. Syed Noh
Dr. Saifuddin Rehan
Dr. Hilmi Hela Ladin
Ts. Dr. Hanafi Azami
Ts. Dr. Nur Azam Abdullah
Dr. Tengku Nordayana Binti Kamaruddin
Dr. Alia Abdul Ghaffar
Dr. Amelda Dianne Andan
Dr. Wan Luqman Hakim
Dr. Norfazrina Mohd Yatim
Dr. Mohammed Aldheeb
Dr. Muhammad Abdullah
Dr. Adib Hamdani Bin Rosli
Dr. Mohd Azan Mohammed Sapardi
Ir. Dr. Mohd Farid Bin Aladdin
Ts. Dr. Fatimah Dzaharudin

Keynote/Invited speakers/Advisory board
Prof. Dr. Waqar Asrar (Chairman)
Dr. Hanafi Azami
Dr. Hilmi Hela Ladin
Dr. Tengku Nordayana Binti Kamaruddin

Financial Sponsorship, Food and Souvenirs
Assoc. Prof. Dr. Salmiah Ahmad (Chair)
Dr. Saifuddin Rehan
Dr. Nabilah Ramli
Azweeda Dahlan

Web, Apps and Social Media
Assoc. Prof. Dr. Fadly J Darsivan (Chairman)
Dr. Norfazrina Mohd Yatim
Dr. Mohammed Aldheeb

Publication in Journals
Assoc. Prof. Dr. Moumen Idres (Chairman)
Prof. Dr. Sher Afghan Khan
Assoc. Prof. Dr. Erwin Sulaeman
Prof. Dr. Waleed Faris

Proceedings and Programme book
Assoc. Prof. Dr. Mohd Sultan Ibrahim (Chairman)
Dr. Wan Luqman Hakim
Ts. Dr. Fatimah Dzaharudin

Registration
Dr. Syed Noh (Chairman)
Assoc. Prof. Dr. Sany I Ihsan
Zamilah Abdulaziz
Mohammad Azizudeen
Syazwani Mahmoodin
Nooryani Haji Ali

Hospitality and Transportation
Dr. Saifuddin Rehan (Chairman)
Firdauz Hakem
Nor Hafez Adnan
M Farid

Publicity
Dr. Hanafi Azami (Chairman)
Dr. Nur Azam Abdullah
Dr. Tengku Nordayana Binti Kamaruddin
Dr. Mohd Azan Sapardi
Prof. Dr. Sher Afghan Khan
Prof. Dr. Waleed Faris
Prof. Dr. Masjuki Hassan
Prof. Dr. Ahmad Faris Ismail
Dr. Adib Hamdani Bin Rosli

Advisory Board
Prof. Dr. A K Ghosh, IIT Kanpur, India
Prof. Dr. Ashraf A Omar, UIR, Morocco.
Prof. Dr. K. J. Kwon, KARI, Korea
Prof. Dr. Mazharul Islam, AUST, Bangladesh
Prof. Dr. Mehdi Ahmadian, Virginia Tech., USA
Prof. Dr. Mirza Faisal Baig, AMU, India
Prof. Dr. Mohamad ElSayed, EMU, USA
Prof. Dr. Mohamad Razi Nalim, IUPUI, USA
Prof. Dr. Mohamed Hafez, UC Davis, USA
Datuk Naguib Mohd Nor, STRAND, Malaysia
Prof. Dr. Tarek Abdel Salam, ECU, USA
Prof. Dr. Yulfian Aminanda, UTB, Brunei Darussalam

Conference Organizer
Department of Mechanical and Aerospace Engineering
Faculty of Engineering
International Islamic University Malaysia
Jalan Gombak, 53100 Kuala Lumpur, Malaysia

IOP Conf. Series: Materials Science and Engineering 1244 (2022) 011002 doi:10.1088/1757-899X/1244/1/011002

Peer Review Statement

All papers published in this volume have been reviewed through processes administered by the Editors. Reviews were conducted by expert referees to the professional and scientific standards expected of a proceedings journal published by IOP Publishing.

- **Type of peer review:** Double Anonymous
- **Conference submission management system:** easychair
- **Number of submissions received:** 100
- **Number of submissions sent for review:** 100
- **Number of submissions accepted:** 80
- **Acceptance Rate (Submissions Accepted / Submissions Received × 100):** 80
- **Average number of reviews per paper:** 1.99
- **Total number of reviewers involved:** 40
- **Contact person for queries:**
  Name: moumen idres
  Email: midres@iium.edu.my
  Affiliation: international islamic university malaysia

Tensile strength of single banana fibers (SBFs) improved by novel alkaline treatment

Mohd Khairul Efni Norman[1] and Mohd Yussni Hashim[1]

[1]Faculty of Mechanical and Manufacturing Engineering,
Universiti Tun Hussein Onn Malaysia, 86400 Parit Raja, Johor, Malaysia

E-mail: yusni@uthm.edu.my

Abstract. Environmental concern on global warming controversy has raised the issue that forced the researchers to work on green materials. Among them, plant fibers are recognized as environmentally friendly with a high potential to explore their adaptability and utilization for many engineering applications. This paper aims to examine the banana fiber behavior of various alkaline treatment conditions and compare the results with untreated fiber conducted on the same tests. Both the tests were subjected to physical and tensile load, then each condition was tested individually to achieve a valid average. The Paper frame technique was adopted for this single fiber tensile test (SFTT) and corresponding to the stress-strain curves. Surface morphology was done using scanning electron microscopy (SEM) to compare and find the effect of alkaline treatment on surface fiber modification.
Keywords: Banana fiber, Single fiber, Alkaline treatment and Tensile strength

1. Introduction

The utilization of conventional materials such as artificial mineral-based fiber composites can harm the health of consumers and the environment. Several limitations affect these products leading to skin irritation, harmful chemical materials, costly operation and non-biodegradable nature that contribute to the environmental deterioration [1]. Indirectly, the release of pollutants and ecological issues significantly affects the natural ecosystem and will continue in the future as the use of these materials in composite manufacturing is unavoidable due to the high demand for lightweight products. Several attempts have been made commonly practiced where potential alternatives based on industrial by-products are a possible substitute for supplemental filler materials to replace synthetic fibers in composite materials. Natural abundant sources such as cellulose fibers (plant fibers) in partial replacement for petroleum-based plastic materials help reduce combined weight and cost, increase productivity and improve mechanical properties of high-end products [2].

Natural fibers have been fully utilized in the automotive and construction industries due to their advantages in excellent price-performance, low weight and environmentally friendly character [3]. However, the application of natural fibers is still a challenge and mostly the issue lies in preparing high-quality output as a reinforcement filler in composite materials Thereby there is still some difficulty in ensuring their mechanical properties capability and ability. For this reason, many articles reference in the literature were searched further for the additional relevant reviewed and data will also be gathered from multiple sources at various time points. Much of the current literature on surface modification pays particular attention by conducting various chemical treatments (commonly used alkaline treatment of

NaOH) to reduce the hydrophilic nature of fibers (moisture absorption) and to enhance the fiber surface adhesion with matrix (interfacial interaction) [4]–[6].

A great deal of previous research (Sanjay et al. [7]) into natural fiber modification has been conclusively shown that using different alkaline treatment concentrations for different time periods was a common fiber treatment to enhance optimal conditions. However, higher alkaline treatment conditions (up to 8% NaOH) significantly contributed to the deteriorating effect by yielding the lowest tensile strength and excessed delignification on the fiber surface [8]. Following the studies by Jamal et al. [9], the optimization method was applied to investigate the conditions of alkaline treatment resulting from the concentration, time and temperature. Together, these studies indicate that the process parameters used to quantify mechanical properties are specifically toward varies response at the low, medium or high conditions. Therefore, the objective of the current study is to improve the bond strength on the single banana fibers (SBFs) influenced by varies alkaline treatment for use as reinforcing materials.

2. Experimental Work

The overview of methodology is summarized in Figure 1.

2.1 Materials

Banana fibers were obtained from a local plantation and sodium hydroxide (NaOH) was provided by a local supplier. Initially, raw banana stems were extracted into stem cutting, fiber extracting and straightening techniques. The fibers were then manually extracted after subjected to the effective retting process for about 28 hours at a controlled pH level (between 6.8 and 7.4) with sodium carbonate at room temperature conditions. Figure 2(a) shows the banana stems were soaked into a water tank filled with tap water and some chemical additive (Na_2CO_3).

2.2 Surface modification

A 2g banana fibers sample was treated in an alkaline treatment with conditions being specified at the various concentration (4, 6, 8% NaOH), time (6, 12, 24 hours) and temperature (27, 60, 100°C). Upon completing the treatment process as presented in Figure 2(b), the soaked fibers were rinsed thoroughly with clean water flow and neutralized with 1% acetic acid to maintain alkalinity of pH level (7 to 9). Then, the banana fibers were dried in a hot air oven at 100°C for 1 hour and eventually for next analysis.

2.3 Optical microscope (OM)

Single banana fibers for both untreated and treated fiber were randomly selected and observed using OM as shown in Figure 2(c). Observation data were collected from an image analyzer with code model Olympus SHZ10 Research Stereo equipment. The visualization procedure was done according to ASTM E2015-04(2014) standard.

2.4 Tensile test

A standard low modulus fiber was tested through a single fiber tensile test (SFTT) per ASTM C1557-03 standard as set out in Figure 2(d). Specially selected fibers were mounted on a paper frame and further glued to maintain consistency and prevent slipperiness. The tensile testing was carried out in the Lloyd Universal Testing Machine with a 10 N load cell and 20 mm gauge length. A constant cross speed was adopted at 1.0 mm/min and maintained throughout the testing. Totally five samples of each treatment condition were tested and the average values of tensile strength were recorded once the fiber break or fracture occurs.

2.5 Scanning electron microscope (SEM)

The high resolution of fiber morphology images was observed using SEM with a model of the JEOL machine. The analysis was focused mainly on the longitudinal and transversal surfaces to identify the changes in fiber surface roughness and fractures specimen. Images of damaged specimens were taken by subjected to magnify at a range of 200 to 500 times magnification.

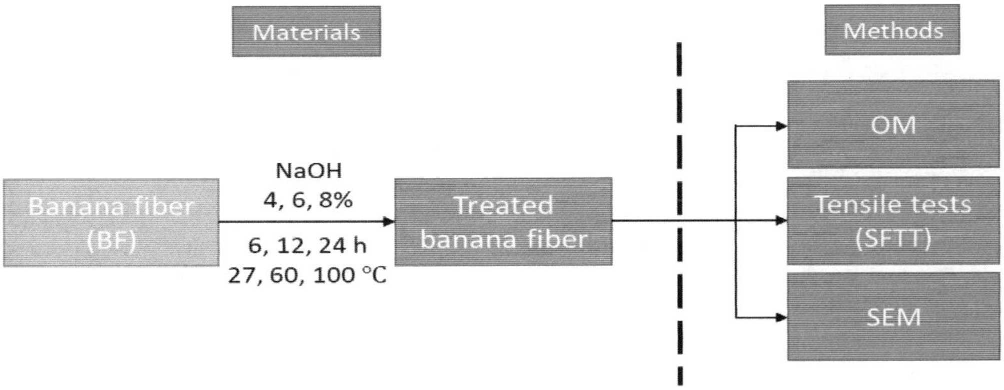

Figure 1. Schematic process of study for overall raw materials and test methods.

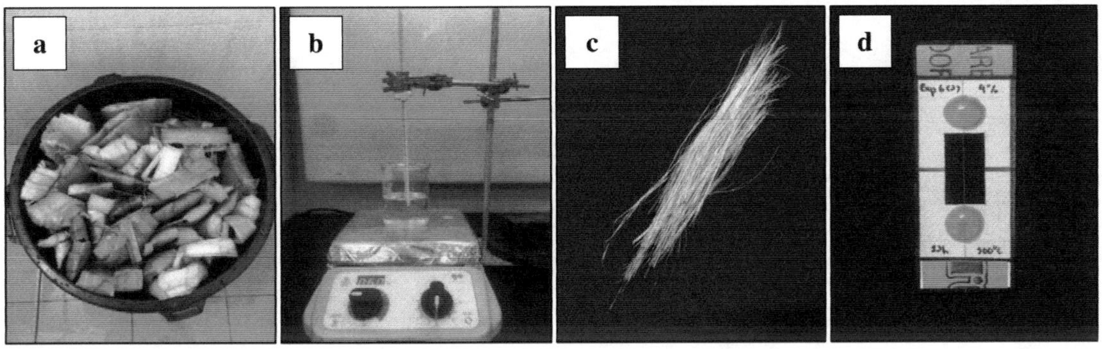

Figure 2. Experimental setup (a) water retting process, (b) chemically treated fiber (NaOH), (c) short random banana fibers, (d) SFTT samples.

3. Results and Discussion

3.1 Influence of alkaline treatment on physical fiber

The average of three measurements along the fiber was observed at three different regions (bottom, centre and top) and considered the diameter of that particular single fiber. In total, ten sets of average fiber were performed and the results of 27 tests as shown in Table 1. The tested process parameters, including alkaline concentration, time and temperature levels are listed in Table 1. The mean diameter of banana fiber showed that with the addition of alkaline concentration to treatment processes, the diameter decreased. In this case, the treated fibers exhibited an increment in average diameter at 47.01% compared to untreated fibers with 212.57 µm and 7.78 standard deviation value.

The attributes of alkaline treatment are primarily dependent on the process parameters of fiber enhancement. This process of improving surface fiber with higher alkaline treatment condition is true particularly at a higher average diameter. Based on data presented, it explained that the single fiber measurements at high-levels (8% NaOH and 24 hours) showed the highest mean diameter reading (216.02 µm) followed by low-levels (112.65 µm) for the lowest reading. Physically, the fibers treatment subjected to a high-level soaking times exposure at 18 and 24 hours resulted in severe surface damage of the fiber textures [4]. Significant reduction of the cross-sectional area is observed with the increase of long-term alkaline treatment conditions as shown in Figure 3. A similar trend of decline in area measurement has been obtained to that of a diameter trend. On the other hand, SEM images showed a clean and smooth surface chemically treated for all banana fibers, unlike untreated ones. This analysis confirmed that proper surface modification could help retain good mechanical properties.

IOP Conf. Series: Materials Science and Engineering 1244 (2022) 012001 doi:10.1088/1757-899X/1244/1/012001

Table 1. Reported result of banana fiber on diameter changes after alkaline treatment.

Time (h)	Temperature (°C)								
	27			60			100		
	Alkaline concentration (%)								
	4	6	8	4	6	8	4	6	8
6	155.02	112.65	164.68	177.23	193.83	115.30	133.72	187.90	198.10
12	137.77	188.68	192.63	187.65	163.72	135.97	152.12	157.43	161.95
24	115.43	152.83	132.13	131.32	173.15	216.02	164.45	135.12	181.27

**Note*: The unit of diameter changes were measured in μm.

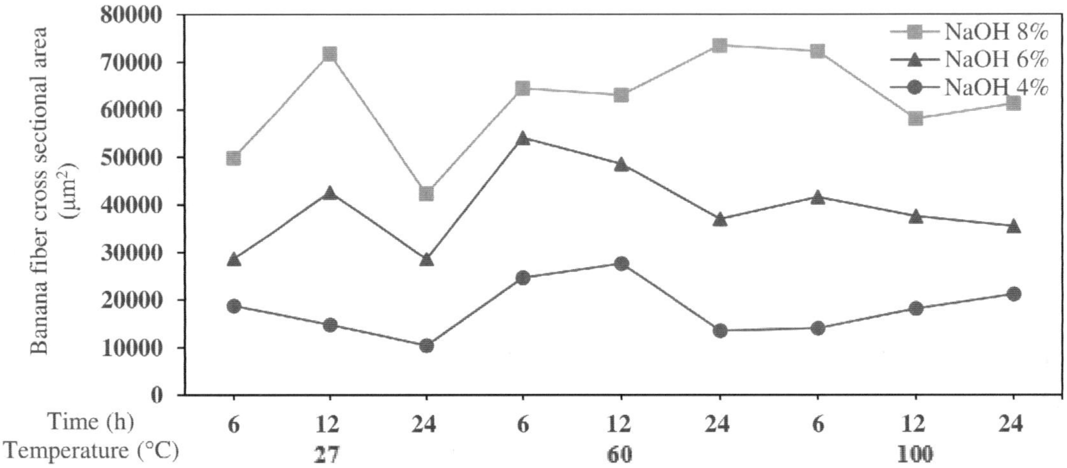

Figure 3. Cross-sectional area of chemically treated SBFs.

In this connection, hemicellulose and lignin was removed by making the interfibrillar region less dense and less rigid [10]. The relation between cellulose and lignin compound presents non-uniform properties and imperfections in parameter control of average diameter and cross-section along the length of fibers. A good example is that natural fiber varies from one fiber to another due to different physiological, morphological and structural function in determining its properties [11]. Recent evidence reports that the qualities of single or bundle fiber characters related to some genetic factors (growing conditions) such as perimeter or diameter and amount of cellulose deposition order [12]. Therefore, the diameters of the individual fibers are mostly not constant and not easy to measure caused by a tangle of strands and bundles of elementary fibers that held together [13].

3.2 Influence of alkaline treatment on tensile strength

Tensile strength of both raw and treated single banana fibers was obtained from the stress/strain curves graph. Table 2 and Figure 4 summarizes the tensile strength of SBFs after alkaline treatment. In order to get more precious data, a shorter span length (gauge) is preferred over other lengths in producing an excellent fiber strength during a tensile test. This finding was also in line with previous work done by Hossain et al. [10], who figured out that the tensile strength decreased when the addition of fiber length leads to the stressed fiber segment by weakening the structure. Studies such as that conducted by Arafa [12] have shown that all the characters under single fiber seemed to have a bit lower strength than bundle fiber. Throughout this study, the strong relationship between single fiber and maturity level from environmental conditions significantly affected the strength readings either improved or vice versa.

From the data obtained, the average tensile strength of banana fibers with surface modification is about 55.76 MPa to 328.05 MPa, which means that treated fibers have particularly improved at 132.46%.

Table 2. Reported result of banana fiber on tensile strength after alkaline treatment.

Time (h)	Temperature (°C)								
	27			60			100		
	Alkaline concentration (%)								
	4	6	8	4	6	8	4	6	8
6	319.92	205.19	202.40	184.45	202.60	96.96	328.05	194.36	67.47
12	278.42	258.20	214.87	114.00	111.44	134.83	165.72	115.98	70.71
24	130.62	140.72	143.75	87.52	219.94	157.89	55.76	106.42	73.67

**Note*: The unit of tensile strength was measured in MPa.

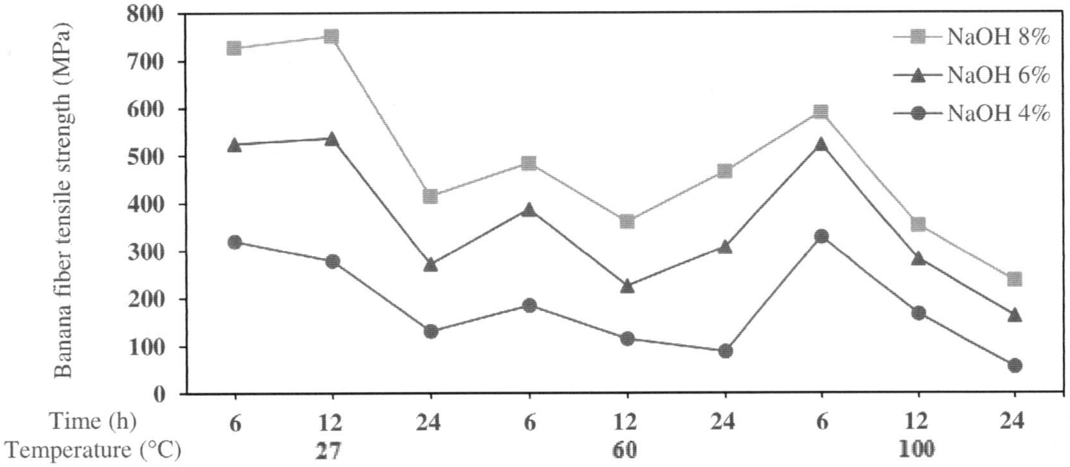

Figure 4. Tensile strength of chemically treated SBFs.

For untreated fiber case, the average tensile strength is 141.12 MPa which is relatively lower than the other average of treated fiber. The failure mechanism mainly depends on process parameters with a longer treatment level, resulting in a higher chance of fiber fracture and reduced tensile strength. As argued by Hashim et al. [14], the negative results were due to the increase in immersion temperature and alkaline concentration affect fiber strength behavior. In a follow-up study by the same author [15], however, the tensile strength of kenaf fiber revealed that the elementary fiber changes after various alkaline treatments and reduced compared to those untreated fibers. An insight into the fiber characteristic strength has confirmed that the effectiveness of surface modifications indirectly correlated to the alkaline treatment setting used.

Similarly, prior studies have been able to show that the fiber strength was somewhat encouraging results convincingly. One interesting finding is consistent with Ibrahim et al. [16], who confirms that 6% NaOH offered a better tensile strength. From a mechanical point of view, the treatment at 4% NaOH was quite satisfied as the tensile strength improves with a more excellent value. An experimental approach purely controls this improvement criterion. Thus, increase the response levels up and partially increase the fiber strength, contributing to the material's overall mechanical properties. Compare to a study by Sair et al. [17], the treated hemp fibers were exhibited maximum improvement on tensile strength especially reaches at 8% NaOH up to 39% increment and notices the decrement results after this treatment. This result contradicts the findings presented here, in which the higher treatment response mostly exposed to poor tensile strength. A possible explanation for this might be increasing alkaline treatment resulting in a weaker and damaged fiber, then decreasing the fiber strength [18], [19]. On the other hand, recent evidence suggests that single fiber characteristics are not homogenous and structural integrity of tensile tested after fracture can be analyzed using SEM equipment [20].

3.3 Influence of alkaline treatment on surface morphology

Each fiber can be observed under the image analyzer after successfully removing some cement components as displayed in Figure 5. Surface morphology of banana fiber with different conditions reveals continuous surface and a few fibrillations along the fiber as indicated by an arrow in the microstructural analysis. Untreated fiber (UT) surface shows sufficiently or fully covered by impurities revealing that poor stress transfer in tensile loading during the pulled-out test as shown in Figure 5(a). These surface materials contain a combination of dry and liquid substances such as oil, wax and many lump materials [21]. A relatively binding material such as impurities exists in this stage, allowing in the form of lumps on the outer layer to influence the accuracy of tensile data analysis. Solid lumps of crystal-shaped materials can be observed in rectangular order with uniformly arranged on the external surface.

In Figure 5(b) shows a clear image of the uneven structure, which caused a selective reduction in fiber diameter and area. From the tensile test, treated fiber at 4% NaOH or low-levels yielded the highest strength compared to those raw banana fibers with longer fiber pull-out. Therefore, it can be assumed that the removed materials were eventually dissolved during treatment, thereby increasing surface roughness and cleaner. The rougher fiber surface was exhibited at 6% NaOH or common levels, yet forming grove structure. It reveals the surface irregularities of banana fiber in Figure 5(c), showing the rough surface that enhances its mechanical performance. The formation of rougher surface occurred to a greater extent causing the topographical changes from removed materials [15]. Figure 5(d) shows fiber delamination at 8% NaOH or high-levels presented over the entire fiber surface, reducing the fiber characteristics. Further, surface fiber resulting in several damages such as severe fiber fibrillation and surface separation could impact fiber integrity from excessive removal of binding materials. These results match those observed by Zhang et al. [22], who found the excessive alkaline treatment into high NaOH expected to deteriorate the quality of fiber surface and unsuitable for enhancing the mechanical properties.

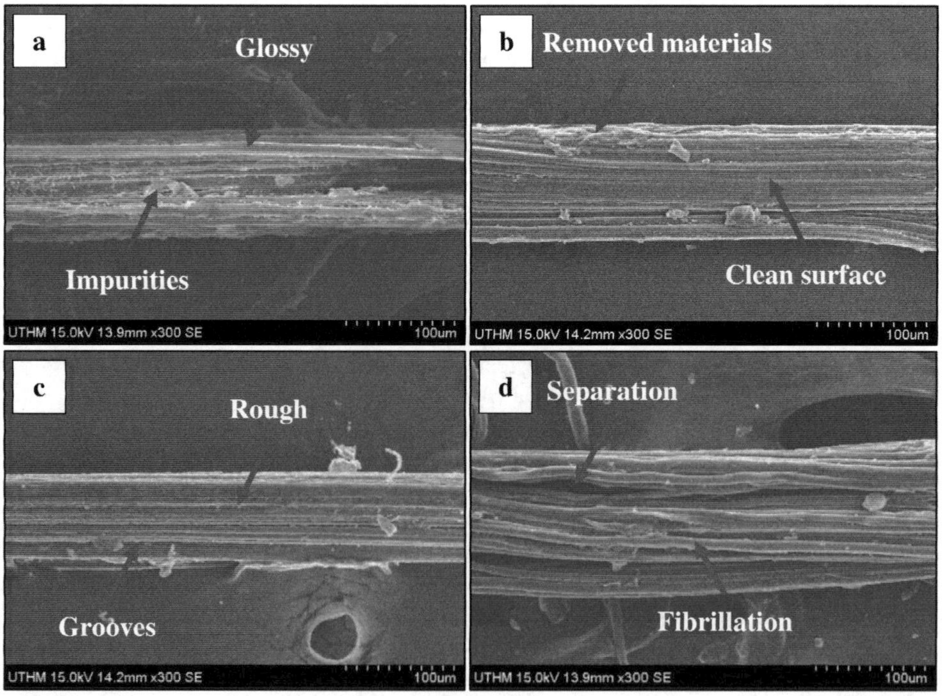

Figure 5. SEM images (a) UT, (b) 4%-6h-27°C, (c) 6%-12h-60°C, (d) 8%-24h-100°C.

4. Conclusion

The present study demonstrated that alkaline treatment affected the physical fiber characteristics and tensile strength of fiber obtained from single banana fibers (SBFs). The diameter and cross-sectional area of treated fibers decreased with an increase in the alkaline treatment process at 47.01%. At higher level conditions, both fiber diameter and cross-sectional area were observed increment due to increased exposure to long-term alkaline treatment conditions. Thus, a significant difference in surface morphology of chemically treated fiber was found rougher than the raw fiber due to removing some impurities from the outer layer of the fiber surface. Accordingly, the surface roughness seemed to be reduced after high alkaline treatment and tends to be damaged by forming a deformed structure (elementary fiber) as can be seen at 8% NaOH or high-level conditions. The data presented here would suggest fiber modification offer an advantage over the raw condition. It provided more effective fiber strength with a higher tensile strength at 328.05 MPa similar to commercial fiber. This advantage significantly improves the tensile strength after alkaline treatment with an improvement of 132.46%. Further concluded that the novel alkaline treatment revealed the great potential and useful for numerous engineering fiber-based materials and applications.

Acknowledgement

Communication of this research is made possible through monetary assistance by Universiti Tun Hussein Onn Malaysia and the UTHM Publisher's Office via GPPS grant code H455 and Publication Fund E15216 respectively.

References

[1] L. O. Afolabi, P. S. M. Megat-Yusoff, Z. M. Ariff, and M. S. Hamizol, "Fabrication of Pandanus Tectorius (Screw-pine) Natural Fiber Using Vacuum Resin Infusion for Polymer Composite Application," *J. Mater. Res. Technol.*, vol. 8, no. 3, pp. 3102–3113, 2019, doi: 10.1016/j.jmrt.2017.05.021.

[2] K. Panyasart, N. Chaiyut, T. Amornsakchai, and O. Santawitee, "Effect of Surface Treatment on the Properties of Pineapple Leaf Fibers Reinforced Polyamide 6 Composites," *Energy Procedia*, vol. 56, no. C, pp. 406–413, 2014, doi: 10.1016/j.egypro.2014.07.173.

[3] C. C. Ihueze, C. E. Okafor, and C. I. Okoye, "Natural Fiber Composite Design and Characterization for Limit Stress Prediction in Multiaxial Stress State," *J. King Saud Univ. - Eng. Sci.*, vol. 27, no. 2, pp. 193–206, 2015, doi: 10.1016/j.jksues.2013.08.002.

[4] M. J. M. Ridzuan, M. S. A. Majid, M. Afendi, K. Azduwin, S. N. A. Kanafiah, and Y. Dan-mallam, "The Effects of the Alkaline Treatment's Soaking Exposure on the Tensile Strength of Napier Fibre," *Procedia Manuf.*, vol. 2, no. February, pp. 353–358, 2015, doi: 10.1016/j.promfg.2015.07.062.

[5] J. A. Halip, L. S. Hua, Z. Ashaari, P. M. Tahir, L. W. Chen, and M. K. Anwar Uyup, *Effect of Treatment on Water Absorption Behavior of Natural Fiber–Reinforced Polymer Composites*. Elsevier Ltd, 2019.

[6] N. A. Latip, A. H. Sofian, M. F. Ali, S. N. Ismail, and D. M. N. D. Idris, "Structural and Morphological Studies on Alkaline Pre-Treatment of Oil Palm Empty Fruit Bunch (OPEFB) Fiber for Composite Production," *Mater. Today Proc.*, vol. 17, pp. 1105–1111, 2019, doi: 10.1016/j.matpr.2019.06.529.

[7] M. R. Sanjay, S. Siengchin, J. Parameswaranpillai, M. Jawaid, C. I. Pruncu, and A. Khan, "A Comprehensive Review of Techniques for Natural Fibers as Reinforcement in Composites: Preparation, Processing and Characterization," *Carbohydr. Polym.*, vol. 207, no. November 2018, pp. 108–121, 2019, doi: 10.1016/j.carbpol.2018.11.083.

[8] M. H. Zin, K. Abdan, N. Mazlan, E. S. Zainudin, and K. E. Liew, "The Effects of Alkali Treatment on the Mechanical and Chemical Properties of Pineapple Leaf Fibres (PALF) and Adhesion to Epoxy Resin," *IOP Conf. Ser. Mater. Sci. Eng.*, vol. 368, no. 1, 2018, doi: 10.1088/1757-899X/368/1/012035.

[9] E. F. A. Jamal *et al.*, "Optimization of Alkali Treatment Condition on Tensile Properties of Kenaf Reinforced Polyester Composite Using Response Surface Method," *Int. J. Integr. Eng.*, vol. 10, no. 1, pp. 40–46, 2018, doi: 10.30880/ijie.2018.10.01.007.

[10] S. I. Hossain, M. Hasan, M. N. Hasan, and A. Hassan, "Effect of Chemical Treatment on Physical, Mechanical and Thermal Properties of Ladies Finger Natural Fiber," *Adv. Mater. Sci. Eng.*, vol. 2013, no. 10, pp. 1–7, 2013, doi: 10.1155/2013/824274.

[11] M. M. E. Costa *et al.*, "Influence of Physical and Chemical Treatments on the Mechanical Properties of Bamboo Fibers," *Procedia Eng.*, vol. 200, pp. 457–464, 2017, doi: 10.1016/j.proeng.2017.07.064.

[12] A. S. Arafa, "Alternate Relationship between Single Fiber Properties and Both of Fiber Microscopic and Physical Properties," *J. Text. Sci. Eng.*, vol. 04, no. 06, 2016, doi: 10.4172/2165-8064.1000175.

[13] V. Mazzanti, M. S. De Luna, R. Pariante, F. Mollica, and G. Filippone, "Natural Fiber-induced Degradation in PLA-hemp Biocomposites in the Molten State," *Compos. Part A*, vol. 137, no. May, p. 105990, 2020, doi: 10.1016/j.compositesa.2020.105990.

[14] M. Y. Hashim, A. M. Amin, O. M. F. Marwah, M. H. Othman, N. H. Hanizan, and M. K. E. Norman, "Two Parameters Weibull Analysis on Mechanical Properties of Kenaf Fiber Under Various Conditions of Alkali Treatment," *Int. J. Integr. Eng.*, vol. 12, no. 3, pp. 245–252, 2020, doi: 10.30880/ijie.2020.12.03.028.

[15] M. Y. Hashim *et al.*, "Interfacial Shear, Tensile Strength and Young Modulus of Kenaf Fiber: The Effects From Different Settings of Alkali Treatment," *Key Eng. Mater.*, vol. 791, no. March, pp. 10–18, 2018, doi: 10.4028/www.scientific.net/KEM.791.10.

[16] M. I. Ibrahim, M. Z. Hassan, R. Dolah, M. Z. M. Yusoff, and M. S. Salit, "Tensile Behaviour for Mercerization of Single Kenaf Fiber," *Malaysian J. Fundam. Appl. Sci.*, vol. 14, no. 4, pp. 437–439, 2018, doi: 10.11113/mjfas.v14n4.1099.

[17] S. Sair *et al.*, "Effect of Surface Modification on Morphological, Mechanical and Thermal Conductivity of Hemp Fiber: Characterization of the Interface of Hemp -Polyurethane Composite," *Case Stud. Therm. Eng.*, vol. 10, no. October, pp. 550–559, 2017, doi: 10.1016/j.csite.2017.10.012.

[18] R. Gunti, A. V. R. Prasad, and A. V. S. S. K. S. Gupta, "Preparation and Properties of Successive Alkali Treated Completely Biodegradable Short Jute Fiber Reinforced PLA Composites," *Polym. Compos.*, no. January, 2015, doi: 10.1002/pc.23395.

[19] M. H. Hamidon, M. T. H. Sultan, A. H. Ariffin, and A. U. M. Shah, "Effects of Fibre Treatment on Mechanical Properties of Kenaf Fibre Reinforced Composites: A Review," *J. Mater. Res. Technol.*, vol. 8, no. 3, pp. 3327–3337, 2019, doi: 10.1016/j.jmrt.2019.04.012.

[20] R. B. Adusumalli, K. C. Venkateshan, C. Kunchi, and S. R. Vadlamani, "Tensile Testing of Single Fibres," *Procedia Struct. Integr.*, vol. 14, no. 2018, pp. 150–157, 2019, doi: 10.1016/j.prostr.2019.05.020.

[21] M. K. Bin Bakri, E. Jayamani, and S. Hamdan, "Processing and Characterization of Banana Fiber/Epoxy Composites: Effect of Alkaline Treatment," *Mater. Today Proc.*, vol. 4, no. 2, pp. 2871–2878, 2017, doi: 10.1016/j.matpr.2017.02.167.

[22] K. Zhang, F. Wang, W. Liang, Z. Wang, Z. Duan, and B. Yang, "Thermal and Mechanical Properties of Bamboo Fiber Reinforced Epoxy Composites," *Polymers (Basel).*, vol. 8, no. 6, 2018, doi: 10.3390/polym10060608.

Flat plate and evacuated tube collectors solar water heater system – A comparative study

Faleh F.F. Hujailan[1], Md Abdul Maleque[1]* and Md Mustafizur[2]
[1]Department of Manufacturing and Material Engineering, International Islamic University Malaysia, P.O. Box 10, 50728 Kuala Lumpur, Malaysia
[2]College of Engineering, Universiti Malaysia Pahang, 26300 Pahang

*Corresponding author's email address: maleque@iium.edu.my

Abstract. Renewable energy for the sustainable energy application in the buildings are increasing day-by-day. The aim of this study is to compare flat plate and evacuated tube collectors system in solar water heating of urban building. The two types of solar water heater systems were designed and evaluated based on the requirements of the urban building dwellers. The F-chart method along with commercial software was used to study the performance efficiency of the solar water heater system at an operating temperature of 65 °C. An economic analysis was also carried out using the same tool for both types of collectors with the desired designed parameters. The performance efficiency results of flat plate and evacuated tube collector are 77 % and 67.3 %, respectively. The economic analysis showed that flat plate collector is cost-effective for the solar water heater system due to its steadiness and lower maintenance costs.

Keywords: Thermal efficiency; Flat plate collector, evacuated tube collector; economic analysis; solar water heater system

1. Introduction

Solar thermal water heating systems in buildings are well-established technologies in many countries around the world. It helps to protect the environment by eliminating the need for heat generated by the combustion of fossil fuels and mitigating greenhouse gas emissions. Solar energy is one of the key sources of renewable energy for building heating, cooling, hot water and generation of electricity. For decades, solar thermal energy has been used by ancient people to heat and dry solar energy. For several nations, it has become a big industry. It is well known that the construction sector contributes 40% of the total global primary energy consumption [1]. In Spain, about 16% of total electricity consumption is related to the construction industry, which is why many of the energy policies and strategies for reducing CO_2 emissions from climate change are targeted at reducing building energy consumption. A large amount of the overall energy production may be provided by solar thermal technologies at an acceptable price [2]. In recent times, solar energy has been developed in a broad range of thermal processes for power generation, water heating, and drying and water purification. There is a recent tendency in the construction industry to implement solar thermal systems with the idea of building-integrated solar thermal systems (BIST). The "multifunctional energy facade" is known as BIST, which differs from traditional solar panels in which it offers a broad range of architectural design solutions (i.e., colour, texture, and shape) [3]. BIST is thus a promising option for enhancing energy efficiency and reducing maintenance costs in contemporary build environments with excellent applicability and building protection along with increased energy production.

In contrast to the configurations of building-added, BIST provides a range of advantages by higher esthetic benefit. It has versatile functions including heating/cooling of the building, the availability of hot water, energy generation as well as enhancing ventilation and the overall appearance of buildings as an ideal solution [4]. However, this is a key element in the development of renewable energy that could cover, in whole or in part, the energy needs of a building [5, 6]. Designers take into consideration solely energetic roles and an excess of distinct qualities. A typical solar water heater (SWH) is made up of a glass cover to encapsulate the absorber and the insulator for solar irradiation [7]. The collector system often becomes a part of the facade, which is more about functionality than the aesthetic, an issue with conventional collectors. Because solar energy is typically not enough to satisfy demand, solar heat must also be stored in a thermal unit [8]. These steps include coloured collectors, configurations that make it possible to achieve optimum inclinations of the solar collector on vertical facades, the inclusion of solar collection in cladding systems for the architecture, and the inclusion of solar collectors in fixed shading devices and glazing systems. A few studies were based on overcoming the obstacles of conventional collectors of solar systems by evaluating the operating effectiveness, the practicality of construction, economic feasibility, and implementation opportunities of novel modern architecture interventions that easily incorporate solar thermal systems into buildings. It has resulted in greater lobbying power and an increasing impact of the solar water heater industry on policy-making and increased collaboration between SWH companies and the real estate market, allowing SWH systems to be more competitive in the building environment [9]. It is incredibly challenging to identifying practical and appropriate design solutions for cost-effective solar thermal collectors and solar thermal systems that are highly desired. Therefore, the purpose of this study is to compare the performance of flat plate and evacuated tube collector's solar water heating systems in the facade of the building on the basis of the urban house occupants.

2. Experimental works

This solar thermal system work aims to emphasize key aspects of system architecture and the incorporation of collectors in the design and collector design of the building facade. The electrical heater backup is used to fulfil the hot water requirements in the system of both collectors with storage tank. The system performance measurement of collectors was conducted at an operational temperature of 65 °C and with a tilt angle of 40 ° by the process of F-chart and the T*SOL software simulation technique which is a dedicated software for solar thermal heating systems to determine the optimum solution to be used. The F-chart model relates annual solar energy fraction to the heat output. In the context for advancing with the new SWH schemes, the restriction and feasibility review were also carried out with the principle of financial analysis using the same software module.

The solar thermal water heater needs to be integrated to determine the size of the storage tank. A solar thermal collector collects heat by direct absorption of sunlight that converts energy into a more usable form of energy. Depending on the occupancy and consumption profile of the chosen building, the number of collectors is needed. Fig. 1 displays the design simulation of the flat plate collector and the evacuated tube collector.

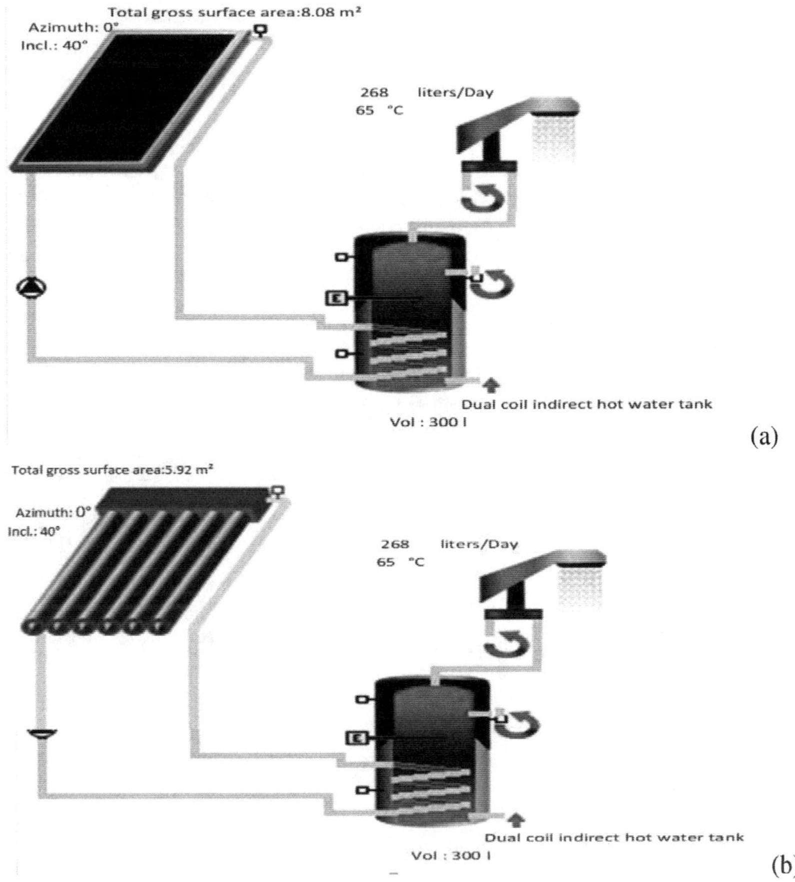

Figure 1. (a) Flat-plate and (b) Evacuated tube collector collectors

3. Results and discussions

3.1 Thermal Performance of solar heater

Detail simulation results of SWH flat-plate and evacuated tube collectors which comprises collector power, installed surface area, maximum energy delivered by the collector, irradiation on the collector, the solar contribution and auxiliary heating share is shown in Table 1.

Table 1 Flat plate and evacuated tube collector annual energy output

Description	Flat plate	Evacuated tube
Installed collector power (kW)	5.66	4.441
Installed solar surface area (gross), (m^2)	8.08	5.92
Irradiation on to collector surface (active), (MJ/m^2)	8,154.31	8,154.31
Max energy delivered by collector (MJ/m^2)	3,297.50	3,435.29
Energy delivered by collector loop (MJ/m^2)	2,958.10	3,102.87
DHW heating energy supply (MWh)	5.33	5.32
Solar contribution to DHW (MWh)	5.87	4.99
Energy from auxiliary heating (kWh)	1,757.90	2,458.20
Natural gas (H) savings (m^3)	722.50	614.00
CO_2 emissions avoided (kg)	1,527.90	1,298.29

DHW solar fraction (%)	77.00	76.00
Fractional energy savings (%)	**77.00**	**67.30**
System efficiency (%)	35.40	37.20

As solar energy is the fastest-growing electricity power source, its contribution is very high over the total energy consumption. From Table 1 it can be seen that the maximum annual energy output per collector area was 3,297.50 MJ/m² and 3,435.29 MJ/m² for flat-plate and evacuated tube collector, respectively. It was also noticed that fractional energy savings for flat-plate exhibited higher energy savings (with 77.0%) compared to evaluated tube collector (with 67.3%).

3.2 Economic Analysis

The financial analysis results of the flat-plate and evacuated tube solar collector system, including system specification, comprehensive economic and financial analysis parameters, are summarized in Table 2. The economic analysis showed that the flat-plate collector is better than the evacuated tube due to its stability and low maintenance costs. It has also been observed that the collector for flat plate has more surface area to have the same amount of energy as an evacuated tube, but the evacuated tube is likely to quickly reach a stagnation stage. Owing to its lower installation area, the evacuated tube collector would have a lower initial and running costs.

Table 2 Financial analysis for flat-plate and evacuated-tube collector system

System Overall	**Flat-plate system**	**Evacuated-tube system**
Active surface area (m^2)	7.32	5.92
System yield (MWh)	5.62	4.99
Annual fuel/energy savings (m^3)	691.00	614.00
Financial Analysis Parameters		
Life span (Years)	20.00	20.00
Interest on capital (%)	2.50	2.50
Reinvestment return (%)	2.50	2.50
Energy cost escalation rate (%)	3.00	3.00
Running cost escalation rate (%)	1.50	1.50
Financing		
Total investments ($)	2,562.00	2,072.00
Subsidies	None	None
Loan capital	None	None
Remaining investment ($)	2,562.00	2,072.00
Running costs in the first year ($)	84.00	88.00
Savings in the first year ($)	691.00	614.00
Financial Analysis		
Cost of solar energy ($/Kwh)	0.046	0.047
Capital return time (Years)	4.00	3.80
Payback period (Years)	4.30	4.00
Profitability		
Return on assets (%)	649.40	698.10
Return on equity (%)	649.40	698.10
Internal rate of return rate (%)	26.50	28.30
Net present value ($)	10,078.00	8,914.00
Reinvestment Premise		
Profit ($)	18,149.00	15,930.00
Modified internal rate of return (%)	11.01	11.42

4. Conclusion

The integrated solar water heater systems for flat plate and evacuated tube collectors have been designed and assessed the performance for comparison on the basis of the requirements of the inhabitants of the urban house. Based on the findings, the maximum annual energy output per collector area for flat-plate and evaluated collector are 3,297.50 MJ/m2 and 3,435.29 MJ/m2 respectively. A flat-plate collector requires more surface than an evacuated tube to deliver the same amount of electricity, but the evacuated tube quickly hit a stagnation stage during operation. In conclusion, due to its stability and low maintenance costs, flat plate type collector is highly advisable to implementation in the building on the basis of the performance. The economic analysis of flat plate collector demonstrated a cost-effective design for the solar water heater system due to its steadiness and lower maintenance costs.

Acknowledgement

The authors greatly acknowledge Department of Manufacturing & Materials Engineering of International Islamic University Malaysia (IIUM) and Universiti Malaysia Pahang for technical support.

References

[1] Agathokleous, R., Barone, G., Buonomano, A., Forzano, C., Kalogirou, S., & Palombo, A. (2019). ''Building façade integrated solar thermal collectors for air heating: experimentation, modelling and applications.'' *Applied Energy*, 239, 658-679.

[2] Rosiek, S., & Batlles, F. J. (2009). Integration of the solar thermal energy in the construction: Analysis of the solar-assisted air-conditioning system installed in CIESOL building. *Renewable Energy*, 34(6), 1423–1431.

[3] Quijera, J., Alriols, M. and Labidi, J., (2014). Integration of a solar thermal system in canned fish factory. *Applied Thermal Engineering*, 70(2), pp.1062-1072

[4] Shen, X. Z., Wang, Z. H., Tang, T. Y., & Shi, Y. W. (2015). Building Integrated Solar Thermal (BIST) Technologies and Their Applications: A Review of Structural Design and Architectural Integration. *Journal of Fundamentals of Renewable Energy and Applications,* 05(05).

[5] Renewable energy for passive house heating: II. Model. (2004). *Fuel and Energy* Abstracts, 45(3), 198.

[6] McClure, B. (2020). The sun's shift on your horizon | EarthSky.org. Retrieved 2 December 2020, from https://earthsky.org/astronomy-essentials/how-far-does-sun-move-on-your-horizon-each-day

[7] Savvides, A., Vassiliades, C., Michael, A., & Kalogirou, S. (2019). Siting and building-massing considerations for the urban integration of active solar energy systems. *Renewable Energy*, 135, 963-974.

[8] Faddouli, A., Fadili, S., Labrim, H., Hartiti, B., Ertugrul, M., Belouaggadia, N., & Stitou, M. (2020). Comparative study of a normal solar water heater and smart thermal/thermoelectric hybrid systems. *Materials Today: Proceedings*, 30, 1039-1042.

[9] Souliotis, M., Panaras, G., Fokaides, P., Papaefthimiou, S., & Kalogirou, S. (2018). Solar water heating for social housing: Energy analysis and Life Cycle Assessment. *Energy and Buildings*, 169, 157-171.

Influence of Surface Discontinuity on Sound Absorption Coefficient of Vehicle Interior Material

Mohamed [1], Erwin Sulaeman [1*], Amelda Dianne [1] Adri Zainal Abidin [2]

[1] Department of Mechanical Engineering, Kulliyyah of Engineering, International Islamic University Malaysia, Kuala Lumpur, 53100, Malaysia

*Corresponding author: esulaeman@iium.edu.my

Abstract. There has been considerable effort in the automotive industry to reduce and control the noise in the car cabin. Selecting or designing a better absorber will enhance the comfort of the passengers in the car. There are many techniques to reduce or control the noise in the car cabin. One of the techniques is to place specific material in order to reduce its acoustical noise. This paper presents a study on sound absorption of the felt, polyurethane foam and fabric which are commonly used as car interior material. Furthermore, the influence of surface discontinuity created between two joined material parts is also investigated. The sound absorption coefficient is measured using two-microphone impedance tube. This tube is used to provide noise source frequency ranging from 50 Hz to 10 kHz. Using the dBAlpha software that is connected to the microphone via Symphonie Analyzer, the sound absorption coefficient of the material for every frequency is presented in tabular as well as in graphical form. The test results show that material selection is significantly important for the absorption of noise. Besides, it is found that the surface discontinuity gives different effect depending on the material. For the felt material, the surface discontinuity has insignificant effect. However, the surface discontinuity may give a moderate effect for other materials such as polyurethane foam and fabric.

1. Introduction

With the government regulations and policies to reduce the noise generated in the cabin intensified, improvement in the design and selection of a better sound absorber has become a more important part in the automotive industry. The sound absorbers convert the sound into heat or transmit it away using different techniques. The sound absorption property of a material is mostly defined and presented using the term sound absorption coefficient α, which is the ratio of absorbed sound intensity in an actual material to the incident sound intensity.

Many researchers and engineers are exploring new ways to enhance the sound absorption coefficient of the materials. For instance, changing the configuration of the materials or introducing new technologies on the material surface might improve its absorption capability. These developments of the absorbing materials are in accordance with the fact that they are essential in the design of the automotive interiors as eliminating the noise in the cabin enhances the comfortability of the passengers.

Majority of the research papers focused on the design and modification of the sound absorbers as well as the parameters that affects the absorption performance of the materials. Moreover, some of the studies explored the limitations as well as the improvement of the impedance tube [1-10].

The study conducted by Ismail *et al.* [1] discovered that synthetic fibers mostly used are hazardous to human as well as the environment and thus alternative absorbing materials such as textile, foam, rubber or plastic are becoming of greater interest to the researchers because there are recyclable and hence environmental friendly. Moreover, another research conducted by Jayamani and Hamdan [2] shows that thin, lightweight, and low-cost absorbers that can work efficiently when exposed to both low and high frequency sound waves are desired. Yet another study done by Lu, Cheng and He [3] states that closed-cell materials are poor sound absorbers since sound wave cannot reach easily the interior structure of the materials. This points out a research gap that relates to the porous materials.

Many techniques are used to improve the absorption rate of the materials. For instance, some individuals are trying to introduce porous materials while others are working on the composite materials. The research performed by Lu, Hess, and Ashby [4] elaborates that energy absorbed in the material mainly happens through viscous and thermal dissipation processes. On the other hand, Youn Eung Lee and Chang Whan Joo [5] illustrated that ultrafine fiber contents provide a significant increase in the absorption rate of the material. Therefore, it is clear that the effect of fiber fineness to sound absorption is of greater consideration. Yet another discussion from Lee and Joo [5] shows that surface roughness enhances resonance and thus increases the rate of the absorption, however, the rate of sound absorption depends on the width and the depth. A similar research done by Seddeq [6] supports that creating barrier along the sound propagation will reduce its energy by converting mechanical energy to heat energy. There are many techniques used by researchers to disturb the motion of the sound wave thus reducing its energy such as vortex shedding method, introducing porosity and roughness on the material surfaces. Moreover, Coates and Kierzkowski [7] proclaimed that when thickness is roughly one-tenth the wavelength of the incident sound waves, the absorbers operate much better due to the peak absorption occurs at the resonant frequency of the wavelength. Furthermore, Koizumi, Tsujiuchi and Adachi [8] suggested that placing a material of higher density at the source would lower the noise of the cabin drastically. Consequently, Koizumi, *et al.* [8] stated that increasing the fibers' diameter would increase the absorption rate of the material since the energy loss increases as the surface area increases.

The techniques used to reduce noise in the car cabin vary from one researcher to another. Many researchers show that designing the absorbers appropriately can significantly reduce the noise in the cabin and thus having a very comfortable environment for the passengers. Besides, J. Krüger [9] stated that active noise abatement approach is considered and need to be optimized to have maximum insertion loss.

Some research found that there are few efforts on increasing the absorption coefficient for low frequency range. Liu *et al.* [10] investigated non-woven fabrics, especially kapok fiber, in an attempt to find a high α at low frequency region. They concluded that increasing kapok fiber content and thickness will enhance the absorption characteristics. In a further study, Liu *et al.* [11] suggested to use multi-layer of composite materials using non-woven fabrics with kapok. They found that since kapok fiber has a high degree of hollow and small fiber diameter, it is an excellent sound absorption. Combination of kapok with knitted fabric is also investigated by Makki and Oktariani [12]. Many other researchers, such as Mamtaz *et al.* [13], Or *et al.* [14], and Borell *et al.* [15] also investigated the use of natural fiber that offers good sound absorption and poses environmental-friendly material. The use of waste materials is also investigated by Borell *et al.* [15] and Bujoreanu *et al.* [16]. A comprehensive review of multi-layer fibrous structures toward acoustic absorption is presented by Tang and Yan [17].

However, there are some drawbacks on the use of the natural fiber, for example if the sound absorber is exposed to sources vulnerable to fire such as near automotive engines or genset. For such a case, a metal-based sound absorber may be more appropriate. A multi-layered elastic micro-perforated plates is proposed by Kim *et al.* [18]. A compound micro perforated panel absorber array adopting the series-parallel coupling manner is proposed by Qian *et al.* [19]. Gai *et al.* [20] performed experimental study on sound absorption performance of microperforated panel with membrane cell. Recently, Shen *et al.* [21] develop a thin sound absorber consisted of multilayer compressed porous metal with rear cavity. They produced the best actual average sound absorption coefficient of 0.51 in the 100–6000 Hz range obtained by a 4-layer compressed porous metal with the total thickness of 5 mm. To obtain this result, Shen *et al.* [21] performed theoretical modeling, parameter optimization and finite element simulation Their results is validated by standing wave tube measurement.

All the references above deal with sound absorber having continuous surface exposed to the source of the noise. None of the literatures deal with discontinuous surface which can occur in practical application of the material. In the present work, a study on the effect of surface discontinuity on the sound absorption coefficient (α) is conducted. Comparison of the results between continuous and discontinuous surface is elaborated for some materials and conclusion is made.

2. Experimental Procedures

2.1. Raw Material Preparation

Samples were prepared from a range of commercial absorbents, including felt, fabric and polyurethane foam to study the absorption properties. In addition, sound absorption coefficient of materials was measured on the influences of surface discontinuity using two-microphone tube. For the surface discontinuity case, all the materials were cut into two halves and then combined using glue lightly. The specimens were prepared into two equal halves so that discontinuity will be consistent for all the materials. Most of the car absorbers are not connected throughout the interior surface because of openings such doors and windows. The test samples used are shown in Figures 1-2. The materials were cut down in cylindrical shape according to the impedance tube size thus able to fit the tube appropriately without any leakages. A special cutter machine dedicated to cut the material with a high precision for the impedance tube test is used. Considering the thickness of the material, impedance tube has adjustable knob to control the space between the tube surface and the specimen that allows the samples to fit in the space tightly. The dimensions and the properties of the materials that were used are shown in Table 1.

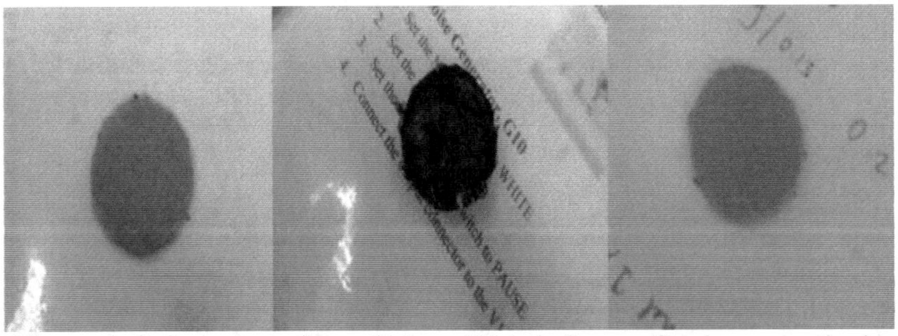

Figure 1. Polyurethane foam, Felt and Fabric for small tube respectively

Table 1. Properties of the sound absorber materials

Test Samples	Mass (gram)	Diameter (cm)	Thickness (cm)
Felt (Large tube)	6.8	10	1.1
Felt (Small tube)	0.4	3	1.1
Polyurethane foam (Large tube)	1.41	10	0.1
Polyurethane foam (Small tube)	0.08	3	0.1
Fabric (Large tube)	2.18	10	0.2
Fabric (Small tube)	0.13	3	0.2

Figure 2. Continuous and Discontinuous material (Large Tube) respectively; (a) Polyurethane foam, (b) Fabric, (c) Felt.

2.2. Experimental Set-up

There are several ways of measuring the absorption coefficient of a material such as a plane wave impedance tube that is equipped with two microphones. Impedance tube, also known as two microphone tube, provides the incidence sound absorption coefficient that determines the sound absorption properties of material used. The tests were conducted according to ISO 10534-2 standard. The impedance tube method is based on a measurement of the transfer function between the two microphones from which various parameters can be calculated, such as the reflection coefficient, impedance and absorption. The two-microphone tube used has two tubes depending on the diameter: a large one with the diameter of 100 mm, covering the frequency range from 50 Hz to 1.6 kHz, and a smaller tube with the diameter of 30 mm, covering the frequency range from 500 Hz to 10 kHz. The tube is connected to the loudspeaker box. The loudspeaker placed at one end generates sound waves which propagate within the tube and strikes the sample at the other end, thus the wave will either be reflected or absorbed depending on the properties of the test sample. In addition, reflected sound resulting in a standing wave. The sound pressure distribution is measured using a long probe microphone attached to the analyser. The long probe microphone is well insulated from airborne noise and impact vibration. Symphonie Analyzer is used to obtain the complex transfer function between the two microphones; the closest microphone to the loudspeaker will be used as the reference. The impedance tubes and dBAlpha software used are shown by Fig. 3. The measurements were conducted at the room temperature of 25°C.

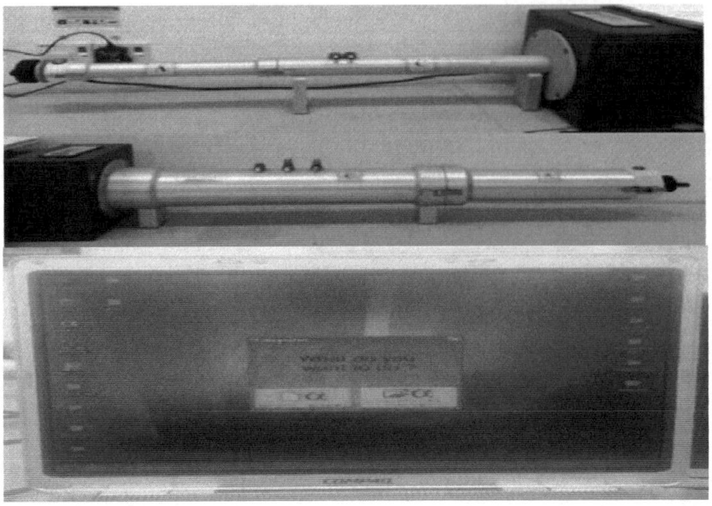

Figure 3. Impedance tubes and dBAlpha Software; (a) Small Tube, (b) Large Tube, (c) dBAlpha Software.

2.3. Mathematical Expressions

The sound pressure distribution, which is measured using a long probe microphone, always has a value greater than or equal to unity. The complex pressure amplitudes ($\hat{p}_1$and $\hat{p}_2$) are sent to the analyser in order to determine the sample's reflection coefficient amplitude R. After analysing the complex transfer function ($\widehat{H}_{12}$) between the two microphones using the pressure distribution data via equation 1, the reflection coefficient R of the material is determined using equation 2.

$$\widehat{H}_{12} = \frac{\hat{p}_1}{\hat{p}_2} \tag{1}$$

$$\hat{R} = \frac{\widehat{H}_{12} - e^{-jks}}{e^{jks} + \widehat{H}_{12}} \tag{2}$$

Where k is the wave number and s is the microphones spacing.

The sound absorption coefficient α is calculated using an octave band frequency analyzer through equation 3.

$$\propto = 1 - \|\hat{R}\|^2 \tag{3}$$

3. Experimental Results and Discussions

There are three test samples used in the present investigations including felt, Polyurethane foam and fabric. The effect of varying the material thickness towards the sound absorption coefficient of the material is investigated. Furthermore, the influence of surface discontinuity created between two joined materials is also presented. Therefore, the sound absorption coefficient is measured based on the variation of thickness as well as introducing surface discontinuity between two joined material parts.

3.1. Surface Discontinuity

Each of the materials was cut into halves and joined together lightly using glue. This is to provide surface discontinuity between the parts. The effect of the discontinuity relative to the continuous surface was studied and evaluated separately for each material since the thickness are different. The experiment was done for both large tube and small tube; this provides the results for the low to high frequency.

Sound absorption coefficient of felt in both cases remains almost same through the frequency as illustrated on Figure 4. In addition, this indicates that leaking of the sound wave through felt is very minimal even when discontinuity is introduced. Therefore, very low to medium frequency has almost insignificant effect towards the absorption property of felt. Furthermore, Figure 5 illustrates that the relative difference between the two cases is somehow significant at a very low frequencies as well as high frequencies. Nevertheless, the relative difference of the discontinuous surface is less than 30% for all frequencies investigated compared to the continuous surface.

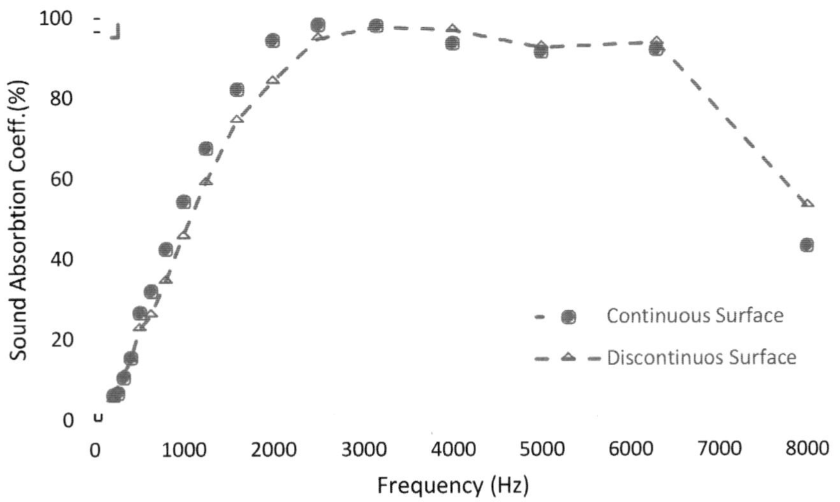

Figure 4. Influence of surface discontinuity on sound absorption coefficient of felt

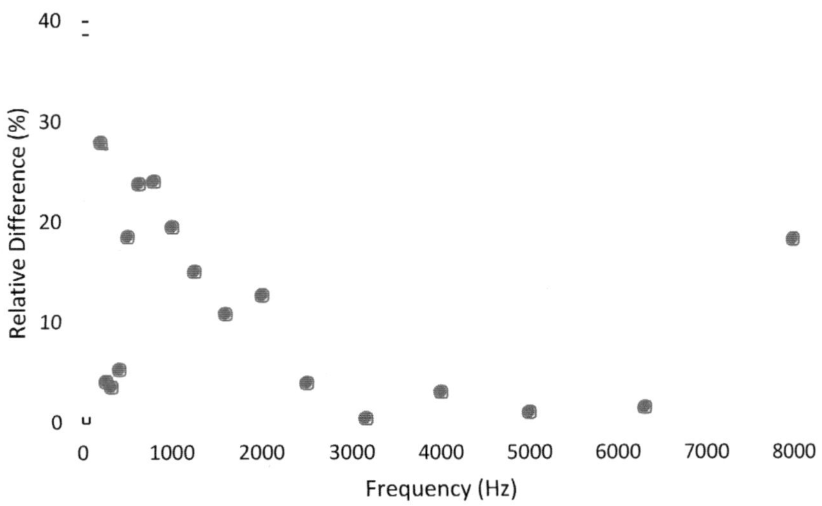

Figure 5. Relative difference of the absorption coefficients between continuous and discontinuous felt

For the polyurethane foam material, the absorption coefficient for both cases increase drastically from 50 Hz to 6300 Hz and then dropped steadily within a short range of frequency as shown in Figure 6. The highest coefficient is seen to be at 6300 Hz. At this maximum point, the continuity surface absorbs more as compared to the discontinuity case. Moreover, in both cases, the material performs effectively at around 4000 Hz. The absorption coefficient over the frequency range is elaborated on Figure 6. The relative difference of the two cases is more profound within the very low frequency region as shown in Figure 7. This illustrates similar behavior with felt at very low frequency. However, the relative difference of the foam is higher than the felt which can reach 50% for the low frequency. The relative difference for the form shows a steady increase at high frequency which can be higher than 90% in the range of frequency investigated in the present work. Therefore, the effect of the discontinuity surface is high at high frequency for the polyurethane foam.

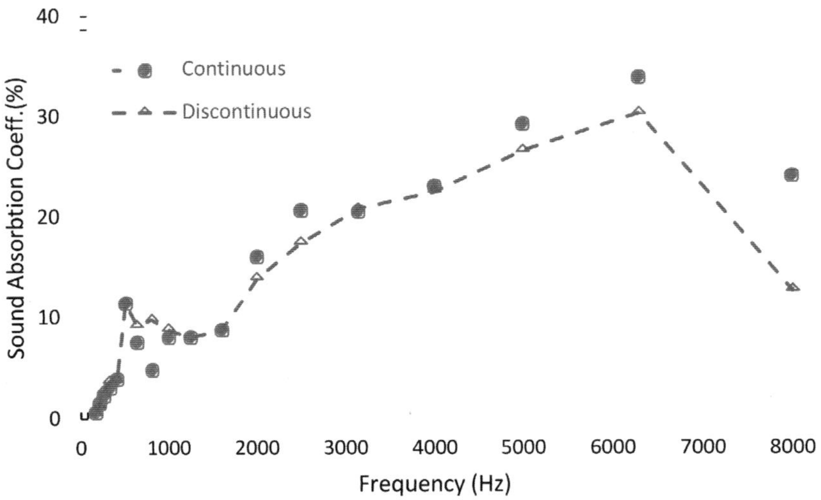

Figure 6. Influence of surface discontinuity on sound absorption coefficient of polyurethane foam

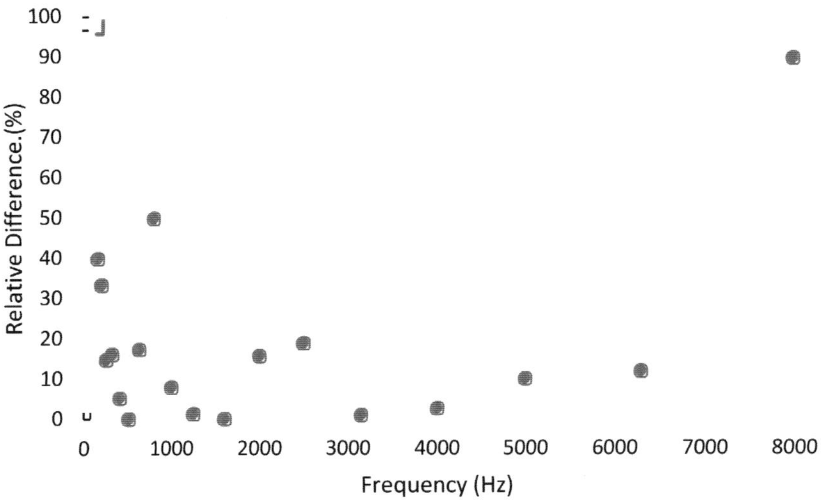

Figure 7. Relative difference of the absorption coefficients between continuous and discontinuous polyurethane foam

Figure 8 shows the absorption coefficient of fabric over given range of frequency. The trend for fabric is seen to have almost the same behaviour as polyurethane foam whereby the coefficient increased until a maximum point and then start dropping. The maximum absorption coefficient of fabric too happens to fall at 6300 Hz and this is the best working frequency range for the material too. On the other hand, the relative difference between continuous and discontinuous surfaces, as shown in Figure 9, illustrates that the material insightful difference of the two cases is seen again at the very low frequency range. For high frequency, the discontinuous surface of the fabric is however performing better than the discontinuous surface of the polyurethane foam. Figure 9 shows that the relative difference of the fabric is lower than 30% in the high frequency region as compared to the foam which can reach 90%.

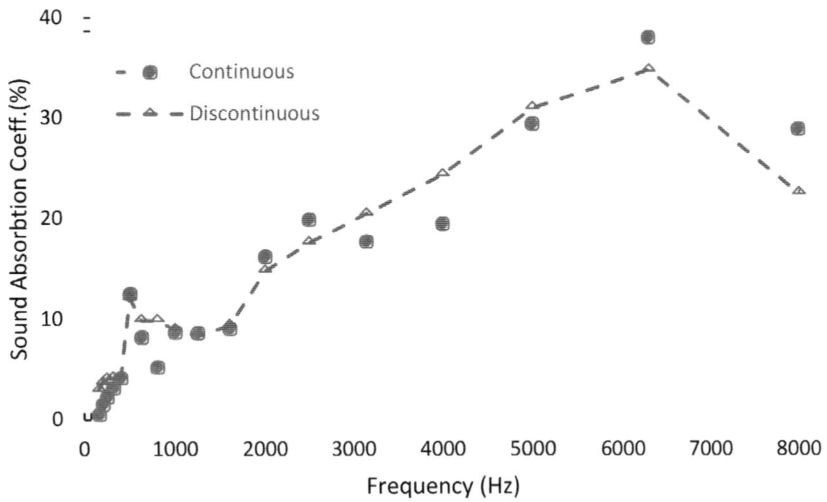

Figure 8. Influence of surface discontinuity on sound absorption coefficient of fabric

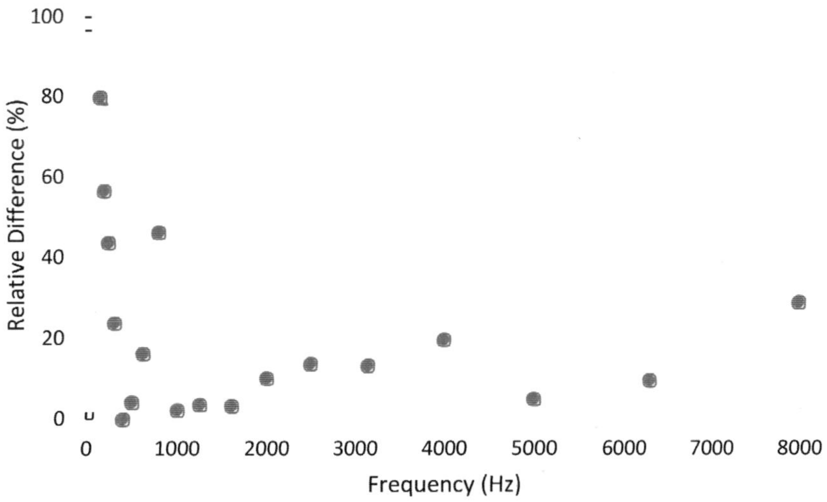

Figure 9. Relative difference of the absorption coefficients between continuous and discontinuous fabric

4. Conclusion

Following the experiments performed, it can be concluded that, in general the effect of the surface discontinuity is depended on the material selection. The effect of the surface discontinuity for the felt is insignificant compared to the polyurethane foam and fabric because of less leakage introduced by the discontinuity at low to high frequency. It is found also that the effect of the surface discontinuity is depended on the frequency regions. The surface discontinuity effect is less than 30% for the three materials on the moderate frequency. For low frequency, the surface discontinuity effect of the fabric shows poor performance. For high frequency, the surface discontinuity effect of the foam is the worst. Among these materials, felt has the highest sound absorption coefficient. From these results, we suggest that felt be used to improve passenger comfort as it has the highest sound absorption coefficient.

Acknowledgement
The support of Malaysian Ministry of Higher Education under the research grant FRGS17-036-0602 is gratefully acknowledged. We would also like to thank financial support by IIUM under the KOE Postgraduate Tuition Fee Waiver Scheme 2019 (TFW2019).

References

[1] Ismail, Lindawati, Mohd Imran Ghazali, Shahruddin Mahzan, and AM Ahmad Zaidi. "Sound absorption of Arenga Pinnata natural fiber." *World Academy of Science, Engineering and Technology* 67 (2010): 804-806.
[2] Jayamani, Elammaran, and Sinin Hamdan. "Sound absorption coefficients natural fibre reinforced composites." In *Advanced Materials Research*, vol. 701, pp. 53-58. Trans Tech Publications Ltd, 2013.
[3] Lu, Tian J., Feng Chen, and Deping He. "Sound absorption of cellular metals with semiopen cells." *The Journal of the Acoustical Society of America* 108, no. 4 (2000): 1697-1709.
[4] Lu, Tiara J., Audrey Hess, and M. F. Ashby. "Sound absorption in metallic foams." *Journal of applied physics* 85, no. 11 (1999): 7528-7539.
[5] Lee, Youn Eung, and Chang Whan Joo. "Sound absorption properties of thermally bonded nonwovens based on composing fibers and production parameters." *Journal of applied polymer science* 92, no. 4 (2004): 2295-2302.
[6] Seddeq, Hoda S. "Factors influencing acoustic performance of sound absorptive materials." *Australian Journal of Basic and Applied Sciences* 3, no. 4 (2009): 4610-4617.
[7] Coates, Michael, and Marek Kierzkowski. "Acoustic textiles: lighter, thinner and more sound absorbent." *Tech Text Int* 11, no. 7 (2002): 15-18.
[8] Koizumi, Tsujiuchi, N. Tsujiuchi, and A. Adachi. "The development of sound absorbing materials using natural bamboo fibers." *WIT Transactions on The Built Environment* 59 (2002).
[9] Krüger, J., and M. Quickert. "Determination of acoustic absorber parameters in impedance tubes." *Applied Acoustics* 50, no. 1 (1997): 79-89.
[10] Liu, Xueting, Xiong Yan, Li Li, and Huiping Zhang. "Sound-absorption properties of kapok fiber nonwoven fabrics at low frequency." *Journal of Natural Fibers* 12, no. 4 (2015): 311-322.
[11] Liu, Xueting, Xiaoning Tang, and Zhongmin Deng. "Sound absorption properties for multi-layer of composite materials using nonwoven fabrics with kapok." *Journal of Industrial Textiles* (2020): 1528083720904926.
[12] Makki, A. I., and E. Oktariani. "Acoustic absorptive properties of Kapok fiber, Kapok fiber layered tricot fabric and Kapok fiber layered double weave fabric." In *Journal of Physics: Conference Series*, vol. 1381, no. 1, p. 012065. IOP Publishing, 2019.
[13] Mamtaz, Hasina, Mohammad Hosseini Fouladi, Mohd Zaki Nuawi, Satesh Narayana Namasivayam, Masomeh Ghassem, and Mushtak Al-Atabi. "Acoustic absorption of fibro-granular composite with cylindrical grains." *Applied Acoustics* 126 (2017): 58-67.
[14] Or, Khai Hee, Azma Putra, and Mohd Zulkefli Selamat. "Oil palm empty fruit bunch fibres as sustainable acoustic absorber." *Applied Acoustics* 119 (2017): 9-16.
[15] Borrell, JM Gadea, E. Juliá Sanchis, J. Segura Alcaraz, and I. Montava Belda. "Sustainable sound absorbers from fruit stones waste." *Applied Acoustics* 161 (2020): 107174.

[16] Bujoreanu, Carmen, Florin Nedeff, Marcelin Benchea, and Maricel Agop. "Experimental and theoretical considerations on sound absorption performance of waste materials including the effect of backing plates." *Applied Acoustics* 119 (2017): 88-93.
[17] Tang, Xiaoning, and Xiong Yan. "Multi-layer fibrous structures for noise reduction." *The Journal of the Textile Institute* 108, no. 12 (2017): 2096-2106.
[18] Kim, Hyun-Sil, Pyung-Sik Ma, Sang-Ryul Kim, Seong-Hyun Lee, and Yun-Ho Seo. "A model for the sound absorption coefficient of multi-layered elastic micro-perforated plates." *Journal of Sound and Vibration* 430 (2018): 75-92.
[19] Qian, Y. J., J. Zhang, N. Sun, D. Y. Kong, and X. X. Zhang. "Pilot study on wideband sound absorber obtained by adopting a serial-parallel coupling manner." *Applied Acoustics* 124 (2017): 48-51.
[20] Gai, Xiao-Ling, Xian-Hui Li, Bin Zhang, Tuo Xing, Jun-Juan Zhao, and Zhi-Hui Ma. "Experimental study on sound absorption performance of microperforated panel with membrane cell." *Applied Acoustics* 110 (2016): 241-247.
[21] Shen, Xinmin, Panfeng Bai, Liang Chen, Sandy To, Fei Yang, Xiaonan Zhang, and Qin Yin. "Development of thin sound absorber by parameter optimization of multilayer compressed porous metal with rear cavity." *Applied Acoustics* 159 (2020): 107071.

IOP Conf. Series: Materials Science and Engineering 1244 (2022) 012004 doi:10.1088/1757-899X/1244/1/012004

A review of effect of higher-order nonlinear theories on analysis of piezoelectric laminates

Adnan Akhlaq[1], M.S.I Shaik Dawood[2], J S Mohamed Ali[3], E Sulaeman[4]

Department of Mechanical Engineering, Kulliyyah of Engineering, International Islamic University Malaysia

E-mail: adnanakhlaq87@gmail.com

Abstract. The paper aims at reviewing the modeling of piezoelectric laminated structures for stress distribution, vibration and noise control, shape control, health monitoring, and energy harvesting. Accurate modeling of piezoelectric laminates using linear and non-linear modeling techniques is essential at the high and low electric fields. The review is done in the field application of (1) different deformation theories (2) geometric non-linear theories (3) material non-linear models for the analysis of piezoelectric laminates. The literature review emphasizes using a higher-order non-linear model to analyze piezoelectric laminates and their effect on the composite structure.

1. Introduction

Piezoelectric materials are the most widely used form of smart materials. When an electric field is applied across them, they undergo deformation and produce an electric field when strained, thus can act as actuator and sensor. Piezoelectric actuators and sensors can be embedded or surface bonded to the laminated composite materials for control applications. The piezoelectric materials are applied on flexible structures for application to stress distribution, vibration and noise control, shape control, health monitoring, and energy harvesting [1]. Piezoelectric materials are often applied to the elastic substrate and even with composites to utilize the piezoelectric effect for high load applications [2]. Piezoelectric materials combined with beams, plates, and shell structures for application in aerospace and automotive engineering. Therefore, many researchers have attempted to model piezoelectric laminated structures accurately using different deformation theories, non-linear geometric theories, and non-linear material theories.

When applied as layers or patches embedded or surface bonded with composite laminates, Piezoelectric materials combine the better mechanical properties of laminated structure for superior static and dynamic response. Laminated structures having one dimension much larger than the other two are modeled as beam and small thickness compared with the two-dimensional modeled plates and shells. The structures can be modeled using classical laminate theory, first-order shear deformation theory, and higher-order shear deformation theory. Geometric non-linearity and material non-linearity of piezoelectric laminates affect their response. The review is done considering all aspects of modeling theories used for static and dynamic analysis of piezoelectric laminates. However, specific emphasis is given to literature dealing with non-linear material theories of piezoelectric materials. For general studying about the piezoelectric materials and different deformation theories, books like K. Bhaskar [3], Chopra [4] and Wang [5] are suitable for knowledge and reference.

The review work is done for the static, buckling, dynamic, and vibration control analysis using piezoelectric laminates, which are further categorized under the different deformation theories used for modeling, different geometric approaches applied, and the basis of material non-linearity of piezoelectric materials. The review aims at identifying the advantages and limitations of various shear deformation theories, non-linear geometric theories, and non-linear material theories.

2. Modeling of laminated structures using different deformation theories

A lot of work is done on employing a geometrically linear model to model laminated structures activated using piezoelectric materials for use in aerospace, civil structures, vibration control, form control, noise, acoustic control, energy harvesting, and health monitoring. The analytical approach can be used to model laminated structures. That gives accurate results, but for different end conditions or different orientations of laminated structures, it becomes difficult to model using an analytical method; thus finite element approach will be helpful. A finite element technique with large degree of freedom or nodal displacement requires substantial amount of computation time. Sze & Yao [6, 7] developed a finite element approach for modeling solid shells with piezoelectric materials. Achuthan et al [8] studied use of finite element method to investigate the shape control of piezoelectric beams. Perel & Palazotto [9] formulated a finite element for the dynamics of a composite beam with piezoelectric actuators. S. Y. Wang [10] proposed a finite element approach for investigating static and dynamic performance of piezoelectric bimorph. Kapuria et al [11] gave a finite element model for analysing hybrid piezoelectric beams under electro-mechanical loads. Kapuria & Kumari [12] used finite elements analysed a simply-supported cross-ply laminated circular cylindrical shells with monolithic piezoelectric fiber-reinforced composite (PFRC) sensory/actuator layers for free vibration and steady-state forced vibration response. Bendary et al [13] developed a simple finite element model to describe advanced smart beam behaviour. Adnan Elshafei & Alraiess [14] investigated static deformations, natural frequencies, and mode shapes of isotropic and orthotropic beams with dispersed piezoelectric actuators were computed using the same finite element model.

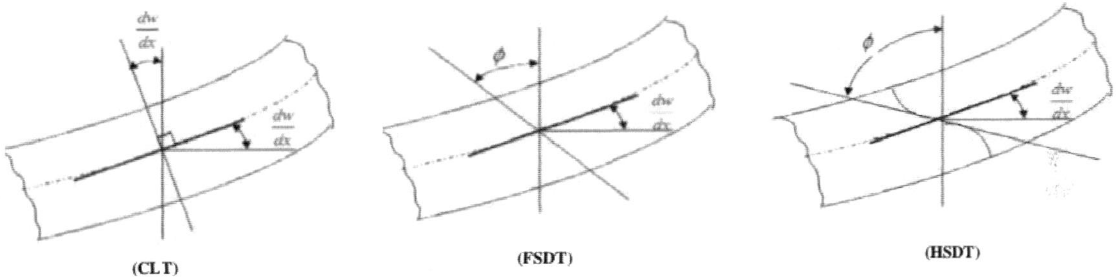

Figure 1. Laminate Deformation based on Different Deformation Hypothesis

Modeling the response of moderately thick to thin laminates accurately where thickness is minimal in comparison to in-plane dimensions a two-dimensional (2D) laminate theories for electro-mechanical loading is used. The different computationally efficient theories where displacement is assumed to be a single global variation throughout the thickness of laminate are classical laminate theory (CLT), First-order shear deformation theory (FSDT), and Higher-order shear deformation theory (HSDT). The different theories are developed by Taylor's series expansion of in-plane displacement w.r.t thickness coordinates.

2.1. Classical theory

Classical theory is the most straightforward theory based on the Kirchhoff-Love assumption, which states straight lines normal to the mid-plane remain straight and normal before and after deformation. For bending analysis, the displacement distribution according to classical theory is given by:

$$u = u_0 - z\frac{\partial w_0}{\partial x} \tag{1}$$

$$v = v_0 - z\frac{\partial w_0}{\partial y} \tag{2}$$

$$w = w_0 \tag{3}$$

Where (u, v, w) are the displacements along the (x, y, z) coordinates respectively and u_0, v_0, w_0 are the longitudinal, lateral and transverse deflection respectively of a point at the mid-plane $(i.e. z = 0)$ and z is the distance in the thickness direction. We consider transverse shear strains to be zero in this theory, which yields accurate findings when the thickness to in-plane dimension ratio is small. This hypothesis holds good for extremely thin structures with zero transverse shear strain.

Researchers have used classical Kirchhoff theory for developing coupled piezoelectric finite element model. Moita et al [15] applied finite element theory for vibration control of thin plates based on classical laminate theory. Moita *et al.* [17] used CLT to optimal vibration control of reinforced composite structure. Erturk [18] analytically analyzed for power output generated by piezoelectric patch based on classical Kirchhoff theory. U. Aridogan et al [19, 20] used Kirchhoff's theory for modeling analytically piezoelectric energy harvester structurally integrated on a thin plate. Yoon et al [21] developed an electro-mechanical analytical model for estimating power generation by energy harvester for the surface-bonded piezoelectric patch based on Kirchhoff theory. The classical theory is applied for analyzing electro-mechanical performance [22]. Further, Chattaraj & Ganguli [23] used the theory to introduce the effect of self-induced effect in the performance of the piezoelectric actuator. Przybylski & Gasiorski determined the non-linear vibration behavior of beam elements.

2.2. *First-order shear deformation theory*

The classical theory is generally considered for thin structures where shear strain deformation is not considered along the thickness direction. For moderately thick laminates transverse shear effect is to be taken into consideration. This theory is based on Reissner–Mindlin hypothesis, which considers transverse shear strains. It is based on the assumption that the displacement along thickness is constant, whereas the in-plane displacements vary linearly through the thickness. In contrast, straight lines transverse and normal to the mid-surface before deformation remain straight after deformation but not normal to mid-surface [25]. A shear correction factor is necessary to meet the boundary condition at the structure's upper and lower surfaces, the accuracy of this theory is dependent on the choice of appropriate shear correction factors [26, 27].

The displacement field according to FSDT is given by

$$u = u_0 + z\phi_x \tag{4}$$

$$v = v_0 + z\phi_y \tag{5}$$

$$w = w_0 \tag{6}$$

where ϕ_x and ϕ_y represent rotation about y and x-axis, respectively. The FSDT is used in papers for static and dynamic analysis by Cen *et al.* [28], Kapuria & Dumir [30], Krommer [31], Marinković *et al.* [32], and vibration control of piezoelectric laminated structure by Balamurugan & Narayanan [33], Wang *et al.* [34]. Based on FSDT, a meshless approach for static and vibration control of piezoelectric laminated structures is developed by Liu *et al.* [35] and Wang [10], developed an efficient FEM model for static and dynamic analysis of piezoelectric bimorph. Using FSDT, the static analysis and vibration control of functionally graded piezoelectric structures are investigated by Nguyen-Quang *et al.* [36], Xue *et al.* [37], Mallek *et al.* [38].

2.3. Higher-order shear deformation theory

Thin and moderately thick structures are subjected to classical and first-order shear deformation theory, which assumes that in-plane deformations vary linearly across the thickness direction. To perform analysis on thick structures and to deal with the limitations of classical and FSDT, the higher-order shear deformation theories are developed where the transverse shear strains vary along with the thickness and also provide transverse shear stress-free conditions at outside surfaces. As a result, no shear correction factor is required. In this theory, the in-plane displacement is supposed to be a cubic variation of the thickness coordinate whereas the out-of-plane displacement is kept constant. The displacement field is expressed here as

$$u = u_0 + z\phi_x + z^2\theta_x + z^3\lambda_x \quad (7)$$

$$v = v_0 + z\phi_y + z^2\theta_y + z^3\lambda_y \quad (8)$$

$$w = w_0 \quad (9)$$

where $\boldsymbol{\phi_x}$ and $\boldsymbol{\phi_y}$ represent the normal to mid-surface rotation about the y and x-axis, respectively, while the functions $\boldsymbol{\theta_x}$, $\boldsymbol{\theta_y}, \boldsymbol{\lambda_x}$ ***and*** $\boldsymbol{\lambda_y}$ are determined by applying stress-free boundary conditions to the top and bottom surfaces. By defining $\boldsymbol{h}$ as the displacement function thus reduces to

$$u = u_0 + z\phi_x - \frac{4z^3}{3h^2}(\phi_x + \frac{\partial w_0}{\partial x}) \quad (10)$$

$$v = v_0 + z\phi_y - \frac{4z^3}{3h^2}(\phi_y + \frac{\partial w_0}{\partial y}) \quad (11)$$

$$w = w_0 \quad (12)$$

Zhou *et al.* [39], Correia *et al.* [40, 41], Moita *et al.* [17], Selim *et al.* [42, 43] applied this hypothesis for applications to shape control, vibration control, noise attenuation, and precision positioning using piezoelectric laminated structures.

2.4. Layerwise theory

Laminated theories have considered the same displacement fields for different layers of laminates, assuming a single layer. Layerwise theory with varying fields of displacement, layers equilibrium, and continuity equations at the interface of layers are considered to overcome the limitations of single layer theory and improve accuracy in results. Further, the stress-free condition at the top and bottom layer is applied to reduce the number of unknowns. Layerwise theory having displacement field in each discrete layer a function of in-plane displacement can be used to first-order or higher-order theories. Vasques & Dias Rodrigues [44], Ray & Reddy [45] applied layerwise first-order theory, and Kapuria [46], Kapuria *et al.* [30], Yaqoob Yasin *et al.* [47], Rahman & Alam [48], Kapuria & Ahmed [49] applied layerwise third-order theory.

3. Analysis of piezoelectric laminates using geometrically non-linear theories

Thin Piezoelectric structures undergoing small displacements on application weak electric field belong to the geometrically linear model. However, researchers observed large displacement or rotations of piezoelectric structures using a weak electric field through the thickness. A linearly distributed electric potential through the thickness is assumed, the piezoelectric structures are first analyzed using the von Karman type of non-linearity. The square and product of derivatives of transverse deflection are considered in in-plane normal and shear strain components for moderately large deflection and slight rotation. Von Karman type of non-linearity is used in buckling analysis of piezoelectric structure by Kapuria & Dumir [50] using classical theory, Varelis & Saravanos [51] using FSDT, Kapuria &

Alam [52] using third-order zig-zag theory. For the static analysis using Von Karman non-linearity, Varelis *et al.* [53] and Panda *et al.* [54] applied a finite element model based on FSDT. For dynamic analysis of piezoelectric structures using von Karman, non-linearity is used by Mukherjee & Chaudhuri [55] using FSDT. Schmidt & Vu [56] based on first and third-order theory, and Sarangi & Ray [57] applied first-order zig-zag theory based on von Karman non-linearity for active control of vibration of composite laminates.

More geometric non-linear factors are introduced to the strain displacement relations for structures subjected to moderate rotations. Lentzen *et al.* [58] proposed Finite Element models for static and dynamic analysis of piezo-laminated structures under moderate rotation. Zhang *et al.* [60] and Hanen Mallek *et al.* [61] implemented the non-linear reasonable rotation model based on FSDT for static analysis of piezo-laminated structure.

Further considering fully geometrically non-linear effects with moderate rotations, Kundu *et al.* [62] did the post-buckling analysis using FSDT. Gao & Shen [63] developed vibration control using FSDT, Dash & Singh [64] performed dynamic analysis using HSDT, Amabili [65] included entirely geometric non-linear effect accounting to geometric imperfections in HSDT for static and dynamic analysis. Amabili [66] used HSDT for geometric non-linearity in thickness stretching, and Alijani & Amabili [67] used both geometric imperfection and thickness stretching for dynamic analysis based on HSDT.

Since the von Karman, moderate rotation and fully geometric non-linear theories are limited to moderate displacement and small rotations or large displacement and moderate rotations. Therefore, large displcement with large rotations is fully geometric non-linearity strain displacement relation with significant rotation is to used to analyze the piezo-laminated structures accurately. For static and dynamic study of piezoelectric laminated structures, a geometric non-linear finite element model based on FSDT was applied by Zhang *et al.* [68–70] and Rao *et al.* [71].

4. Analysis of piezoelectric laminates considering material non-linear theories

Analysis of Piezolaminated structure using a linear constitutive equation is commonly used for cases where small electric field producing little strain. The electric potential is expected to vary linearly over the thickness of the material, resulting in a constant electric field. If a substantial deformation is obtained when the piezoelectric patch is bent by applying electric field, a higher-order function of electric potential in terms of thickness is considered using a linear piezoelectric constitutive equation. Considering electric potential as a functional relation of thickness, Ahmad *et al.* [72], Varelis & Saravanos [51], Marinković *et al.* [32], Bendary *et al.* [13], Adnan Elshafei & Alraiess [14] performed the static and dynamic analysis of piezoelectric laminates. Marinković *et al.* [73, 74] investigated geometric non-linearity, including functional dependent electric potential. In addition, Pandey & Arockiarajan [75] studied the performance of MFC under thermal conditions using thickness-dependent electric potential function.

Piezo-ceramics being brittle, produce minor strains in the structure before the fracture. However, on applying a strong electric field, piezoelectric layers yield an electro-elastic material non-linear effect. Therefore, the non-linear constitutive equation for piezoelectric materials is to be included in electro-elastic analysis. Broadly two approaches describe piezoelectric non-linearity mathematically. First is reversible non-linearity that represents piezoelectric constitutive equation by having piezoelectric constant as a higher-order function of the electric field. This is described experimentally by Masys *et al.* [76]. The second is the irreversible constitutive model taking ferroelectric hysteresis and repolarization effect into account for numerical treatment as presented by Kamlah & Böhle [77], Zhou & Chattopadhyay [78], Landis [79], Klinkel [80], Linnemann *et al.* [81]. Achuthan *et al.* [8] studied the non-linear behavior of piezoelectric material for shape control using HSDT. Tan & Tong [82] investigated the non-linear behavior of piezoelectric material using the non-linear quadratic equation. Yao *et al.* [83, 84] studied the static and dynamic characteristics of bending actuators for the unimorph and bimorph cantilever using the strain function of the higher-order electric field based on the Kirchhoff hypothesis. Elhadrouz *et al.* [85], Kusculuoglu & Royston [86] compared the experimental and numerically found non-linear characteristics considering hysteresis using the finite element model. On application of high electrical loads, non-linear electro-mechanical characteristics are observed in piezoelectric ceramics. Wischke *et al.* [2] evaluated the electrostriction and the elastostriction non-linear parameters for piezoelectric ceramics. Kapuria & Yasin [87, 88] using HSDT developed a coupled layerwise finite element model for static and active vibration control of piezo-laminated composite structures under strong electric fields.

Further, Yasin & Kapuria [89] applied the model under a high electric field to investigate the effect of piezoelectric non-linearity on vibration suppression of piezoelectric laminated shells. Using the Euler-Bernoulli beam model, an analytical investigation was performed to analyze the performance of rectangular bimorph cantilevers considering the second-order electric field dependent equation assuming small strain under a large electric field is presented Chattaraj & Ganguli [22]. Rao *et al.* [90] presented finite element analysis for non-linear electro-mechanical behavior of piezoelectric laminated structures using second order non-linear constitutive equation assuming small strain under large electric field based on FSDT. In addition, the electro-mechanical coupling coefficient affects the self-induced electric displacement field. Chattaraj & Ganguli [23] analysed the performance of the piezoelectric bimorph actuator considering the self-induced electric displacement field at high electric field. Ahoor *et al.* [91] performed dynamic analysis of bimorph actuators considering strain as a function of higher order electric field to express peizoelectric material non-linearity. Yao *et al.* [84] proposed a model for static analysis, including both non-linear piezoelectric and von karman type non-linear geometric effect considering a strong electric field based on classical theory. Under strong electric field application, a geometrically non-linear finite element model based on FSDT was developed by Zhang *et al.* [92] to analyse piezoelectric laminate, taking into account both fully geometrical non-linearity and material non-linearity.

5. Conclusion and future scope

In this review, work analysis of piezoelectric laminates has been classified based on applying different deformation theories, non-linear geometric theories, and non-linear material models. This work mainly focuses on using efficient deformation theory with the geometric non-linear or non-linear material model to accurately model piezoelectric laminates for static or dynamic analysis depending on the advantages and limitations of various theories discussed literature review. A clear description is provided that classical theory and first-order shear deformation theory are both unable to effectively predict the response of moderately thick and thick laminates. For moderately thick or thick piezoelectric laminates, higher-order shear deformation should be used to improve static and dynamic response.

Application of Von Karman type geometric non-linearity is limited to structures with moderate deformation and small rotation. Large deformation structures under a significant rotation of an entirely geometric non-linear model with large rotation are applied. For accurate modeling of piezoelectric laminated structures fully geometric non-linear theory with large rotation is to be applied. The piezoelectric linear model is applicable for small strain conditions under a weak electric field. The material non-linearity of piezoelectric ceramics is to be considered when a large electric field is applied. Numerical and experimental analyses have shown that considering material non-linearity gives the better response of piezoelectric laminated structures.

Researchers are interested in piezoelectric materials because of its use in a variety of engineering domains, including aeronautical, mechanical, civil, micro- and nano-engineering, and so on. Piezoelectric materials are ideal for applications such as noise, vibration, shape and position control, non-destructive testing, and health monitoring systems because they can operate as sensors or actuators. There is a lot of application where piezoelectric materials are to be operated at a high electrical load. It becomes increasingly important to predict the displacement under actuation mode accurately. Researchers have generally considered electrostriction in non-linear material analysis. Considering the elastostriction coefficient in actuation mode will give a more precise displacement of actuators which will be helpful in further analysis of piezoelectric laminates. The use of higher-order

IOP Conf. Series: Materials Science and Engineering 1244 (2022) 012004 doi:10.1088/1757-899X/1244/1/012004

theory along-with material non-linearity is to be further investigated in order to accurately model piezoelectric laminates for static and dynamic analysis.

References

[1] Chopra I 2002 Review of State of Art of Smart Structures **40** 16–9

[2] Wischke M, Haller D, Goldschmidtboeing F and Woias P 2010 Assessing the elastostriction and the electrostriction parameter of bulk PZT ceramics *Smart Mater. Struct.* **19**

[3] K. Bhaskar T K V 2014 *Plates Theories and Applications* vol 28

[4] C.M. Wang K H L 2000 *Shear Deformable Beams and Plates*

[5] Inderjit Chopra J S 2014 *Smart Structures Theory* (Cambridge University Press)

[6] Sze K Y R, Yao L Q S and Yi S 2000 A hybrid stress ANS solid-shell element and its generalization for smart structure modelling . Part II * smart structure modelling 565–82

[7] Sze K Y R and Yao L Q S 2000 A hybrid stress ANS solid-shell element and its generalization for smart structure modelling . Part I * solid-shell element formulation 545–64

[8] Achuthan A, Keng A K and Ming W C 2001 Shape control of coupled nonlinear piezoelectric beams *Smart Mater. Struct.* **10** 914–24

[9] Perel V Y and Palazotto A N 2002 Finite element formulation for dynamics of delaminated composite beams with piezoelectric actuators *Int. J. Solids Struct.* **39** 4457–83

[10] Wang S Y 2004 A finite element model for the static and dynamic analysis of a piezoelectric bimorph *Int. J. Solids Struct.* **41** 4075–96

[11] Kapuria S, Dumir P C, Ahmed A and Alam N 2004 Finite element model of efficient zig-zag theory for static analysis of hybrid piezoelectric beams *Comput. Mech.* **34** 475–83

[12] Kapuria S and Kumari P 2010 Three-dimensional piezoelasticity solution for dynamics of cross-ply cylindrical shells integrated with piezoelectric fiber reinforced composite actuators and sensors *Compos. Struct.* **92** 2431–44

[13] Bendary I M, Elshafei M A and Riad A M 2010 Finite element model of smart beams with distributed piezoelectric actuators *J. Intell. Mater. Syst. Struct.* **21** 747–58

[14] Adnan Elshafei M and Alraiess F 2013 Modeling and analysis of smart piezoelectric beams using simple higher order shear deformation theory *Smart Mater. Struct.* **22**

[15] Moita J M S, Correia I F P, Soares C M M and Soares C A M 2004 Active control of adaptive laminated structures with bonded piezoelectric sensors and actuators *Comput. Struct.* **82** 1349–58

[16] Simões Moita J M, Franco Correia V M, Martins P G, Mota Soares C M and Mota Soares C A 2006 Optimal design in vibration control of adaptive structures using a simulated annealing algorithm *Compos. Struct.* **75** 79–87

[17] Moita J S, Martins P G, Mota Soares C M and Mota Soares C A 2008 Optimal dynamic control of laminated adaptive structures using a higher order model and a genetic algorithm *Comput. Struct.* **86** 198–206

[18] Erturk A 2011 Piezoelectric energy harvesting for civil infrastructure system applications: Moving loads and surface strain fluctuations *J. Intell. Mater. Syst. Struct.* **22** 1959–73

[19] Aridogan U, Basdogan I and Erturk A 2014 Analytical modeling and experimental validation of a structurally integrated piezoelectric energy harvester on a thin plate *Smart Mater. Struct.* **23**

[20] Aridogan U, Basdogan I and Erturk A 2014 Multiple patch-based broadband piezoelectric energy harvesting on plate-based structures *J. Intell. Mater. Syst. Struct.* **25** 1664–80

[21] Yoon H, Youn B D and Kim H S 2016 Kirchhoff plate theory-based electromechanically-coupled analytical model considering inertia and stiffness effects of a surface-bonded piezoelectric patch *Smart Mater. Struct.* **25** 25017

[22] Chattaraj N and Ganguli R 2016 Electromechanical analysis of piezoelectric bimorph actuator in static state considering the nonlinearity at high electric field *Mech. Adv. Mater. Struct.* **23** 802–10

[23] Chattaraj N and Ganguli R 2017 Effect of self-induced electric displacement field on the response of a piezo-bimorph actuator at high electric field *Int. J. Mech. Sci.* **120** 341–8

[24] Przybylski J and Gasiorski G 2018 Nonlinear vibrations of elastic beam with piezoelectric actuators *J. Sound Vib.* **437** 150–65

[25] Zhang S Q, Li Y X and Schmidt R 2015 Modeling and simulation of macro-fiber composite layered smart structures *Compos. Struct.* **126** 89–100

[26] Wu Z, Chen R and Chen W 2005 Refined laminated composite plate element based on global-local higher-order shear deformation theory *Compos. Struct.* **70** 135–52

[27] Aydogdu M 2009 A new shear deformation theory for laminated composite plates *Compos. Struct.* **89** 94–101

[28] Cen S, Soh A K, Long Y Q and Yao Z H 2002 A new 4-node quadrilateral FE model with variable electrical degrees of freedom for the analysis of piezoelectric laminated composite plates *Compos. Struct.* **58** 583–99

[29] Kapuria S and Dumir P C 2002 First order shear deformation theory for hybrid cylindrical panel in cylindrical bending considering electrothermomechanical coupling effects *ZAMM Zeitschrift fur Angew. Math. und Mech.* **82** 461–71

[30] Kapuria S, Dumir P C and Ahmed A 2003 An efficient coupled layerwise theory for dynamic analysis of piezoelectric composite beams **261** 927–44

[31] Krommer M 2003 Piezoelastic vibrations of composite Reissner-Mindlin-type plates *J. Sound Vib.* **263** 871–91

[32] Marinković D, Köppe H and Gabbert U 2007 Accurate modeling of the electric field within piezoelectric layers for active composite structures *J. Intell. Mater. Syst. Struct.* **18** 503–13

[33] Balamurugan V and Narayanan S 2001 Shell finite element for smart piezoelectric composite plate/shell structures and its application to the study of active vibration control *Finite Elem. Anal. Des.* **37** 713–38

[34] Wang S Y, Quek S T and Ang K K 2001 Vibration control of smart piezoelectric composite plates *Smart Mater. Struct.* **10** 637–44

[35] Liu G R, Dai K Y and Lim K M 2004 Static and vibration control of composite laminates integrated with piezoelectric sensors and actuators using the radial point interpolation method *Smart Mater. Struct.* **13** 1438–47

[36] Nguyen-Quang K, Dang-Trung H, Ho-Huu V, Luong-Van H and Nguyen-Thoi T 2017 Analysis and control of FGM plates integrated with piezoelectric sensors and actuators using cell-based smoothed discrete shear gap method (CS-DSG3) *Compos. Struct.* **165** 115–29

[37] Xue Y, Li J, Li F and Song Z 2019 Active Control of Plates Made of Functionally Graded Piezoelectric Material Subjected to Thermo-Electro-Mechanical Loads *Int. J. Struct. Stab. Dyn.*

19 1–23

[38] Mallek H, Jrad H, Wali M and Dammak F 2019 Nonlinear dynamic analysis of piezoelectric-bonded FG-CNTR composite structures using an improved FSDT theory *Eng. Comput.*

[39] Zhou X, Chattopadhyay A and Thornburgh R 2000 Analysis of piezoelectric smart composites using a coupled piezoelectric-mechanical model *J. Intell. Mater. Syst. Struct.* **11** 169–79

[40] Franco Correia V M, Aguiar Gomes M A, Suleman A, Mota Soares C M and Mota Soares C A 2000 Modelling and design of adaptive composite structures *Comput. Methods Appl. Mech. Eng.* **185** 325–46

[41] Pinto Correia I F, Mota Soares C M, Mota Soares C A and Herskovits J 2002 Active control of axisymmetric shells with piezoelectric layers: A mixed laminated theory with a high order displacement field *Comput. Struct.* **80** 2265–75

[42] Selim B A, Zhang L W and Liew K M 2016 Active vibration control of FGM plates with piezoelectric layers based on Reddy's higher-order shear deformation theory *Compos. Struct.* **155** 118–34

[43] Selim B A, Liu Z and Liew K M 2019 Active vibration control of functionally graded graphene nanoplatelets reinforced composite plates integrated with piezoelectric layers *Thin-Walled Struct.* **145** 106372

[44] Vasques C M A and Dias Rodrigues J 2006 Active vibration control of smart piezoelectric beams: Comparison of classical and optimal feedback control strategies *Comput. Struct.* **84** 1402–14

[45] Ray M C and Reddy J N 2005 Active control of laminated cylindrical shells using piezoelectric fiber reinforced composites *Compos. Sci. Technol.* **65** 1226–36

[46] Kapuria S 2001 An efficient coupled theory for multilayered beams with embedded piezoelectric sensory and active layers *Int. J. Solids Struct.* **38** 9179–99

[47] Yaqoob Yasin M, Ahmad N and Naushad Alam M 2010 Finite element analysis of actively controlled smart plate with *Lat. Am. J. Solids Struct.* **7** 227–47

[48] Rahman N U and Alam M N 2015 Structural control of piezoelectric laminated beams under thermal load *J. Therm. Stress.* **38** 69–95

[49] Kapuria S and Ahmed A 2021 A coupled efficient layerwise finite element model for free vibration analysis of smart piezo-bonded laminated shells featuring delaminations and transducer debonding *Int. J. Mech. Sci.* **194** 106195

[50] Kapuria S and Dumir P C 2005 Geometrically nonlinear axisymmetric response of thin circular plate under piezoelectric actuation *Commun. Nonlinear Sci. Numer. Simul.* **10** 411–23

[51] Varelis D and Saravanos D A 2004 Coupled buckling and postbuckling analysis of active laminated piezoelectric composite plates *Int. J. Solids Struct.* **41** 1519–38

[52] Kapuria S and Alam N 2004 Zigzag theory for buckling of hybrid piezoelectric beams under electromechanical loads *Int. J. Mech. Sci.* **46** 1–25

[53] Varelis D and Saravanos D A 2004 Mechanics and finite element for nonlinear response of active laminated piezoelectric plates *AIAA J.* **42** 1227–35

[54] Panda S and Ray M C 2008 Nonlinear finite element analysis of functionally graded plates integrated with patches of piezoelectric fiber reinforced composite *Finite Elem. Anal. Des.* **44** 493–504

[55] Mukherjee A and Chaudhuri A S 2005 Nonlinear dynamic response of piezolaminated smart

beams *Comput. Struct.* **83** 1298–304

[56] Schmidt R and Vu T D 2009 Nonlinear dynamic FE simulation of smart piezolaminated structures based on first- and third-order transverse shear deformation theory *Adv. Mater. Res.* **79–82** 1313–6

[57] Sarangi S K and Ray M C 2011 Active damping of geometrically nonlinear vibrations of laminated composite plates using vertically reinforced 1-3 piezoelectric composites *Acta Mech.* **222** 363–80

[58] Lentzen S, Kłosowski P and Schmidt R 2007 Geometrically nonlinear finite element simulation of smart piezolaminated plates and shells *Smart Mater. Struct.* **16** 2265–74

[59] Zhang S Q and Schmidt R 2014 Large rotation theory for static analysis of composite and piezoelectric laminated thin-walled structures *Thin-Walled Struct.* **78** 16–25

[60] Zhang S Q and Schmidt R 2014 Static and dynamic FE analysis of piezoelectric integrated thin-walled composite structures with large rotations *Compos. Struct.* **112** 345–57

[61] Mallek H, Jrad H, Wali M and Dammak F 2019 Geometrically nonlinear finite element simulation of smart laminated shells using a modified first-order shear deformation theory *J. Intell. Mater. Syst. Struct.* **30** 517–35

[62] Kundu C K, Maiti D K and Sinha P K 2007 Post buckling analysis of smart laminated doubly curved shells *Compos. Struct.* **81** 314–22

[63] Gao J X and Shen Y P 2003 Active control of geometrically nonlinear transient vibration of composite plates with piezoelectric actuators *J. Sound Vib.* **264** 911–28

[64] Dash P and Singh B N 2009 Nonlinear free vibration of piezoelectric laminated composite plate *Finite Elem. Anal. Des.* **45** 686–94

[65] Amabili M 2015 A new third-order shear deformation theory with non-linearities in shear for static and dynamic analysis of laminated doubly curved shells *Compos. Struct.* **128** 260–73

[66] Amabili M 2014 A non-linear higher-order thickness stretching and shear deformation theory for large-amplitude vibrations of laminated doubly curved shells *Int. J. Non. Linear. Mech.* **58** 57–75

[67] Alijani F and Amabili M 2014 Effect of thickness deformation on large-amplitude vibrations of functionally graded rectangular plates *Compos. Struct.* **113** 89–107

[68] Zhang S Q, Li Y X and Schmidt R 2015 Active shape and vibration control for piezoelectric bonded composite structures using various geometric nonlinearities *Compos. Struct.* **122** 239–49

[69] Zhang S Q, Wang Z X, Qin X S, Zhao G Z and Schmidt R 2016 Geometrically nonlinear analysis of composite laminated structures with multiple macro-fiber composite (MFC) actuators *Compos. Struct.* **150** 62–72

[70] Zhang S Q, Gao Y S, Zhao G Z, Yu Y J, Chen M and Wang X F 2019 Geometrically nonlinear analysis of CNT-reinforced functionally graded composite plates integrated with piezoelectric layers *Compos. Struct.* **234**

[71] Rao M N, Schmidt R and Schröder K U 2015 Geometrically nonlinear static FE-simulation of multilayered magneto-electro-elastic composite structures *Compos. Struct.* **127** 120–31

[72] Ahmad S N, Upadhyay C S and Venkatesan C 2004 Linear and Nonlinear Analysis of a Smart Beam Using General Electrothermoelastic Formulation *AIAA J.* **42** 840–9

[73] Marinković D, Köppe H and Gabbert U 2008 Degenerated shell element for geometrically

nonlinear analysis of thin-walled piezoelectric active structures *Smart Mater. Struct.* **17**

[74] Marinković D, Köppe H and Gabbert U 2009 Aspects of modeling piezoelectric active thin-walled structures *J. Intell. Mater. Syst. Struct.* **20** 1835–44

[75] Pandey A and Arockiarajan A 2017 Performance studies on Macro fiber composite (MFC) under thermal condition using Kirchhoff and Mindlin plate theories *Int. J. Mech. Sci.* **130** 416–25

[76] Masys A J, Ren W, Yang G and Mukherjee B K 2003 Piezoelectric strain in lead zirconate titante ceramics as a function of electric field, frequency, and dc bias *J. Appl. Phys.* **94** 1155–62

[77] Kamlah M and Böhle U 2001 Finite element analysis of piezoceramic components taking into account ferroelectric hysteresis behavior *Int. J. Solids Struct.* **38** 605–33

[78] Zhou X and Chattopadhyay A 2002 Nonlinear Piezoelectric Constitutive Relationship and Actuation for Piezoelectric Laminates *43rd AIAA/ASME/ASCE/AHS/ASC Struct. Struct. Dyn. Mater. Con*

[79] Landis C M 2004 Non-linear constitutive modeling of ferroelectrics *Curr. Opin. Solid State Mater. Sci.* **8** 59–69

[80] Klinkel S 2006 A phenomenological constitutive model for ferroelastic and ferroelectric hysteresis effects in ferroelectric ceramics *Int. J. Solids Struct.* **43** 7197–222

[81] Linnemann K, Klinkel S and Wagner W 2009 A constitutive model for magnetostrictive and piezoelectric materials *Int. J. Solids Struct.* **46** 1149–66

[82] Tan P and Tong L 2002 A one-dimensional model for non-linear behaviour of piezoelectric composite materials *Compos. Struct.* **58** 551–61

[83] Yao L Q, Zhang J G, Lu L and Lai M O 2004 Nonlinear dynamic characteristics of piezoelectric bending actuators under strong applied electric field *J. Microelectromechanical Syst.* **13** 645–52

[84] Yao L Q, Zhang J G, Lu L and Lai M O 2004 Nonlinear extension and bending of piezoelectric laminated plate under large applied field actuation *Smart Mater. Struct.* **13** 404–14

[85] Elhadrouz M, Ben Zineb T and Patoor E 2006 Finite element analysis of a multilayer piezoelectric actuator taking into account the ferroelectric and ferroelastic behaviors *Int. J. Eng. Sci.* **44** 996–1006

[86] Kusculuoglu Z K and Royston T J 2008 Nonlinear modeling of composite plates with piezoceramic layers using finite element analysis *J. Sound Vib.* **315** 911–26

[87] Kapuria S and Yasin M Y 2013 A nonlinear efficient layerwise finite element model for smart piezolaminated composites under strong applied electric field *Smart Mater. Struct.* **22**

[88] Kapuria S, Yasin M Y and Hagedorn P 2015 Active vibration control of piezolaminated composite plates considering strong electric field nonlinearity *AIAA J.* **53** 603–16

[89] Yaqoob Yasin M and Kapuria S 2016 Influence of piezoelectric nonlinearity on active vibration suppression of smart laminated shells using strong field actuation *JVC/Journal Vib. Control* **24** 505–26

[90] Rao M N, Tarun S, Schmidt R and Schröder K U 2016 Finite element modeling and analysis of piezo-integrated composite structures under large applied electric fields *Smart Mater. Struct.* **25** 1–12

[91] Ahoor Z H, Ghafarirad H and Zareinejad M 2019 Nonlinear dynamic modeling of bimorph piezoelectric actuators using modified constitutive equations caused by material nonlinearity *Mech. Adv. Mater. Struct.* **0** 1–11

[92] Zhang S Q, Zhao G Z, Zhang S Y, Schmidt R and Qin X S 2017 Geometrically nonlinear FE analysis of piezoelectric laminated composite structures under strong driving electric field *Compos. Struct.* **181** 112–20

Error driven Fuzzy Logic Controller (FLC) for Spherical Mobile Robot: Simulation & Experimental Performance Analysis

Nurul Nafisah Kamis[1], Salmiah Ahmad, Abd. Halim Embong, Erwin Sulaeman

Kulliyyah of Engineering,
International Islamic University Malaysia,
53100 Gombak, Selangor Malaysia.

[1]nafisahkamis@gmail.com

Abstract. Spherical mobile robot is a ball-shaped mobile robot that capable to move from one place to another to perform the desired tasks. The challenge for spherical mobile robot falls within control algorithm aspects, where it is nonlinear and highly under-actuated increases the difficulties to control its motion. In this paper, the potentiality of intelligent controller is explored to control the motion spherical mobile robot. An error driven Fuzzy Logic controller (FLC) which consist of modular of two Proportional-Derivative (PD) type FLC and a Proportional-Integral (PI) type FLC were designed and tested to control the position of the spherical mobile robot, angle of the rotation of the pendulum and the angular velocity of the spherical mobile robot, respectively. The performance of the controller was then optimized by tuning the input and output gain of the controller with Particle Swarm optimization (PSO) method. The performance of the robot without the controller and with optimized controller were compared and analysed. As a result, the performance of the system in the form of rise time and settling time of the robot to control its position managed to be reduced by 86% and 89% respectively with optimized PD-type. For PD-type FLC to control the rotation of the pendulum, it was capable to eliminate the overshoot and reduce it settling time by 35.5%. While the optimized PI-type FLC has managed to eliminate the steady state error and achieved it desired velocity setpoint in 1.2s. Lastly, all the designed controllers were successfully tested on the prototype of the spherical mobile robot to study the performance of the spherical mobile robot. However, the overall performance of the implemented designed controller has shown a minimum effectiveness. The controller managed to achieve the setpoint value but with low stability which result from the less accuracy of the hardware sensors.

1. Introduction

The Spherical Mobile robot (SR) is one of the unique types of the mobile robot that has become an interest topic in robotic fields. Due to its spherical shape, it possesses a characteristic of a ball which can move freely in any type of environments. Compared to the wheeled and legged mobile robot, spherical mobile robot is more efficient to move on soft and bumpy area. Besides, it has low overturning's risk and capable to bounce from one obstacle to another obstacle to keep its mobility. In uneven terrain, the mobility of the robot will be more smoothly as it can roll like a ball. This uneven surface allows it to maneuver comfortably while trying to stabilize itself. Most of the spherical mobile robot come with full cover and seal to give it spherical shape. The cover also can be used as the shield to protect the inner circuitry and all the components in the robot. Overall, the physical structure of the spherical mobile robot can improve the robustness and mobility of the mobile robot to wander in any special environments to perform any tasks. Therefore, the type of mobile robot has been reported in many research papers for varieties of applications like education and entertainment (Edutainment) [1] – [4], navigation and exploration, and surveillance and monitoring [5].

However, the geometrical shape of this mobile robot also makes it difficult to be control. The design and control system of the spherical mobile robot is one of the biggest challenges due to its nonlinearity and highly under-actuated system. The motion of the robot is created by disturbing the equilibrium and cause instability to the system. Yet, unplanned instability will cause unpredicted change in direction. As the spherical mobile robot is round and spherical in shaped, has caused it to easily move in any direction, where this movement can be advantages to the robot to move around and also may create problem if it needs to follow any trajectory.

In the early stages, spherical mobile robot was designed with a single-wheel, installed in the spherical shell. The motion of the spherical mobile robot is with respected to the motion of the wheel. As a result, the open loop control system is designed based on kinematic and dynamic motion of the robot which is deduced from the basic equation of motion analysis of the ball with respect to the motion of the wheel inside the robot [6]. Likewise, the longitudinal, and lateral motion control with open loop system had also been developed for Spherobot in [7] and spherical rolling robot [8]. The robot was controlled using open loop system with no feedback of the actual speed of the rotors. This open loop control has also been used as the preliminary control analysis for the rotor-type spherical robot to plan its trajectory [9]. Furthermore, four motion control modes; heading, steering, jumping and zero-radius turning motion for spherical robot have been tested with an open-loop control system using simulation and experimental where the desired velocity was set to analyze the output for each mode independently [10].

Technically, open-loop control system has shown that the controller managed to control the system accordingly, but the design is meticulous and suitable for simulation environment only. Any add-on data or disturbance on the actual environment can affect the result significantly if open loop is applied. In [11], open loop control system was experimented with BHQ-2 – a pendulum driven spherical robot that caused oscillation when it moves on uneven terrain which has proven the inability of this type of controller for complex system. So even though the construction of open loop control system is simple and economical and generally stable but, it is inaccurate and unreliable to a nonlinear system like the spherical robot. Besides, feedback controller based on Lyapunov control with feedback controller for the dynamic model based on back-stepping technique was designed for spherical robot with gyro as its internal driver unit (IDU) [11]. The controller was designed to control the position and attitude of the rolling robot for trajectory tracking with Lyapunov stability theory was used to prove the convergence of the controller. The simulation to track the desired trajectories for linear and circular motion has been successful run to prove the capability of the spherical robot. However, no experimental result was reported for this controller.

Next, nonlinear sliding mode controller was implemented in BYQ-III to tackle the nonlinearity of the spherical mobile robot. Here, constraint Lagrangian method was used for dynamics model and it was then tested with simulation and experiment platforms to control a simple motion of latitudinal locomotion. The results have shown an effective motion controller but producing a bad grace position accuracy. While in [12], they designed a PD-controller and sliding mode controller with adaptive parameter to solve for the path following problem and to compensate for the uncertainties of the system. Hierarchical sliding mode controller has also been used in [13] to control the longitudinal motion of the pendulum driven spherical robot, where double layer structures of the controller was required to guarantee the stability of the whole system.

An intelligent controller was suggested to handle the uncertainty and improve the robustness of the spherical robot's controller. In [13], they have simulated a hierarchical sliding mode control for longitudinal motion while in [14], they have integrated the hierarchical mode controller with fuzzy guidance scheme through a back-stepping technique for path tracking control. Both controllers were capable to tackle the control problem despite any input disturbance in the system. Besides, Fuzzy logic controller (FLC) has been successfully used in many other applications, vehicle suspension system [15] and for the energy management system in battery electrical vehicle (BEV) [16].

PD-controller with PD-type fuzzy control was designed for the pendulum driven spherical robot in [17], where the conventional PD-controller was used to improve the transient response of the spherical robot while fuzzy was used to increase the robustness to the disturbance. Performance of PID-controller with neuro-fuzzy controller and sliding mode control was also being compared to the stand-alone conventional PID [18]. The hybrid controller has shown tackled the robustness and the steady-state error issues of the output response of the spherical robot. Most of the previous intelligent controllers in the literature, basically were designed with the combination of several other controllers like PD controller or PID controller. So far there are limited findings in the literature to analyze the performance of different PID-type of FLC based controllers to control the rolling position of the spherical robot in details steps with analysis and justifications, for rolling position control of SR. Analyzing the performance of different types of intelligent controller like FLC with variations of parameters is beneficial to find the most suitable controller for this application since the performance of a controller is not standard for all types of systems and needs to be customized from the analysis.

Therefore, in this paper, two Proportional Derivative (PD)- type FLC and Proportional – Integral (PI)- type FLC have been designed to control the rolling position of the robot, angle of rotation of the pendulum and angular velocity of the spherical robot, respectively. The model of spherical robot is explained in section 2 while the design of error based Fuzzy Logic Control is in section 3. In section 4, brief discussion of the selection of optimization method and the discussion of the result implementation of the controller in prototype of the spherical robot is justify in section 5. Lastly, overall finding is concluded in conclusion which in in section 6.

2. Model of Spherical Mobile Robot

The spherical robot system has been modelled as a Pendulum-driven spherical mobile robot as in [17]. Assumptions have been made for the modelling; The robot rolls without slipping; The geometric center of the spherical shell is the geometric center of the system; When the spherical in static condition, the pendulum is assumed to be in vertical downward.

2.1. *Kinematic Modelling*

The kinematic motion of the spherical mobile robot is represented on reference frame, R_0 that fixed on the ground; X_0 and Y_0 as in Figure 1. While X_1, and Y_1 and X_2, and Y_2 are the moving frames R_1 and R_2 are attached to the center of the spherical mobile robot and thus allowing translation only with respect to R_0 and R_1. Ideally, changes in the angle of rotation of the pendulum, θ_p will cause the change in angle of rotation of the spherical body, θ_s in the opposite direction. Position of the spherical mobile robot is donated as p_s So, the equation of motion of the robot can also be represented based on angular velocity, ω_s and linear velocity, v_s of the spherical shell, the angular velocity ω_p, and linear velocity, v_p of the pendulum. The kinematic equation is described in Eq. (1) - (3) with R as radius of spherical mobile robot.

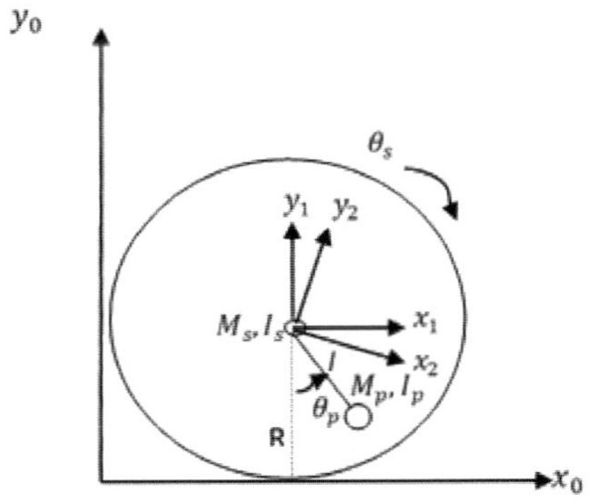

Figure 1: Geometrical diagram of spherical robot

$$\omega_s = -\dot{\theta}_s \tag{1}$$

$$v_s = -R\dot{\theta}_s \tag{2}$$

$$p_s = R\omega_s \tag{3}$$

$$\omega_P = (\dot{\theta}_p - \dot{\theta}_s) \tag{4}$$

$$v_p = (-R\dot{\theta}_s + (\dot{\theta}_p - \dot{\theta}_s)lcos(\theta_p - \theta_s) + ((\dot{\theta}_p - \dot{\theta}_s)lsin(\theta_p - \theta_s)) \tag{5}$$

2.2. Dynamic Modelling

Lagrangian function, L is used to derive the dynamic model of the rolling spherical mobile robot which is the subtraction of kinetic energy of linear and rotation system with the potential energy of the system as in Eq. (6).

$$L = K - P$$
$$= \frac{1}{2}M_S||v_s||^2 + \frac{1}{2}I_S||\omega_s||^2 + \frac{1}{2}m_p||v_p||^2 + \frac{1}{2}I_p||\omega_p||^2 - m_p g r_{p-z}$$
$$= \frac{1}{2}M_S(R\dot{\theta}_s)^2 + \frac{1}{2}I_S(-\dot{\theta}_s)^2 + \frac{1}{2}m_p\left(-R\dot{\theta} + (\dot{\theta}_p - \dot{\theta}_s)lcos(\theta_p - \theta_s) + \left((\dot{\theta}_p - \dot{\theta}_s)\, lsin(\theta_p - \theta_s)\right)\right)^2 + \frac{1}{2}I_p(\dot{\theta}_p - \dot{\theta}_s)^2 - m_p glcos(\theta_p - \theta_s) \tag{6}$$

The calculation for kinetic energy involves both the rigid bodies of the shell and pendulum while only pendulum that has potential energy as the system is hanging at the center of spherical mobile robot. The mass of pendulum and the spherical shell is represented by M_s and m_p respectively. The equation also includes the moment of inertia in the pendulum, I_p and the spherical shell, I_s that can be calculated with Eq. (7) and Eq. (8).

$$I_s = \frac{2}{3}M_s R^2 \tag{7}$$

$$I_p = \frac{1}{12}m_p l^2 + m_p\left(\frac{l}{2}\right)^2 \tag{8}$$

Viscous friction, S is assumed to be acting between the spherical shell and the surface where it can be modelled with an energy dissipation function of the velocity of the system $\dot{q}_i$ and damping constant, ξ as in Eq. (9).

$$S=\frac{1}{2}\xi\dot{q}_i^{\,2}=\frac{1}{2}\xi\left(\dot{\theta}_s^{\,2}+\dot{\theta}_p^{\,2}\right) \tag{9}$$

From the following equations, generalized forced, Q_i can be derived from Euler-Langrage equation using Hamilton's principle as in Eq. (10).

$$\frac{d}{dt}\left(\frac{\partial L}{\partial \dot{q}_i}\right)-\frac{\partial L}{\partial \dot{q}_i}+\frac{\partial S}{\partial \dot{q}_i}=Q_i \tag{10}$$

Where the generalized coordinates used are θ_s and θ_p for q_1 and q_2. The generalized force in the equation is the input torque the rotated the pendulum and reaction torque about the shaft will be rotated in the opposite. Thus, the equation of motion can be rewritten as Eq. (11) and Eq. (12).

$$M_{11}\ddot{\theta}_s+M_{12}\ddot{\theta}_p+C_{11}+G_{11}=\tau \; M_{11}\ddot{\theta}_s+M_{12}\ddot{\theta}_p+C_{11}+G_{11}=\tau \tag{11}$$

$$M_{22}\ddot{\theta}_p+M_{21}\ddot{\theta}_s+C_{21}+G_{21}=\tau \; M_{22}\ddot{\theta}_p+M_{21}\ddot{\theta}_s+C_{21}+G_{21}=\tau \tag{12}$$

where

$$M_{11}=M_SR^2+m_pR^2+m_pl^2+I_s+I_p+2m_pRl\cos(\theta_p\text{-}\theta_s)$$
$$M_{12}=M_{21}=-m_pl^2\text{-}I_p\text{-}m_pRl\cos(\theta_p\text{-}\theta_s)$$
$$M_{22}=m_pl^2+I_p$$
$$C_{11}=m_pRl\sin(\theta_p\text{-}\theta_s)\left(\dot{\theta}_p\text{-}\dot{\theta}_s\right)^2+\xi\dot{\theta}_s$$
$$C_{21}=\xi\dot{\theta}_p$$
$$G_{11}=\text{-}G_{21}=\text{-}m_pgl\sin(\theta_p\text{-}\theta_s)$$

The model of DC motor is described as in Eq. (13) and Eq. (14), where it is used to rotate the hanging internal mechanism that acts like a pendulum to the system.

$$\ddot{\theta}_m=\frac{1}{J}(k_t-b\dot{\theta}_m) \tag{13}$$

$$\frac{di}{dt}=\frac{1}{L}(-R_mi+V_m-k_e\dot{\theta}_m) \tag{14}$$

The θ_m, k_e, k_t, b, R_m, V_m, L and J are used to represent the rotation of the motor, moment of inertia of the motor, electromotive force constant, motor torque constant, motor's viscous friction constant, electrical resistance, voltage armature, electrical inductance and moment of inertia of the motor respectively. The parameters are set as in Table 1 and 2 to analyse the rolling motion of the spherical mobile robot. The overall block diagram of the system is shown in Figure 2 consisting of DC motor model, dynamic and kinematic model of the spherical mobile robot.

Table 1: Parameters of the spherical mobile robot.

Parameters	Value
Mass sphere, M_S	0.5 kg
Mass Pendulum, m_p	0.1 kg
Radius, R	0.135 kg
Length, l	0.12 m
Gravitational acceleration, g	9.81 N.m
Damping coefficient, ξ	0.2

Table 2-Electrical parameters of DC Motor

Parameters	Value
Back emf constant, K	0.01 V/(rad/s)
Rotor damping, b	0.1 N.m /(rad/s)
Armature resistance, R_m,	1 ohm
Armature inductance, L	0.5 H
Rotor inertia, J	0.01 kg.m^2

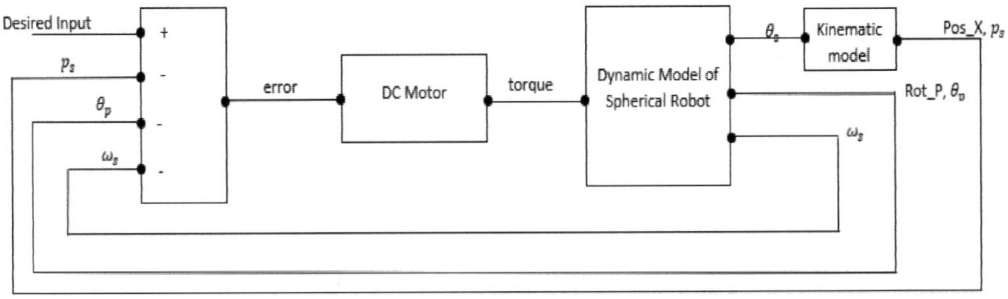

Figure 2: Schematic Diagram of the spherical robot on three axis planes.

3. Error based Fuzzy Logic Controller

Error based Fuzzy Logic Control (FLC) has been proposed to control the position of the rolling mechanism of spherical mobile robot. FLC is an intelligent controller that uses reasoning and human knowledge as a main element in control system. The system used a linguistic language to for a knowledge-based to determine the control action of the spherical mobile robot. The significant components in the FLC system are fuzzification, fuzzy inference system (FIS) and defuzzification, as can be depicted in Figure *3* .

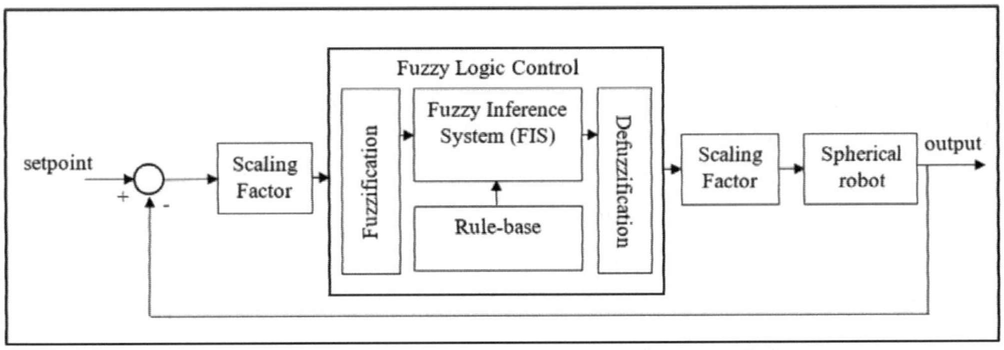

Figure 3: Configuration of Fuzzy Logic Control System (FLC).

Since FLC is nonlinear in nature, its effectiveness varies from one application to another. Therefore, in this paper, two Proportional-Derivative (PD)-type and a Proportional-Integral (PI)-type of FLC has been designed and investigated to control the rolling position of the Spherical mobile robot, position of the rotation of the pendulum and the angular velocity of the spherical mobile robot.

3.1. Proportional-Derivative (PD)-type Fuzzy Logic Controller

The Proportional-Derivative (PD)-type FLC is a multi-input single-output (MISO) system that receives error, e and derivative of error, $\dot{e}$, and supplies a single output voltage to the actuator. Two set of PD-type FLC were designed to control the rolling position of the spherical mobile robot and the angle of rotation of the pendulum in the spherical mobile robot. Five triangular membership function for to represent the input and output of the FLC has been used for PD-type FLC to control the rolling position of the spherical robot. While a combination of trapezoidal and triangular membership function has been used to represent the input of the FLC and five triangular membership function is used to represent the output of the FLC has been designed to control the angle of pendulum's rotation of the spherical robot.

Then, total of 25 rule-based were constructed based on the five variables from each input per controller. The overall FIS of the rule-based is listed in Table 3. The example of if-else statement for PD-type FLC is, "*If e is NB, AND ė is NB, then V is NB*".

Table 3: 25 Rules-based for PD-type FLC

Input		**Derivative of error ($\dot{e}$)**				
		NB	**NS**	**Z**	**PS**	**PB**
Error	**NB**	NB	NB	NS	NS	Z
(e)	**NS**	NB	NS	NS	Z	PS
	Z	NS	NS	Z	PS	PS
	PS	NS	Z	PS	PB	PB
	PB	Z	PS	PS	PB	PB

The similar pattern of rules is used for both controllers. The rules represent the relationship between input error and derivative of error with the output of the controller as the rotation of the pendulum also effect the position of the spherical mobile robot. For instance, if the error e and derivative of error, $\dot{e}$ are NB mean that the spherical mobile robot moves fast and moving away from the desired output then, the voltage of the DC motor should be NB which means that the robot will move in the opposite direction to the desired distance. However, the rotation of the pendulum must be controlled to not exceed 45 degrees from the downward position as it might degrade the rolling motion of the spherical mobile robot.

3.2. Proportional-Integral (PI)-type Fuzzy Logic Controller

The Proportional-Integral (PI)-type FLC is also a multi-input single output (MISO) system which receives error, e and integral of error, $\int e$. It is designed to control the angular velocity of the spherical mobile robot. It gives an output in term of voltage of the motor used to rotate the rolling mechanism of the spherical mobile robot. The combination of sigmoidal and triangular membership function has been used to represent the input error and integral error of the FLC. While five triangular membership function is assign for output of the FLC.

Like PD-type FLC, PI-type FLC also consists of 25 rule-based, as shown in Table 4. The example of if-else statement for PI-type FLC is, "If e is NB, AND $\int e$ is NB, then V is NB".

Table 4: Rules-based for PI-type FLC

Input		**Integral of error (*fe*)**				
		NB	***NS***	***Z***	***PS***	***PB***
Error (e)	***NB***	NB	NB	NS	NS	Z
	NS	NB	NS	NS	Z	PS
	Z	NS	NS	Z	PS	PS
	PS	NS	Z	PS	PB	PB
	PB	Z	PS	PS	PB	PB

Rules above are constructed to control the angular velocity of the spherical mobile robot so that it will not move too fast which cause it to be unstable and move randomly. It spherical shaped is one of the factors that can cause it velocity to be uncontrollable which will then affect it final desired position. The rules are constructed to control the input of the dc motor which rotate the pendulum which than rotate the spherical shell. So, to increase or decrease the angular velocity of the robot, angle of the pendulum needs to be control. The defuzzification process for all the controllers is designed using center of gravity method.

4. Particle Swarm Optimization (PSO)

To improve the designed controller, the scaling factor of input and output gain of the controllers have been optimized with Particle Swarm Optimization (PSO) method. The optimization is done with objectives to improve response time of each controller. The PSO has been designed to minimize the integral absolute error (IAE) of the system response by varying its input and output gain to the possible solution. The IAE has been used as the objective function to optimise this error-driven controller as it also improved the output response of the system. The parameters of the PSO used are listed as Table 5.

Table 5: Parameters of the PSO method.

PSO Parameter	**Value**
Population size, npop	20
Number of iterations	100
Weight, w	0.721
Acceleration coefficient, c_1, c_2	0.193
Stopping Criteria	Maximum number of iterations

With parameter mention above, the PSO was used to optimize the input and output gain of the error driven controller by iteratively testing the possible solution to minimize the objective function set for the PSO. During the optimization process, The optimization was run until it reach the number of iteration which is 100.

5. Discussion

The error-based controllers have been designed independently to control position, rotation of the pendulum and the angular velocity of the spherical robot before it being tested in MATLAB/Simulink environment. The system is simulated with step input to observe it ability to reach its desired setpoint. The output response of the system which consist of its rise time, t_s, settling time, t_s, steady-state error, e_{ss}, and percentage overshoot, P_o (if available) is recorded. Figure 4 and Table 6 show the output response of the spherical robot from Matlab / Simulink environment in three conditions: system without controller, with controller and system with optimized controller (FLC controller with PSO).

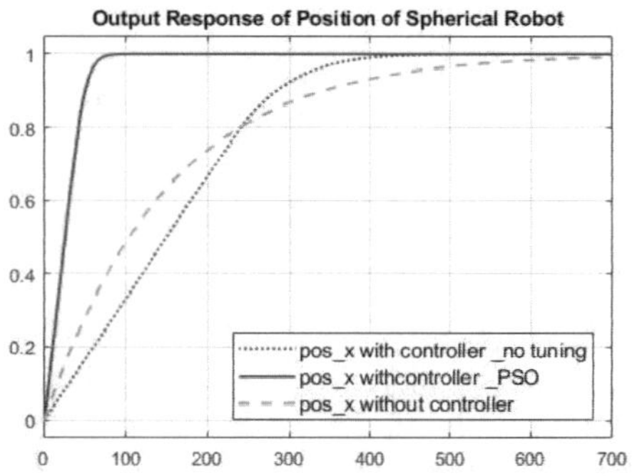

(a)

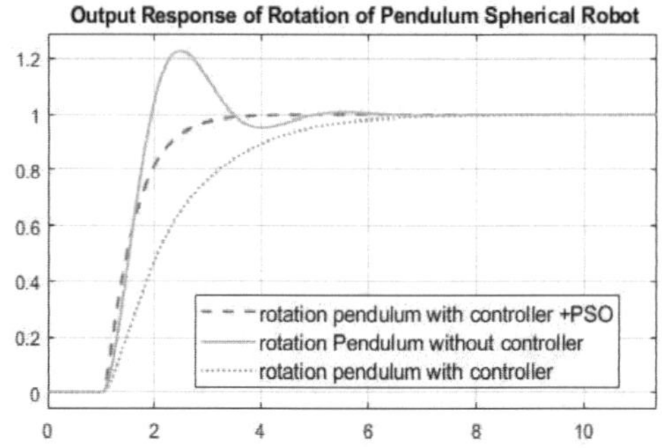

(b)

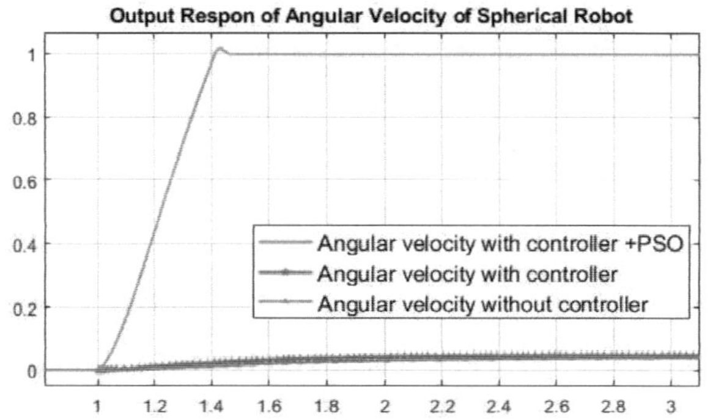

(c)

Figure 4: Output Response of Spherical Robot without Controller, with controller without tuning and with optimized controller (controller with PSO) for (a) control rolling position of spherical robot, (b) control rotation pendulum of spherical robot and (c) control angular velocity of the spherical robot.

Table 6: Output Time response of Spherical mobile Robot Simulation in Matlab/Simulink.

	Without controller	With controller	With optimized Controller
PD-type FLC to control rolling position of Spherical robot	T_r: 330.1 s T_s: 596.5 s e_{ss}: 0	T_r: 235.4 s T_s: 367.5 s e_{ss}: 0	T_r: 43.9 s T_s: 63.9 e_{ss}: 0
PD-type FLC to control angle of rotation of the pendulum	T_r: 0.639 s T_s: 4.654 s P_o: 22.8 %	T_r:2.8 s T_s: 5.7 s e_{ss}: 0	T_r: 1.2 s T_s: 3.0 s e_{ss}: 0
PI-type FLC to control angular velocity of spherical robot	T_r: no T_s: no e_{ss}: 0.9524	T_r: no T_s: no e_{ss}: 0.940	T_r: 0.3 s T_s: 1.2 s P_o : 2 %

Based on the simulation data, it shows that error-driven Fuzzy Logic controller manage to improve the performance of the spherical mobile robot. With optimized PD-type FLC, time response to control spherical mobile robot's rolling position can be reduced for almost 86% and 89% for it rise time and settling time, respectively. In meanwhile, percentage overshoot from time response of the spherical mobile robot without controller can be eliminate but it increases the risetime by 50%. However, it does not increase the settling time but still able to reduce it by 35.5%. For the PI-type

FLC, simulation show the ability of the error driven controller to control the angular velocity of spherical to it setpoint value with 2% of overshoot.

Then, the designed controllers are again being simulated with real set value and experimented in real hardware of spherical mobile robot as shown in Figure 5.

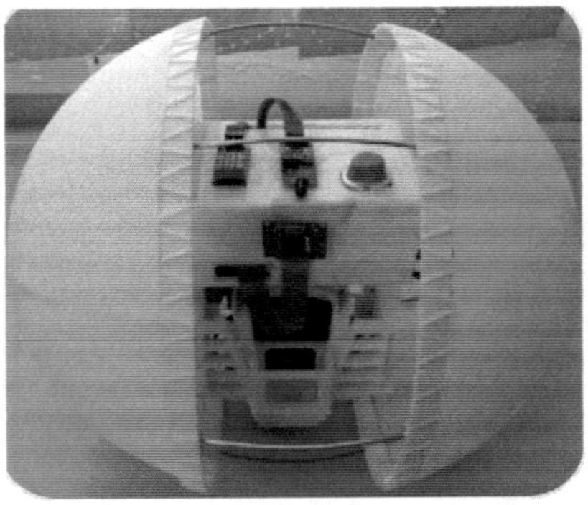

(a)

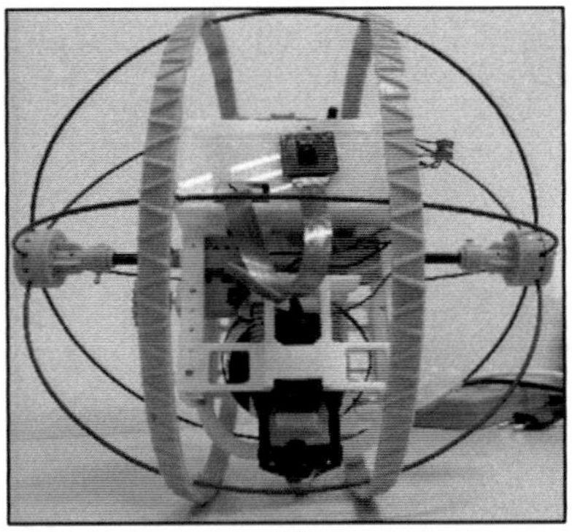

(b)

Figure 5: The prototype of Spherical mobile Robot (a) full body (b) without cover.

The structure of spherical mobile robot is shaped using six spokes which has been setup around the middle rod to give it spherical shaped. The robot also equipped with MPU 6050 and encoder to give feedback of the position and velocity of spherical mobile robot. The two sensors are very important for the robot to allow it to position itself correctly based on the desired input, calculate the angle of pendulum and to measure the velocity of spherical mobile robot.

The result response of the simulation and experimented then being compared. The response time of the controller to control the rotation angle of pendulum is shown as in Figure 6.

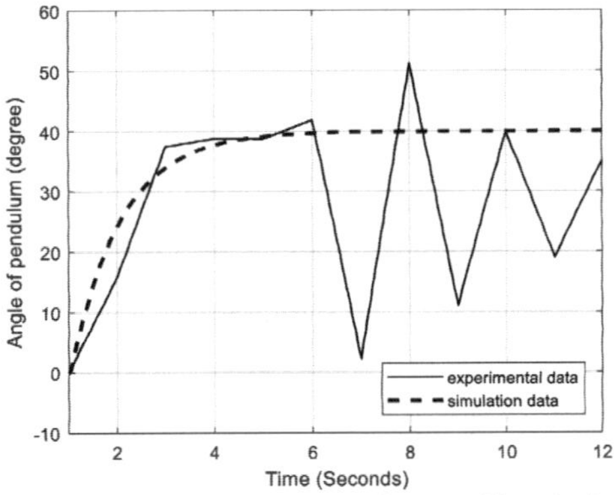

Figure 6:Angle of pendulum's graph when PD-type FLC is implemented in spherical mobile robot (set to 40 degrees).

From the above graph, it is shown that during the experimental tested the controller trying to control the angle as the setpoint. However, when the pendulum achieved the desired setpoint the controller drops the voltage and causes the pendulum to back to the vertical point. Besides, as the pendulum rotates more than 40 degrees, the spherical start to roll and make the pendulum back to its original position. As compared to the simulation, during the experimental the rise time is slower but achieved it desired setpoint faster by 1s. Other than that, during the experiment process, it observed that the power of DC motor needs to be amplified as at some output supplied the DC motor cannot rotate the pendulum.

The PD-type FLC has also implemented in the spherical mobile robot to control the position of the spherical mobile robot. The position of the spherical mobile robot is measured by the encoder. The position of the spherical mobile robot is plot as in Figure 7.

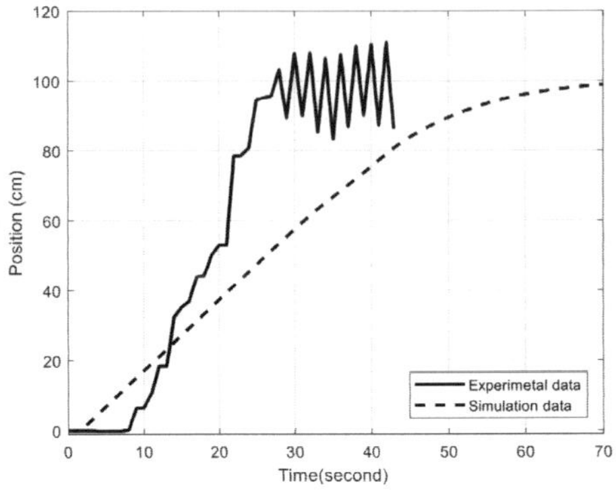

Figure 7:Output response of the PD-type FLC to control position of the spherical mobile robot.

The controller was set to roll for 100 cm. Based on the graph, the spherical mobile robot took less than 30s which is 60% faster as compared to simulation that took 75s for achieved its desired point. But the spherical mobile robot still trying to stop itself in the exact position. The position of the spherical mobile robot is measured by the encoder of the pendulum. As the pendulum is still moving the encoder will calculate the distance even though the small movement by the pendulum does not really roll the spherical mobile robot. This reduces the efficiency to find the exact position of the spherical mobile robot. However, based on the moving average of the data it shows that spherical robot manages to roll to its desired position. The input and output gain of the controller needs to be optimized again for the hardware implementation.

PI-type FLC is implemented in the spherical mobile robot to control the angular velocity of the robot. The experimental result is graph as in Figure 8.

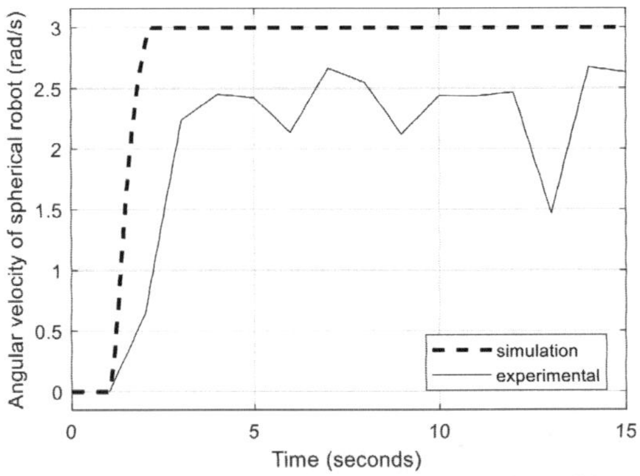

Figure 8: Angular velocity of the spherical mobile robot with PI-type FLC

During the experimental the robot is set to 3 rad/s, however the measured angular velocity does not achieve the desired target but hover around 2.5 rad/s. The steady state error of the system to desired input is around 0.5 rad/s. Proper optimization are needed to optimize the input and output of the controller. Besides, dealing with integral is challenging but the performance of the controller with this PI-type FLC is better than when it run without controller as the steady state are higher.

The implementation of the controller becomes more challenging as it is difficult to get accurate feedback data on its position and angular velocity. Here, the position and velocity are estimated based on the movement of the pendulum. Estimation can be incorrect if the spherical mobile robot rolls due to external and unexpected force that does not involve the pendulum's motion. Besides, the performance of the spherical mobile robot also depends on its physical state. As it undergoes many experiments, the 3D plastic ring that holds the spoke started to deform and this effect the rolling motion of the spherical mobile robot. However, the controller will manage to control the spherical mobile robot if it receives smooth and correct feedback.

6. Conclusion

In this paper, an intelligent controller is designed to control the spherical mobile robot. There are three controllers that have been designed to control the position of the spherical robot, the pendulum of the spherical mobile robot, and the angular velocity of the spherical mobile robot. Two PD-type FLC are designed to control the position of spherical mobile robot and the pendulum whereas PI-type FLC is designed to control the velocity of the spherical mobile robot. These controllers were designed, simulated, and analysed in Matlab/ Simulink environment before it is implemented into the hardware system. The simulation has shown an encouraging result as it manages to improve the performance of output response of the system to control these three parameters: rolling position, angle of rotation of the pendulum and the angular velocity of the rolling spherical mobile robot. Tuning the input and output gain of the error driven controller help to reduce the time response of the system specially to control angular velocity of the spherical mobile robot. The optimization improves the ability of the PI-type FLC to reach it set point by eliminate the steady-state error that occur when the robot simulated without the controller and with untuning controller. Meanwhile, with PD-type FLC eliminates the percentages overshoot in controlling the angle of pendulum. In addition, the simulation to control the rolling position of the spherical robot has improved for average of 87% for both the rise time and settling time with PD-type FLC. Next, the designed controllers are implemented in the prototype of spherical mobile robot. Even though designed controller has been successfully implemented in prototype of the spherical robot however, the controller shows a poor overall performance in regulate the response to achieve it input setpoint. The nonlinearity of the spherical mobile robot does affect in designing the controller for the system. In hardware side, the condition of the spherical robot and the efficiency of the sensor's component can limit the implementation performance of the spherical robot. In the future, the designed controller can be improved by optimized different element in FLC and online optimization in the spherical mobile robot.

Acknowledgement

FRGS 2013-2 Novel Modeling and Investigation of Rolling Mechanism with Bouncing Capability using Natural Phenomena-based Algorithm and PRGS15-021-0031 Upscaling, Testing and Optimization of a Rolling Bouncing Spherical Robot for Surveillance and Monitoring.

References

[1] Bicchi, A., Balluchi, A., Prattichizzo, D., & Gorelli, A. (1997, April). Introducing the" Sphericle": an experimental testbed for research and teaching in nonholonomy. In Proceedings of International Conference on Robotics and Automation (Vol. 3, pp. 2620-2625). IEEE.F. Michaud and S. Caron, "Roball-An autonomous toy-rolling robot," in In Proc. of the Workshop on Interactive Robotics and Entertainment, 2000.

[2] Michaud, F., & Caron, S. (2002). Roball, the rolling robot. Autonomous robots, 12(2), 211-222.

[3] Michaud, F., Laplante, J. F., Larouche, H., Duquette, A., Caron, S., Létourneau, D., & Masson,

P. (2005). Autonomous spherical mobile robot for child-development studies. IEEE Transactions on Systems, Man, and Cybernetics-Part A: Systems and Humans, 35(4), 471-480.

[4] Seeman, M., Broxvall, M., Saffiotti, A., & Wide, P. (2006, October). An autonomous spherical robot for security tasks. In 2006 IEEE International Conference on Computational Intelligence for Homeland Security and Personal Safety (pp. 51-55). IEEE.

[5] Mojabi, P. (2002, May). Introducing August: a novel strategy for an omnidirectional spherical rolling robot. In Proceedings 2002 IEEE International Conference on Robotics and Automation (Cat. No. 02CH37292) (Vol. 4, pp. 3527-3533). IEEE.

[6] Halme, A., Schonberg, T., & Wang, Y. (1996, March). Motion control of a spherical mobile robot. In Proceedings of 4th IEEE International Workshop on Advanced Motion Control-AMC'96-MIE (Vol. 1, pp. 259-264). IEEE.

[7] Mukherjee, R., Minor, M. A., & Pukrushpan, J. T. (1999, December). Simple motion planning strategies for spherobot: a spherical mobile robot. In Proceedings of the 38th IEEE Conference on Decision and Control (Cat. No. 99CH36304) (Vol. 3, pp. 2132-2137). IEEE.

[8] Bhattacharya, S., & Agrawal, S. K. (2000). Spherical rolling robot: A design and motion planning studies. IEEE Transactions on Robotics and Automation, 16(6), 835-839.

[9] Zhan, Q., Zhou, T., Chen, M., & Cai, S. (2006, June). Dynamic trajectory planning of a spherical mobile robot. In 2006 IEEE Conference on Robotics, Automation and Mechatronics (pp. 1-6). IEEE.

[10] Mahboubi, S., Fakhrabadi, M. M. S., & Ghanbari, A. (2013). Design and implementation of a novel spherical mobile robot. Journal of Intelligent & Robotic Systems, 71(1), 43-64.

[11] Otani, T., Urakubo, T., Maekawa, S., Tamaki, H., & Tada, Y. (2006, March). Position and attitude control of a spherical rolling robot equipped with a gyro. In 9th IEEE International Workshop on Advanced Motion Control, 2006. (pp. 416-421). IEEE.

[12] Tao, Y., Hanxu, S., & Qingxuan, J. (2012). Variable structure control of pendulum-driven spherical mobile robots. In Proc. 2010 3rd Int. Conf. Comput. Electr. Eng.(ICCEE 2010 no. 2) (Vol. 53, No. 2, pp. 622-626).

[13] Cai, Y., Zhan, Q., & Xi, X. (2012). Path tracking control of a spherical mobile robot. Mechanism and Machine Theory, 51, 58-73.

[14] Hanafi, D. (2020). Improvement the Vehicle Suspension System Performance Using Fuzzy Controller. International Journal of Integrated Engineering, 12(1), 91-101.

[15] Abulifa, A. A., Soh, A. C., Hassan, M. K., Ahmad, R. M. K. R., & Radzi, M. A. M. (2019). Energy Management System in Battery Electric Vehicle Based on Fuzzy Logic Control to Optimize the Energy Consumption in HVAC System. International Journal of Integrated Engineering, 11(4).

[16] Kayacan, E., Bayraktaroglu, Z. Y., & Saeys, W. (2012). Modeling and control of a spherical rolling robot: a decoupled dynamics approach. Robotica, 30(4), 671-680.

[17] Kayacan, E., Kayacan, E., Ramon, H., & Saeys, W. (2012). Adaptive neuro-fuzzy control of a spherical rolling robot using sliding-mode-control-theory-based online learning algorithm. IEEE transactions on cybernetics, 43(1), 170-179.

[18] Engelbrecht, A. P. (2007). "Particle Swarm Optimization," in Computational Intelligence: An Introduction, (Second Eds) (pp. 289–358). John Wiley & Sons.

Tunnel-magnetoresistive-based Pulsed Eddy Current Probe for Inspection of Corrosion under Insulation

Mohamoud Hussein Mohamed[1], Ali Sophian[1*], Nadzril Sulaiman[1], Erry Yulian Triblas Adesta[2]

[1] Department of Mechatronics Engineering, Kulliyyah of Engineering, International Islamic University Malaysia, Selangor, Malaysia

[2] Department of Manufacturing and Materials Engineering, Kulliyyah of Engineering, International Islamic University Malaysia, Selangor, Malaysia

*Corresponding author: ali_sophian@iium.edu.my

Abstract. The problem of corrosion in the industrial oil and gas pipes has been one of the major contributors in catastrophic structural failures. Among the various types of corrosion, corrosion under insulation (CUI) has been known to cause serious problems. Pulsed eddy current (PEC) non-destructive testing has shown its effectiveness in detecting hidden CUI. Most PEC systems have been developed by using an inductive coil as their sensing device, while some use a magnetic sensor which potentially offers better resolution. A new probe design based on a solid-state Tunnel magnetoresistance (TMR) is presented. TMR sensors offer higher sensitivities compared to other commercially available sensors. The performance of the probe is evaluated by using ferromagnetic mild steel plates that have thicknesses in the range of 8mm to 12mm at different stand-offs with a thin aluminium sheet under the probe. The different thicknesses represent different corrosion levels, while the stand-offs and thin aluminium sheet are to mimic the insulation of different thicknesses and the cladding in the real pipeline structure. The results show an overall mean absolute error (MAE) of 0.19 mm, which is better than our existing Hall-device-based PEC probe.

1. Introduction

In oil and gas industry, non-destructive testing (NDT) plays a crucial role in maintaining a safe and continual operation of its infrastructure. Metals are used in various infrastructures and utilities including pipelines and they are potentially subject to corrosion, and it has been reported that about 10 percent of the used metal is lost to corrosion [1].

In pipelines, one of the types of the corrosion that is commonly found is the corrosion under insulation, or CUI for short, which is very hard to detect by NDT techniques as it is hidden under the cladding and thick insulation layer. This corrosion mainly takes place due to the presence of moisture between the pipe and the insulation [2]. Going undetected, the presence of such corrosion in pipes can lead to tragic accidents [3], which unfortunately have happened in the past. For detection of CUI, the ultrasonic NDT techniques can be used, however they need a relatively long time due to the requirements for the insulation removal and surface preparation. These requirements also mean higher costs and, additionally, the removal of the insulation increases the chance of new surface contamination that may lead to corrosion [4] [5]. Therefore, an NDT technique that can detect this corrosion without removing the insulation and cladding will bring benefits to the industry.

PEC NDT uses a transient electromagnetic field to detect and evaluate both surface and sub-surface irregularities in various metal structures, including multi-layered ones. A rectangular or square excitation current is supplied to the excitation coil to generate a transient magnetic field that will, in turn, induce transient eddy currents inside the specimen. A secondary magnetic field is then induced by the eddy currents opposing the primary one. The net field, which is a combination of the field generated by the coil and the field induced by the eddy current, will then be detected by a sensing device [6]. The sensing device is generally located inside the excitation coil. Thanks to its low frequency components, with its high penetration capabilities PEC testing is ideal to be used in detecting CUI overcoming the need for the removal of the insulation. Therefore, this particular NDT technique cuts down the inspection time and costs [7].

The sensors that are commonly used in a PEC probe can generally be categorized into two types, which are inductive coils and magnetic sensors. The drawback for using and inductive coil as the sensor device is that it cannot detect DC magnetic field and it is less sensitive to low frequency fields. Its spatial resolution is also lower than that of a magnetic sensor generally. The inductive coil's output signal is proportional to the rate of change of the magnetic flux. In [8], it was shown that as the optimum sensitivity varies depending on the cross-section area of the coil and it is hard to find the actual sensitivity of an induction coil.

Magnetic sensors offer the benefit of having a better spatial resolution compared to inductive coils generally. There are many types of magnetic sensors, and the most common one among them are the Hall effect sensors [9]. These sensors are mostly used in as sensor to detect the speed in a wheel and the crankshaft detect the magnetic field by the Hall effect that is caused by the Lorenz force. Even Though, hall sensors are cheap and easy to integrate it in a circuit it has a high noise to signal ratio that restricts its sensitivity in a certain level [10].

Magnetoresistance (MR) sensor technologies provide major advantages compared to hall sensors. The flat frequency response of MR sensors, which ranges from DC to hundreds of MHz, makes them especially appealing for low-frequency and multi-frequency applications [11]. MR sensors exploit the magnetoresistive effect that was discovered by Lord Kelvin in 1857 but it was way later that the first MR sensor was introduced.

Magnetoresistance sensors are made up of a magnetic tunneling junction (MTJ) element, which mainly consists of three main layers where two of them are made of ferromagnetic material and the third layer that is an insulator is sandwiched between the other two layers. These sensors use quantum mechanical phenomenon as the electrons can make a quantum leap to tunnel through the insulator layer. The insulating layer is so thin that electrons can tunnel through the insulator layer if voltage is applied between the two metal electrodes across the insulator [12].

Among MR sensors the most common devices are anisotropic magnetoresistance (AMR) and giant magnetoresistance (GMR). They are mostly used in industrial position sensors, automotive sensors, cell phone compasses, and solid-state memories [13]. Nevertheless, the new developed magnetic sensors like tunnelling magnetoresistance (TMR) have been emerging in more applications due to its high sensitivity. TMR is the fourth generation of magnetoresistive made of an Fe/MgO/Fe-type magnetic MTJ [14]. Because of this MJT, TMR can achieve more than 200% of magnetoresistance variation at room temperature as compared to AMR and GMR where only 25% and 70% of magnetoresistance relative are achieved, respectively [15].

As described by Cheng [16], the gradient of the decay part of the signal can be used to infer the thickness of the sample when a magnetic sensor, instead of an inductive coil, is used in a PEC system. A similar signal feature is used in this work to determine the wall thickness, and in turn the performance of the new probe can be evaluated. A finite element model for PEC systems has been previously built to support the work, which has been discussed in our previous publication [17].

The most common sensors used in a PEC system are pickup signal coil and the Hall-Effect sensor, however, the accuracy of these sensors is not sufficient in detecting corrosion, the noise in the signals of the Hall-Effect sensor and the coil sensor are not suitable for industrial applications. Recently, Research towards improving sensitivity as well as analyzing and removing unwanted noise at low frequencies is still ongoing. Therefore, the Application of using TMR sensor inside the PEC probe can detect the thicknesses of ferromagnetic metal will achieve higher accuracy as this sensor compared to order sensors showed promising results with a 200% MR ratio, this new probe makes the PEC test more suitable in detecting CUI.

In the next section, the methodology of the research will be discussed. This includes the design of the probe and the experimental setup. Then, by using the probe's signal data obtained from the experiment, the performance of the probe is assessed. Finally, the conclusion is provided at the end of the paper.

2. Methodology

2.1. Probe design

The designing of a circular probe was carried out by using Solidworks CAD software. The excitation coil has inner and outer diameters and height of 7.73 mm, 10.57 mm, and 12.0 mm, respectively. It has 200 turns. The probe's coil and sensor holder were fabricated by using a 3D printer. In the middle of the probe, the TMR sensor is inserted into a special slot made to hold the sensor and its PCB. Figures 1 and 2 shows the 3D drawings and picture of the probe (without the TMR sensor holder), respectively.

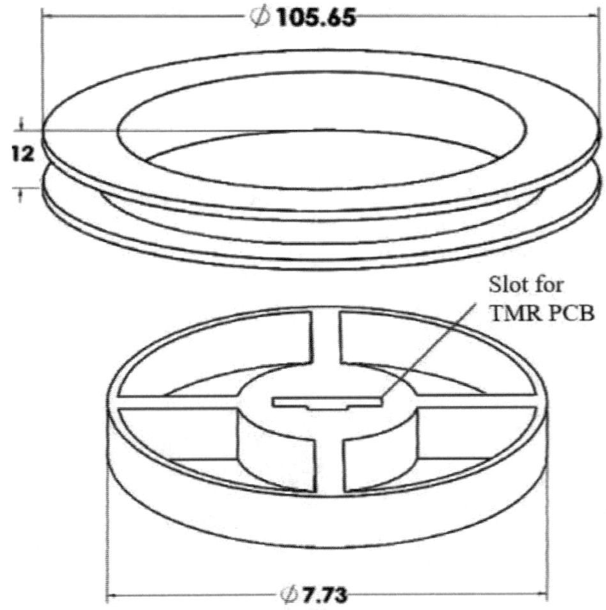

Figure 1. 3D drawing of our TMR-based PEC NDT probe. The top one is the coil former and the bottom one is the holder of the TMR PCB to position the TMR sensor at the centre of the coil. The holder will sit inside the coil former.

Figure 2. The fabricated PEC NDT probe with a slot at the centre where the TMR PCB will be inserted in order to place the sensor as centred as possible.

2.2. Data Acquisition system

A Sony Vaio PCG-41216W portable computer is used to implement the PEC system and to test the performance of the TMR sensor. An NI USB-6218 is used to for data acquisition which captures the TMR sensor's signals and the excitation coil's current at a sampling rate of 100 kS/s.

2.3. TMR SENSOR

TMR 2001 sensor that has a sensitivity of 80mV/V/mT is used in the probe. It comes in a chip size of 3 mm x 3 mm x 1.45 mm. A printed circuit board (PCB) has been designed by using an electronic design automation (EDA) software called Eagle. Figure 3 shows the PCB designed in Eagle. In this project, the PCB is used to mount the TMR sensor. The TMR sensor has a directional sensitivity as shown the figure and it should be vertically placed in the centre of the probe to get a positive reading. A 5V DC supply is used to power up the TMR sensor.

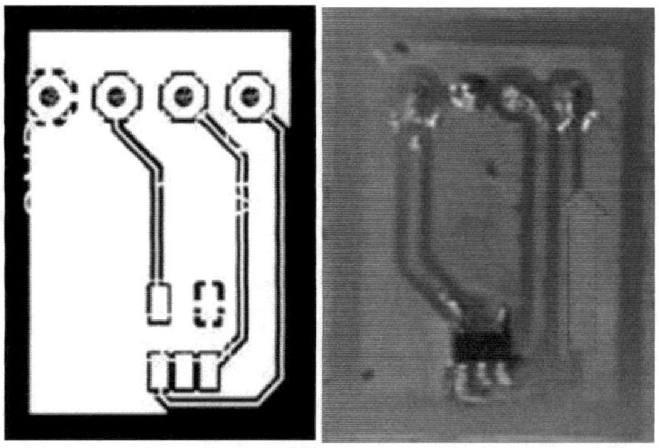

Figure 3. PCB for holding and connecting the TMR sensor was designed in Eagle Software, while the red arrow indicates direction of applied field that generates a positive output; (a) the PCB design and (b) a picture of the PCB

2.4. Experimental set-up of PEC and associated electrical systems

To evaluate the TMR sensor performance an experimental setup is devised. This setup consists of the power source and the excitation circuit, the probe, DAQ device, the specimen. In this experiment the PEC system is used to measure the thickness of ferromagnetic mild steel sheets. Figure 4(a) illustrates the structure of the experimental setup, while figure 4(b) display a picture of the setup. The probe is positioned on the aluminium sheet that represents the aluminium cladding in the real pipe structure. Its thickness is approximately 0.5 mm. Under the aluminium sheet, a Perspex sheet is used of the insulated pipe and how a PEC probe is positioned on the cladding. Perspex sheets have been used to represent the insulator layer in the actual pipe structure. A few Perspex sheets can be used and stacked to create thicknesses of 5 mm, 10 mm, and 15 mm.

Under the Perspex sheet, we have a mild steel sheet. In this experiment, five different mild steel sheets with the thicknesses of 8 mm to 12 mm with an interval of 1 mm have been used to evaluate the performance of this PEC prototype in detecting CUI. The 12-mm thick mild steel sheet is representing the sound pipe wall that has no corrosion. Corrosion will cause the change of the effective thickness and, therefore thinner sheets are used to simulate pipe walls with different levels of corrosion. The worse the corrosion the lower the thickness of the sample.

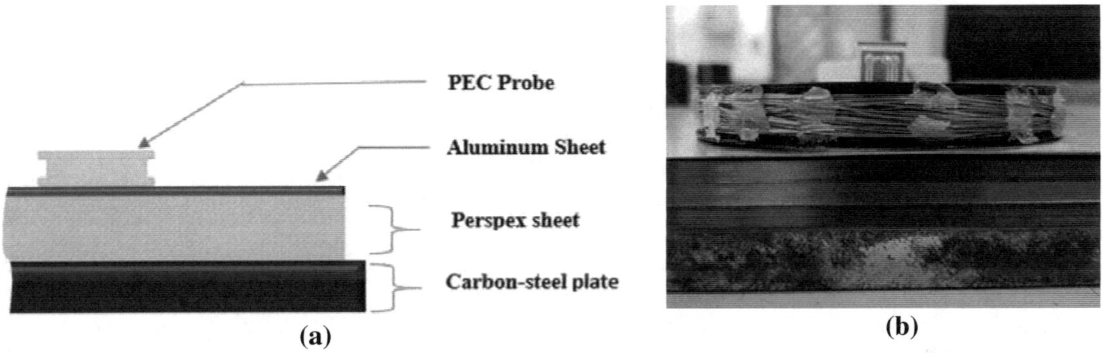

Figure 4. The experimental sample setup: (a) Illustration of the experimental setup. The alumimum sheet simulates the aluminum cladding, the Perspex sheet simulates the insulation, while the carbon-steel plate is representing the wall of the pipe, (b) a picture of the actual setup.

The excitation circuit is used in this project to provide the excitation coil with pulses of excitation current. The external DC power source supplies a continuous 60V DC to the excitation circuit. Capacitors are used to assist the DC power supply to supply the required high current to the excitation coil. To generate the pulsed excitation current, a MOSFET controlled by the USB-6218 is used as a switch. Figures 5(a) and (b) illustrate the system block diagram and the overall experimental setup of the PEC system, respectively.

IOP Conf. Series: Materials Science and Engineering 1244 (2022) 012006 doi:10.1088/1757-899X/1244/1/012006

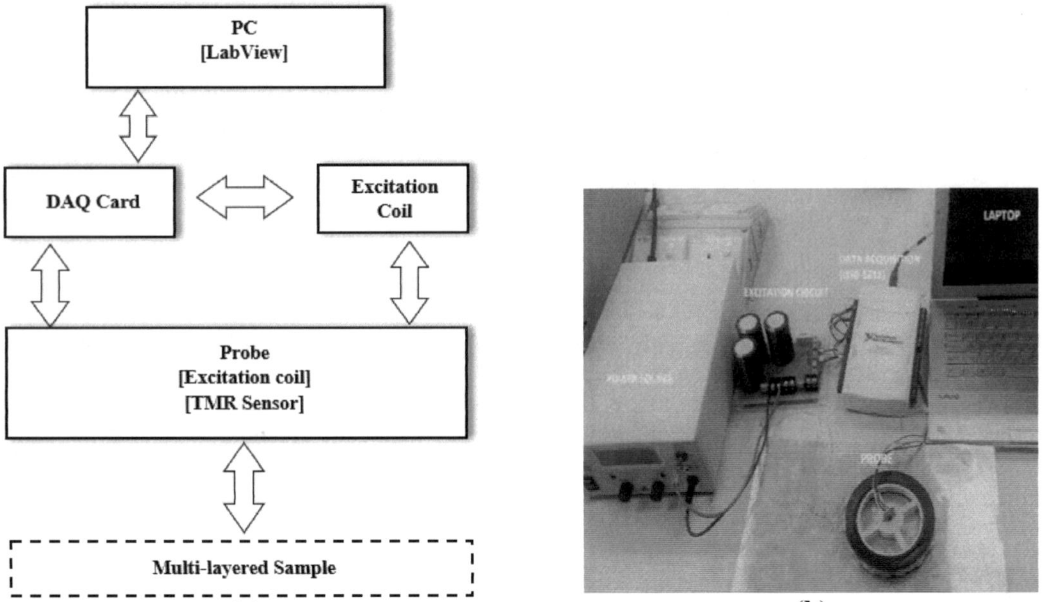

(a) (b)

Figure 5. The experimental setup of the TMR-based PEC NDT system (a) the block diagram, (b) a picture

2.5. Signal processing

The developed LabVIEW code running in the PC controls the data acquisition system to get the sensor's signal data and store them in the hard drive for further off-line analysis. To allow evaluation of the system, 120 readings for each mild steel sheet have been obtained, making a total of 600 readings. The position of the probe was varied randomly between each reading. Later, a signal averaging will be carried out to achieve the prediction of the thicknesses of the mild steel plates. To predict the thicknesses, a feature based on the decay signal, called the decay coefficient, is calculated from each signal. The decay coefficient (λ) of the signal is defined as the reciprocal of the gradient of the signal in the logarithmic scale [17]. This is the signal feature that will be used in this work to infer the sample thickness.

3. Results

Figure 6 shows the comparison of two different thicknesses of ferromagnetic plates, which are 8 mm and the 9 mm, where stand-off was varied at 1, 10 and 15 mm for each ferromagnetic plate. As shown in the graph there are significant differences in the signals between the two thicknesses of the two ferromagnetic plates, while on the other hand, the changes in the lift-off does not have a great effect on the signal of that specific ferromagnetic thickness. The other three mild steel plate thicknesses also show similar patterns. These results underline that this decay property can be used for getting the thickness of the plate relatively consistently despite the variation of the stand-off.

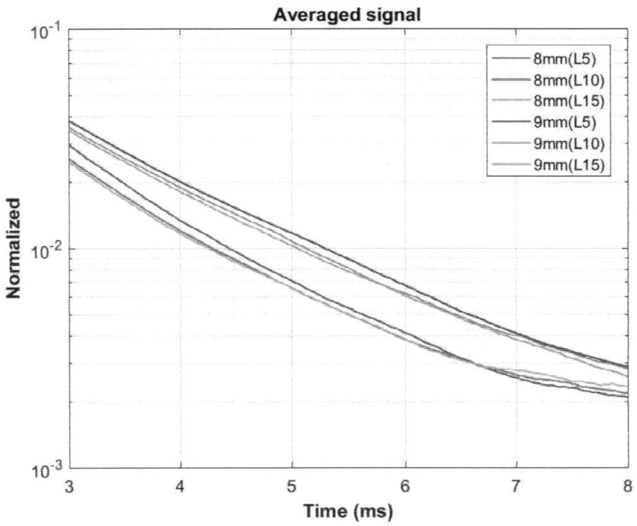

Figure 6. Comparison of the averaged signals from 8-mm and 9-mm-thick plates with stand-offs of 5 mm, 10 mm, and 15 mm.

Figure 7(a) shows the averaged signals in the logarithmic scale, only the falling edge part will be used to compute the thickness of the ferromagnetic sample. As shown in the figure and described in the previous paragraph, there are significant differences between the different thicknesses, on the other hand, the change in the lift-off does not have a great effect on the signal of that specific ferromagnetic thickness. The plots suggest that the gradients in the signal in the time window starting from 2 ms up to 6 ms approximately can be used for good discrimination and determination of the thicknesses. Figure 7(b) shows the zoomed-in of this segment of the signal to show more clearly the differences in the signals and that they demonstrate relatively straight lines.

Figure 8 illustrates the decay coefficients of the averaged signals that were obtained from the sensor against the carbon steel plates thicknesses. Again, the plot shows that the lift-off has negligible effects on the decay coefficient. The relation between the decay coefficient and the thickness can be used to predict the thickness when the decay coefficient of an acquired signal has been obtained.

From the one-to-one mapping of the decay coefficient and thickness of the carbon steel sample, we derived the following quadratic equation:

$$\boldsymbol{T} = 187.192\, \boldsymbol{d}^2 + 128.477\, \boldsymbol{d} + 29.886 \quad (1)$$

where $\boldsymbol{T}$ is the thickness in mm and $\boldsymbol{d}$ is the decay coefficient, which is 1/gradient of the signal's segment.

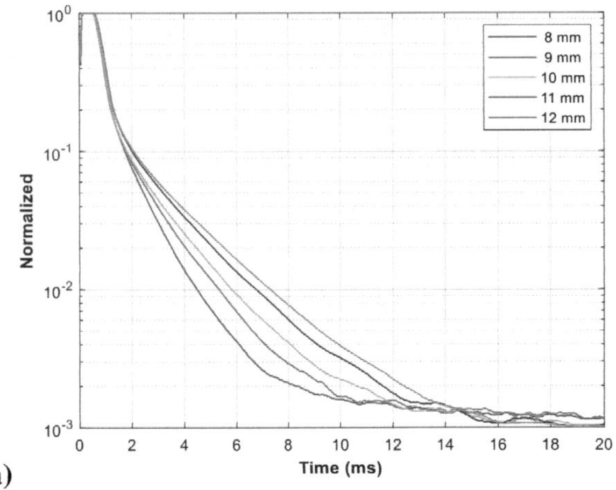

(a)

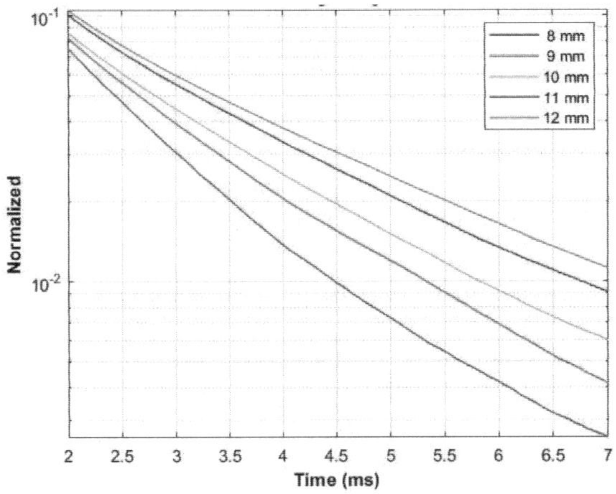

(b)

Figure 7. Average signals of different thicknesses carbon steel plates with different lift-offs: (a) the original signals (b) the segment of the images whose gradients are used for determining the thickness of the sample under test.

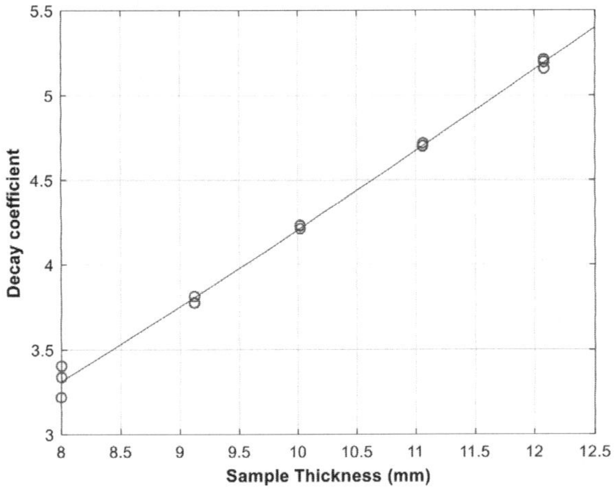

Figure 8. Plot of calculated decay coefficients for different thicknesses and lift-offs vs sample thicknesses

Table 1 and Figure 9 show the performance metrics of the new probe. They show that the probe has been successfully producing signals that can be used to predict the thickness of the sample, which represents a pipe wall with different corrosion levels. The mean absolute error (MAE) and the standard deviation of the errors have been found to be 0.20 mm and 0.25 mm respectively, while the overall MAE is 0.19 mm, which is better than what we have seen before with our Hall-device-based probe that has a MAE of 0.35 mm [17].

Table 1. Performance test results

Actual Thickness (mm)	Mean Predicted Thickness (mm)	Mean Absolute Error (mm)	Standard Deviation of Error (mm)
8.1	7.8	0.26	0.28
9.1	8.4	0.25	0.29
10.0	9.4	0.16	0.19
11.1	10.7	0.14	0.17
12.1	11.6	0.16	0.20
	Overall	0.19	0.25

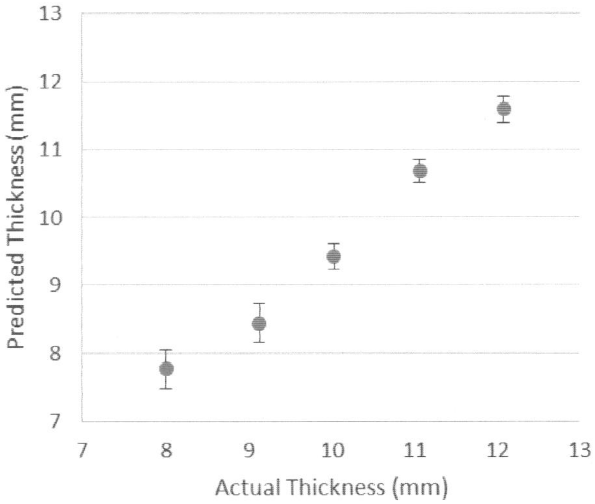

Figure 9. Plot of predicted vs actual thicknesses. The error bar represents the standard deviation of the errors.

4. Conclusion

Corrosion under insulation (CUI) presents a challenging problem for NDT, especially if the test is to be done without removing both the cladding and insulation. Pulsed eddy current (PEC) NDT has been identified the most effective technique todays. In this paper, a new PEC probe design that uses a TMR sensor that has a high sensitivity has been presented. To the authors' knowledge, the use of a TMR sensor in such a PEC NDT system is the first one that has been reported in a research article. The results are very promising, especially when they are compared to those obtained by using Hall devices that we have reported previously. A relatively low mean absolute error (MAE) of 0.20 mm for the TMR- based probe has been found in this study. This is potentially exploited in the future work to implement a PEC NDT system that can be used in the real industrial setting.

References

[1] Fayomi I, Akande I G and Odigie S 2019 *Journal of Physics: Conference Series.* **1378**
[2] Nicola S, Carreto V, Mentzer R A and Sam Mannan M 2013 *CORROSION 2013*
[3] Ayman A, Vincent C, Ali A and Ihsan A 2018 *17th Middle East Corrosion Conference.*
[4] Lukose S 2010 *Inspectioneering Journal* **10**
[5] Towsyfyan H, Biguri A, Boardman R and Blumensath T 2020 *Chinese Journal of Aeronautics* **33**.
[6] Sophian A, Tian G and Fan M 2017 *Chin. J. Mech. Eng.* **30** 500–514.
[7] Lebrun B, Jayet Y and Baboux J-C 1997 *NDT & EInt.* **30** 163-170
[8] Tumanski S 2007 *Measurement Science and Technology* **18**
[9] Li Y, Tian G and Simm A 2018 NDT *E Int.,*vol **41** 477–483.
[10] Herrera-May A, Aguilera-Corte L and Garcı´a-Ramı´rez P 2009 *Sensors (Basel, Switzerland)* **9** 7785-813
[11] Jander A, Carl Smith and Robert Schneider 2005 *Advanced Sensor Technologies for Nondestructive Evaluation and Structural Health Monitoring* **5770**
[12] Tsymbal Y, Mryasov N and LecClair R 2003 *Condensed Matter* **15**.
[13] Deak J G, Zhou Z and Shen W 2017 *AIP Advances* **7** 056676
[14] Chęciński J, Wiśniowski P, Frankowski M and Stobiecki T 2018 *Sensors* **18**
[15] Vidal E G, Munoz D R, Ravelo Arias S I, Moreno J S, Cardoso S, Ferrira R and Freitas P 2017 *Materials* **10** 1134
[16] Cheng W 2012 *J. Nondestruct. Eval.* **31** 215–224
[17] Sophian A, Nafiah F, Gunawan T S, Mohd Yusof N and Al- Kelabi A 2021 *IIUM Engineering Journal* **22** 226-233

Dynamic Fingerprint Benchmarking Model for WiFi Indoor Localization

Yazan Aljeroudi[1] , Erwin Sulaeman[1,*] and Ari Legowo[2]

[1]Department of Mechanical Engineering, International Islamic University Malaysia
[2]Higher Colleges of Technology, United Arab Emirates

*Corresponding author : esulaeman@iium.edu.my

Abstract. WiFi indoor localization is regarded as the most promising reliable technology for indoor localization. Its concept is based on building fingerprint and training a ma-chine learning model on it, then using it for predicting the location based on the measured received signal strength RSS. One aspect of its vulnerability is its time vari-ant change due to the dynamical conditions in the environment and the access points APs. This was the most focus of the researchers in the last decades. However, minori-ty of them have considered the limitations of the existing benchmarking fingerprint in terms of lacking of support to realistic navigations runs when they are used for eval-uating indoor localization models. This article proposes a simple yet novel add-on to existing benchmarking fingerprint that assists in converting it from static fingerprint to consecutive training data with supportability to incremental learning and awareness of various characteristics of indoor navigation runs. The model was evaluated based on two benchmarking fingerprint and the generated incremental data were plotted with two levels of granularity one is corresponding to location and the other to place for referring to high resolution level and low resolution level.

1. Introduction

The tremendous spread of smartphones, portable tablets, and other smart wireless devices in the recent years has enabled the incorporation of various intelligent services and virtual personal assistant [1]. Most intelligent services require the availability of numerous information from users. A part of its intelligent aspect, the service needs to obtain information implicitly with minimum interaction with the user using some models such as human activity recognition [2]. One of the key information for any intelligent service to operate is location of the user or the device [3],[4]. Location is subject to change with time, and many applications or services require precise identification of location [4]. Location estimation or localization has been a research topic from decades [5]. Estimating the location of the user or the device that is carried by the user requires exploiting set of sensors that are carried by the user or embedded in the device. Typically, sensors that are embedded in daily usage device are low cost and subject to errors. This motivates the work on developing intelligent algorithms that can handle the errors and provide relatively accurate estimation of the location exploiting the processor that is deployed in any device and exploiting the remarkable development of intelligent and high-performance algorithms in general and artificial intelligence in particular.

Localization has been categorized under two categories: outdoor [6] and indoor [7, 8]. Outdoor is the less challenging category in general due to the availability of Global Positioning System (GPS) which is considered as the most common localization sensor. GPS has been considered as a main solution for majority of outdoor localization systems [9]. This is because of the wide range of availability, the low cost of accessibility, and the adequate accuracy which is goes to as low as 10 cm for different variant of GPS or DPGS [10]. However, the most challenging category is indoor localization. This is due to the non- accessibility of GPS in indoor environment, and the non-availability of fully equivalent technology in indoor environment. This has motivated researchers to put significant amount of efforts on the problem of indoor localization. Another motivation is the emerging of need for deploying many intelligent services in indoor environment that cannot operate unless a localization algorithm is capable of providing a relatively accurate estimation in an online mode.

WiFi technology is installed in almost any indoor environment nowadays. This has motivated researchers to develop WiFi based indoor positioning solutions. WiFi fingerprinting based localization is considered as one of the most promising approaches for indoor localization [11]. WiFi fingerprinting-based localization is composed of two phases: offline training and online operation. In the offline phase, the fingerprint of the WiFi is measured from set of access points APs distributed in the environment and it is stored with its corresponding locations in a dataset called fingerprint. On the other hand, in the online phase we expect some machine learning algorithm trained in the offline phase to work and to provide the expected location as the location of the best matching record from the WiFi fingerprint. Many researchers have developed various mathematical models and approaches for WiFi fingerprint based indoor localization; however, there is an issue of conducting objective comparative evaluation to them due to the lack of dynamic fingerprint benchmarking for indoor localization. This article aims at providing a novel and simple model that converts an existing benchmarking fingerprint to a sequential dynamic navigation scenario that can be used to evaluate and compare more than one WiFi localization model in an objective way. The remaining of the article is organised as follows. In section 1, we present the literature survey. Section 3 provides the literature survey. The experimental results and analysis are provided in section 4. The conclusion and future work are provided in section 5.

2. Literature Survey

The literature has considered dealing with the vulnerability of the fingerprint to the dynamic conditions of the environment using various models. In the work of [12], the authors have considered fusing multiple fingerprint instead of single one. In order to accomplish it, they developed two-layer fusion profile joint optimization with multiple constraints. Their optimization model is based on fusion profile selection that determines the fusion weights for improving the accuracy of the localization. However, this model concentrates on the dynamic conditions of the fingerprint without taking into consideration the dynamic aspect of the navigation run and its impact on the performance of the model. Other researchers have considered developing dynamic consecutive fingerprint. In the work of [13], the fingerprint was collected by using several consecutive fingerprint while standing at certain position and using certain direction. Collecting the dataset over long period of time (more than 15 months) was regarded as way of countering long term signal change. Similarly, this approach does not enable researchers to evaluate their model on the dynamic of the navigation run which is essential while evaluating models. The focus of countering the dynamic changes of fingerprint was the interest of many other researchers. In the work of [14], a gradient- based fingerprint map (Gmap) was built by comparing absolute RSSI values at nearby positions, and then runs an online extended particle filter (EPF) to localize the user/device. This falls in the same category of aiming at developing an adaptive WiFi localization model to time invariant fingerprints. However, non-of the previous models has considered developing dynamic aware fingerprint for facilitating objective evaluation of developed WiFi models for indoor localization. One of the interesting works that has considered the dynamic of the navigation run as a part of the fingerprint is the work of [15] which has proposed simulation model for generating time series data to support actual conducted navigation scenarios. Their simulator uses marker point provided as a descriptor to the navigation run with adding the velocities and the pauses as elements of describing the fingerprint. This model is one of the novel model that gives an approach of adding dynamic awareness of fingerprint benchmarking tools and datasets. However, this model lacks the incorporation of the essential elements that describes indoor navigation run. The goal of this article is to

IOP Conf. Series: Materials Science and Engineering 1244 (2022) 012007 doi:10.1088/1757-899X/1244/1/012007

propose simple yet novel benchmarking add-on to the existing benchmarking tool to make them capable of generating dynamic and sequential runs for evaluating indoor localization models and approaches.

3. Methodology

Our goal is to develop an algorithm that simulates navigation behavior in indoor navigation environment and use it for converting fingerprint to sequential navigation data, we call it Fingerprint Based Simulator FBS.

3.1 Definition

a. **Location.** It represents the finest resolution of predict visited area by the localization algorithm and it is a part of a place.

b. **Place.** It represents a node in the graph of the topological map that describes the environment

c. **Mobility Vector.** $\boldsymbol{MI} = [\boldsymbol{i_1}\ \boldsymbol{i_2} .. \boldsymbol{i_k} \dots \boldsymbol{i_{last}}]$ denotes the mobility vector $\boldsymbol{i_k} \in \{\mathbf{1, 2} \dots \boldsymbol{l}\}$ where $\boldsymbol{l}$ denotes the number of locations with the constraint of $\boldsymbol{i_k} \in \boldsymbol{Connected}(\boldsymbol{i_{k-1}})$ for any $\boldsymbol{k} > \mathbf{1}$ and $\boldsymbol{i_k} \in \boldsymbol{Connected}(\boldsymbol{i_{k+1}})$ for any $\boldsymbol{k} < \boldsymbol{last}$.

1: Neighbor. $\boldsymbol{Neighbor}(\boldsymbol{i_k})$ denotes the region that is adjacent to $\boldsymbol{i_k}$, however, it does not contain a passage toward $\boldsymbol{i_k}$

2: Connected. $\boldsymbol{Connected}(\boldsymbol{i_k})$ denotes the region that has passage to $\boldsymbol{i_k}$, $\boldsymbol{Connected}(\boldsymbol{i_k}) \subseteq \boldsymbol{Neighbor}(\boldsymbol{i_k})$

3: Before and After. $\boldsymbol{before}(\boldsymbol{i_{k+1}})$ is the region that is visited before $\boldsymbol{i_k}$

4: AP ( ). we define $\boldsymbol{AP}(\boldsymbol{i_k})$ as the access points set that can be reached from inside the region $\boldsymbol{i_k}$. Based on 1

$$AP(i_k) \cap AP\big(Neighbor(i_k)\big) \neq \emptyset \quad (1)$$

And for any MI and i_k in (2)

$$AP(i_k) \cap AP(i_{k+1}) \neq \emptyset \quad (3)$$

3.1.1. Example1. Let us assume that indoor map represented in Figure 1 Map of an indoor environment.

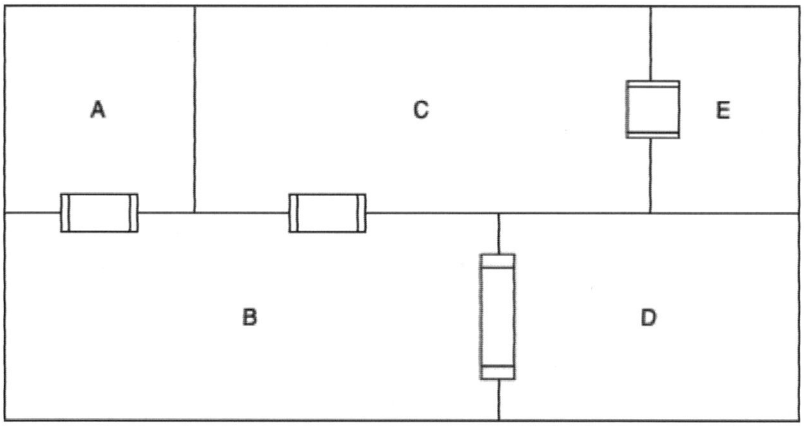

Figure 1. Map of an indoor environment

We can represent the relations between the region with the state diagram shown in Figure 2 the relations between the regions.

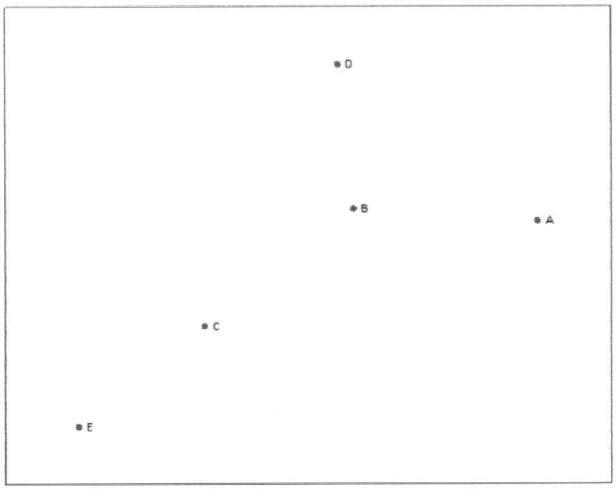

Figure 2. the relations between the places

$Neighbor(A) = \{B, C\}$
$Connected(A) = B$
$Neighbor(D) = \{B, C, E\}$
$Connected(D) = \{B\}$
$MI = (D, B, E, B, D)$ is a not valid mobility vector because $E = before(B)$.
$E \notin Connected(B)$
$MI = (D, B, A, B, C, E)$ is valid mobility vector
The adjacency matrix is of this example is provided in Table 1

Table 1. Corresponding adjacency matrix.

	A	B	C	D	E
A	0	1	0	0	0
B	1	0	1	1	0
C	0	1	0	0	1
D	0	1	0	0	0
E	0	0	1	0	0

3.2. Problem Formulation

Assuming WiFi fingerprint dataset $D(X, Y)$ where $X \in R^{n \times m}$ and $Y \in R^{n \times 1}$ where n denotes the number of records and m denotes the number of AP. The dataset has been captured in an environment that is described by a grid map M that can be converted to topological map described by the graph $G = (V, E)$ where $V = \{P_1, P_2, \dots P_l\}$ and $E = \{e_1, e_2, \dots e_c\}$ where P_i denotes the place i and e_j denotes the connection j. Also, we assume that LAP shows the distribution of set of access points APs using x-y coordinates. Provided that an object carrying WiFi sensor has walked in the environment using trajectory

IOP Conf. Series: Materials Science and Engineering 1244 (2022) 012007 doi:10.1088/1757-899X/1244/1/012007

that has moved between places P_i based on mobility vector MI. The goal is to use MI, G, and LAP in order to generate a scenario with dynamic awareness SD_t.

3.2.1. Algorithm FBS.

The algorithm of fingerprint-based simulator has the role of providing a dynamic sequential data that reflects a realistic indoor navigation behavior instead of the static fingerprint. The process is decomposed into two parts: the first one is generating the mobility index based on the ICG result and it is presented in Table 2 and the second one is generating the sequential fingerprint data and it is presented in Table 3.

Table 2. Pseudocode of generating mobility index using ICG information

```
Input
Label //label of places
revisitingFactor
M    //confusion matrix of ICG
Output
MI   //mobility index
Start
1.Places=[];
2.initialPlace=randi([1 max(Label)]);
3.currentPlace=initialPlace;
4.Places=[Places currentPlace];
5.for i=1:NumberOftransitions
6.  M(currentPlace,currentPlace)=0;
7.  NCIndex=find(M(currentPlace,:)==1);
8.  NIndex=randi([1 length(NCIndex)]);
9.  NextPlace=NCIndex(NIndex);
10.  r=rand;
11.  if(length(Places)>2 && r<revisitingFactor)
12.     NextPlace=Places(end-1);
13.  end
14.  currentPlace=NextPlace;
15.  Places=[Places currentPlace];
16. end
17.  MI = Places
End
```

Table 3. Pseudocode of generating the sequential fingerprint which is the result of FBS

```
Input
MI //mobility index
Label //label of locations
numberOfLocations
Output
SeqData
Start
1.numberOfPlaces=length(MI);
2.for i=1:numberOfPlaces
3.  IndexSet=find(label== MI (i));
4.   for j=1:numberOfLocations
5.     index=randi([1 length(IndexSet)]);
```

```
6.    SeqData= [SeqData ;Data(index,2:end) Label(index) MI (i)];
7.   end
  8. end
End
```

4. Experimental Result and Analysis

4.1. Datasets Presentation, Labeling and ICG

This section presents the quantitative details of the datasets that are used for evaluation. In addition, it presents the application of the ICG on the datasets. We present two datasets: TamperU in subsection 4.1.1 and UJIIndoorLoc in subsection 4.1.2.

4.1.1. TamperU

The dataset covers two buildings in the Tampere University of Technology with 4 and 3 floors respectively. The dataset contains two files for the first building TAMPERE1_training.csv and TAMPERE1_test.csv with 1478 training/reference records and 312 attributes contain the WiFi fingerprint (309 WAPs) and the coordinates (longitude, latitude and height). In the second building there are 583 training/reference records (TAMPERE2_training.csv file) and 175 test ones (TAMPERE2_test.csv).

In order to indicate to the intensity value, negative integer values ranging from -100 dBm to 0 dbM is used. The value -100 dBM indicates to poor signal while the value of 0 dbM indicates to strong signal. For non-active or non-detected AP, a value of 100 was used. In order to summarize, we present the information in Table 4 and a visualization of the fingerprint locations of the building 1 in Figure 3. The classes of the data can be visualized as Figure 4 after applying the labeling algorithm where each point represents a distinct class. Also, we see the results of the ICG from level 15 to 5 in Figure 5. The topological map is provided using level 5 of labeling and it is provided in Figure 6.

Table 4. Summary of the description of TamperU

TamperU	Building 1	Building 2
Number of records	1478	583
Number of attributes	312	357
Number of floors	4	3

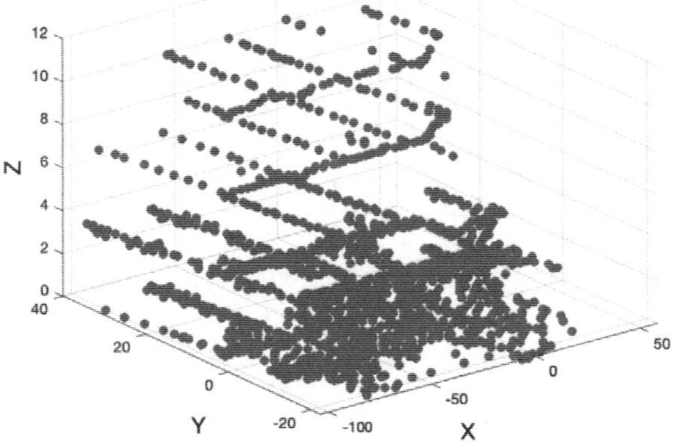

Figure 3. Places of dataset

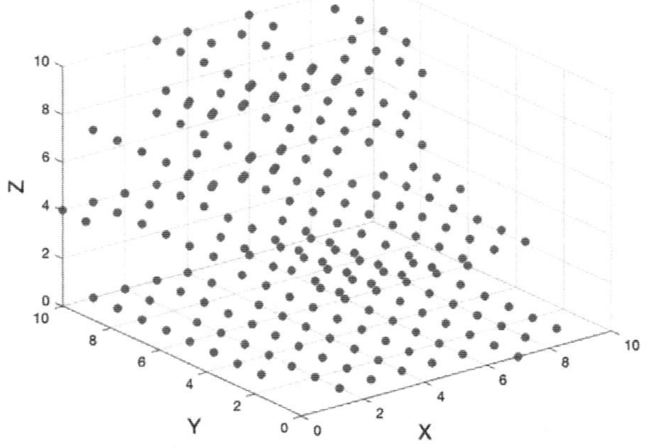

Figure 4. Results of class labeling to TamperU with number of levels =5

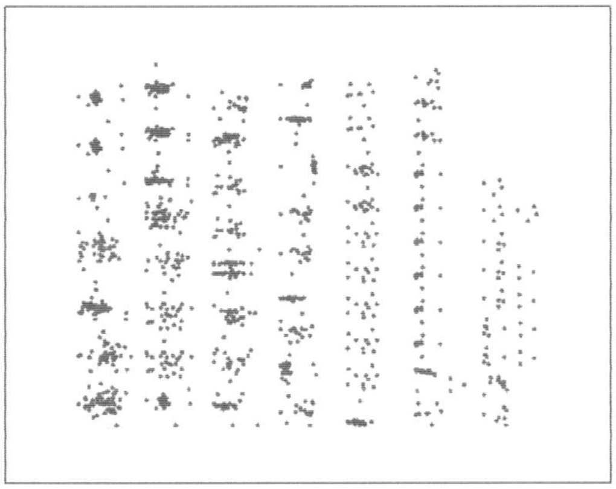

Figure 5. ICG results

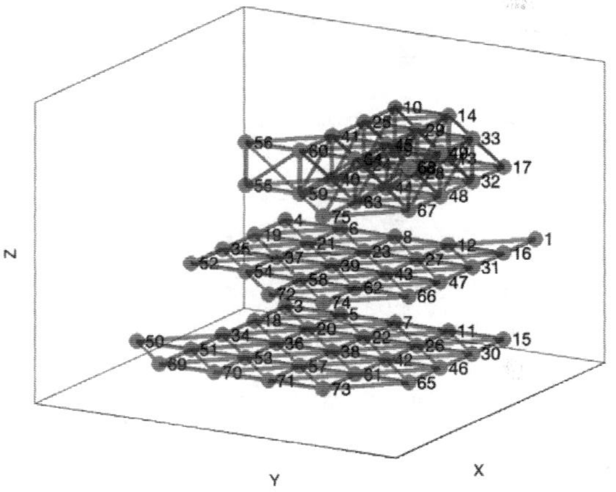

Figure 6. Topological map

4.1.2. UJIInoorLoc.

While TamperU covers only two buildings in Tamper University of Technology, UJIInoorLoc covers three buildings of Universitat Jaume with four flours or more and an area of 110.000m2. The database is composed of 19937 training/reference records (trainingData.csv file) and 1111 validation/test records (validationData.csv file). The number of attributes is 529 and they contain the RSS values of the APs and the coordination of the places. We can visualize the classes as it is given in shown in Figure 7 where each point represents a distinct class and the ICG results in Figure 8. The constraints results are provided in Figure 9. In addition, we present part of the topological map of the dataset in Figure 10.

Table 5. Summary of the description of UJIIndoorLoc

UJIIndoorLoc	Training	Testing
Number of records	19937	1111
Number of attributes	529	529
Number of floors	4	4

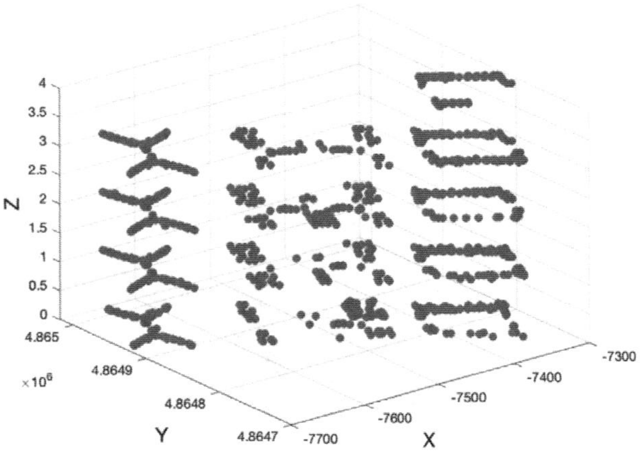

Figure 7. Places of dataset

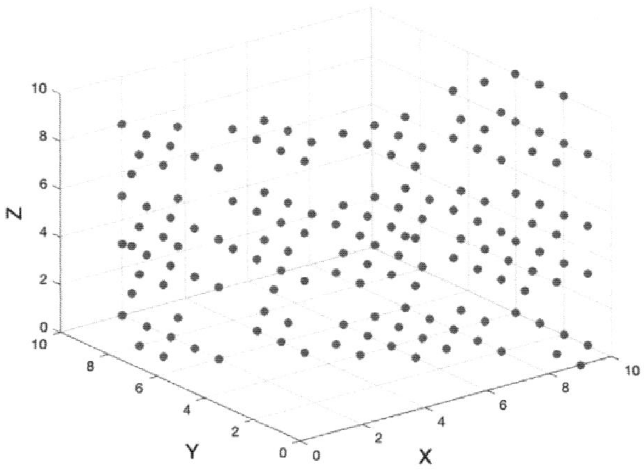

Figure 8. Results of class labeling to UJIIndoorLoc with number of levels =5

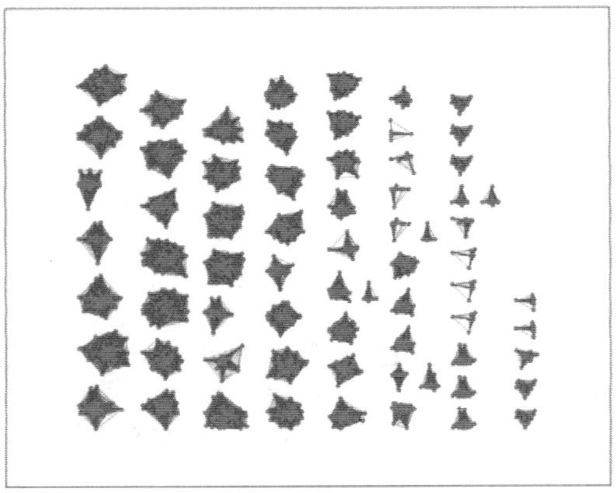

Figure 9. **ICG results**

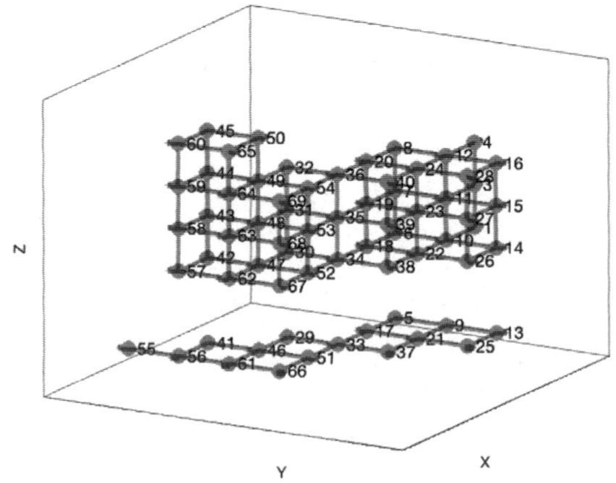

Figure 10. **Topological map**

4.1.3. Resulted Number of Classes for Two Datasets.

This section provides the resulted number of classes after applying our labeling to the two datasets based on the coordination values of the ground truth. We used three levels of granularities: 5, 10 and 15. As we observe in Table 6 the number of the classes increase as the level of the granularity increases. In addition, TamperU includes a greater number of classes comparing with UJI. This is interpreted by larger areas that are covered by TamperU than UJIIndoorLoc.

Table 6. Number of classes for different levels of granularities

Number of Levels	TamperU	UJI
5	75	69
10	227	155
15	355	253

4.2 FBS results

This section will provide set of evaluation scenarios provided from FBS simulator. For evaluation, we generate 8 random scenarios. Each will include different number of transitions. The scenario might include revisiting earlier visited places which will be targeting the knowledge preservation aspect in the classifiers.

Table 7. Description of the generated dynamic navigation runs from the fingerprint

Scenario Number	Number of Transitions	Revisiting Factor	Staying Factor
1	3	0.9	0.5
2	5	0.9	0.5
3	10	0.9	0.5

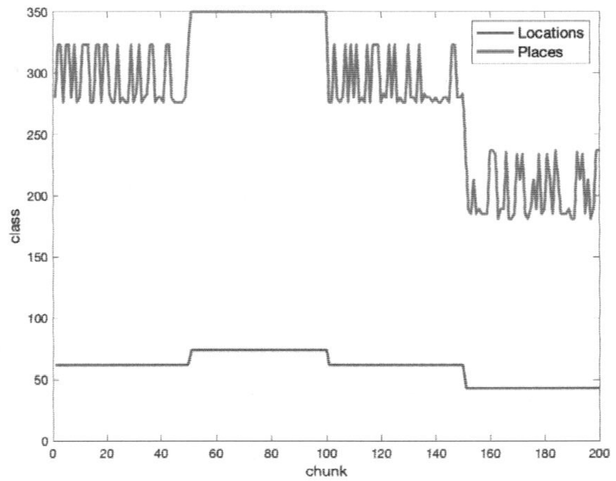

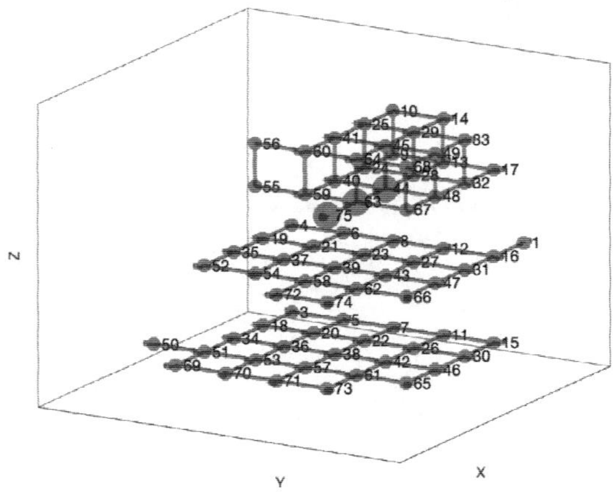

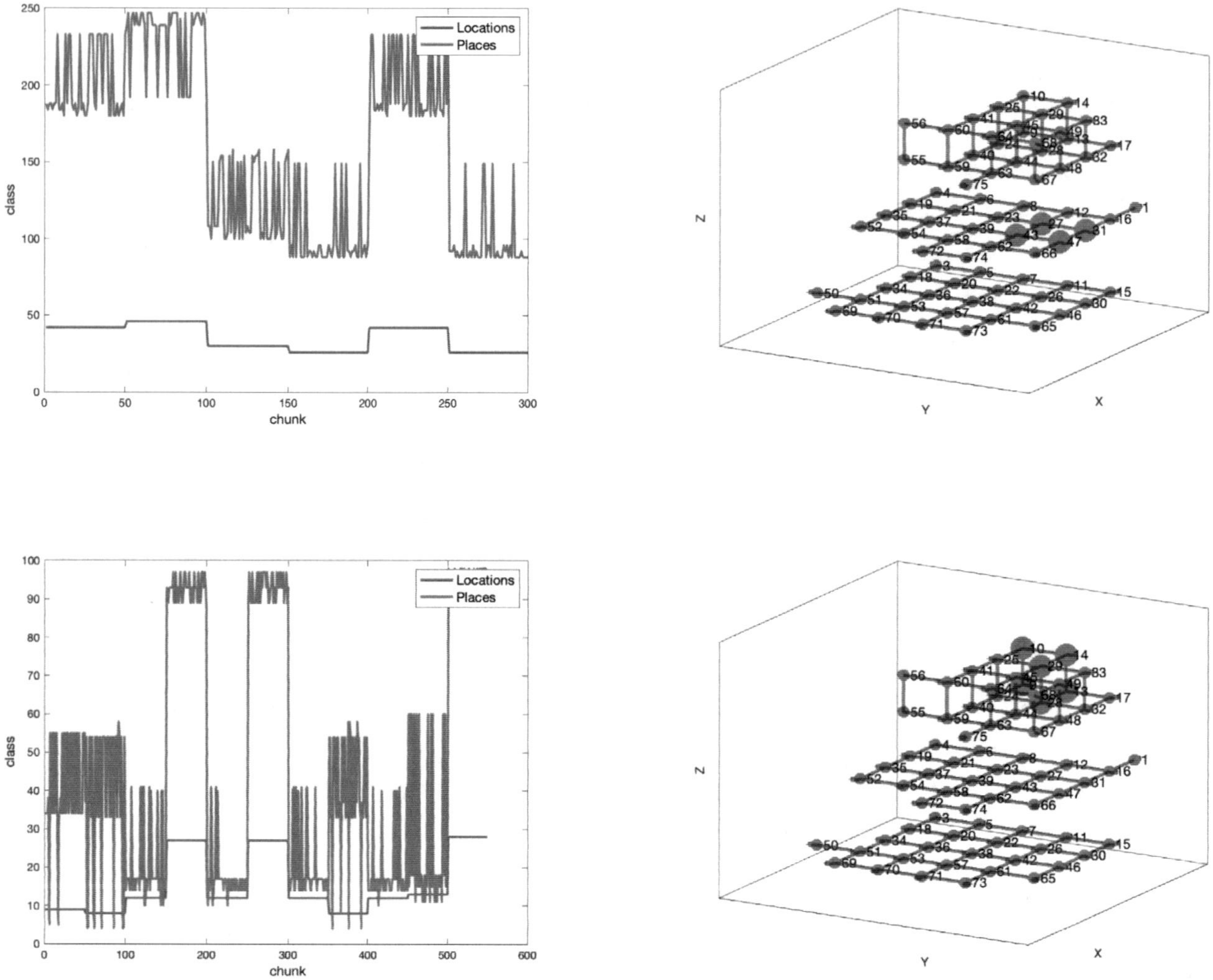

Figure 11 The generated dynamic sequential fingerprint from the provided navigation runs in Table 7 using TamperU dataset

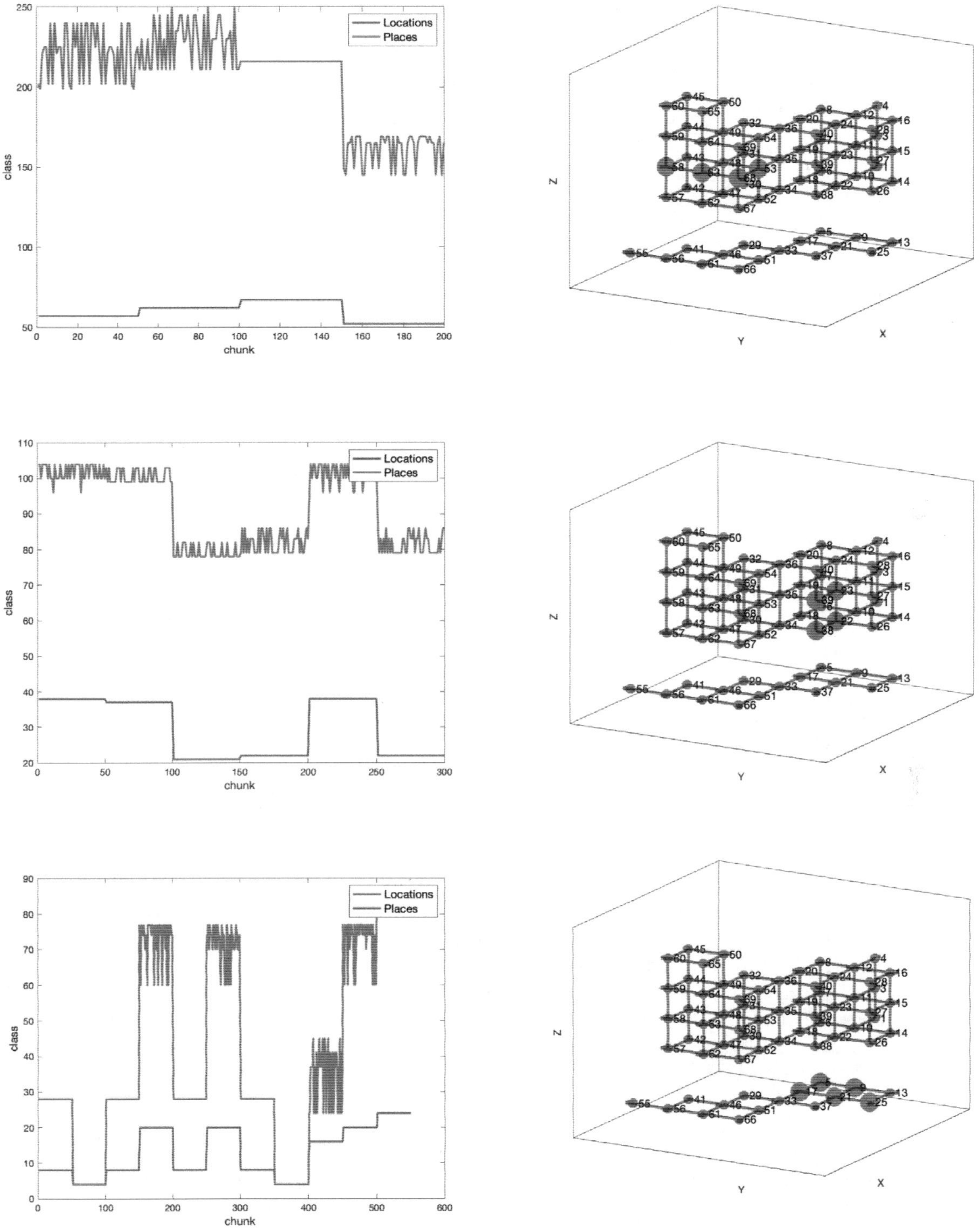

Figure 12. The generated dynamic sequential fingerprint from the provided navigation runs in Table 7 using UJI dataset

5 Conclusion and Future Work

This article has presented a novel model for converting an existing benchmarking fingerprint to dynamical or incremental data for evaluating navigation run. The model can be regarded an add-on to existing benchmarking fingerprint for WiFi localization in the literature to enable objectively evaluating the developed fingerprint based WiFi localization for indoor environment. The objective aspect is enabled by allowing the model to be tested on sequential data generated from the fingerprint and corresponding to the given indoor navigation scenario. This model provides ability to input three parameters: Number of Transitions, Revisiting Factor and Staying Factor. In addition, the model supports more than one level of granularity for labelling. We consider the high resolution level as location and the low resolution level as place. The model has been tested using two benchmarking fingerprint; namely, Tamper and UJI and its results were generated. Future work is to test the developed model on various existing fingerprint based indoor localization approaches and to make an objective quantitative comparison between them. Another future work is to develop finger print based indoor localization that leverages.

References

[1] H. Yang, and H. Lee, "Understanding user behavior of virtual personal assistant devices," *Information Systems e-Business Management*, vol. **17**, no. 1, pp. 65-87, 2019.

[2] J.-L. Reyes-Ortiz, L. Oneto, A. Samà, X. Parra, and D. Anguita, "Transition-aware human activity recognition using smartphones," *Neurocomputing*, vol. **171**, pp. 754-767, 2016.

[3] J.-H. Huh, and T.-J. Kim, "A location-based mobile health care facility search system for senior citizens," *The Journal of Supercomputing*, vol. **75**, no. 4, pp. 1831-1848, 2019.

[4] A. Abdullah, N. Rahman, N. Saad, N. Hashim, and A. Kamaruddin, "Smart Localization and Detection System for School Children," *Journal of Telecommunication, Electronic Computer Engineering*, vol. **10**, no. 1-8, pp. 55-59, 2018.

[5] J. Horn, and G. Schmidt, "Continuous localization of a mobile robot based on 3D-laser-range-data, predicted sensor images, and dead-reckoning," *Robotics Autonomous Systems*, vol. **14**, no. 2-3, pp. 99-118, 1995.

[6] Z. Li, K. Xu, H. Wang, Y. Zhao, X. Wang, and M. Shen, "Machine-learning-based positioning: A survey and future directions," *IEEE Network*, vol. **33**, no. 3, pp. 96-101, 2019.

[7] F. Zafari, A. Gkelias, and K. K. Leung, "A survey of indoor localization systems and technologies," *IEEE Communications Surveys Tutorials*, vol. **21**, no. 3, pp. 2568-2599, 2019.

[8] H. Jiang, C. Cai, X. Ma, Y. Yang, and J. Liu, "Smart home based on WiFi sensing: A survey," *IEEE Access*, vol. **6,** pp. 13317-13325, 2018.

[9] L. N. Balico, A. A. Loureiro, E. F. Nakamura, R. S. Barreto, R. W. Pazzi, and H. A. Oliveira, "Localization prediction in vehicular ad hoc networks," *IEEE Communications Surveys Tutorials*, vol. **20**, no. 4, pp. 2784-2803, 2018.

[10] W. K. Chew, and M. A. Zakaria, "Rover Car Outdoor Localization for Navigation Tracking Using Differential Global Positioning System Estimation." pp. 535-556.

[11] H. Li, L. Sun, H. Zhu, X. Lu, and X. Cheng, "Achieving privacy preservation in WiFi fingerprint-based localization." pp. 2337-2345.

[12] X. Guo, N. R. Elikplim, N. Ansari, L. Li, and L. Wang, "Robust WiFi localization by fusing derivative fingerprints of RSS and multiple classifiers," *IEEE Transactions on Industrial Informatics*, vol. **16,** no. 5, pp. 3177-3186, 2019.

[13] G. M. Mendoza-Silva, P. Richter, J. Torres-Sospedra, E. S. Lohan, and J. Huerta, "Long-term WiFi fingerprinting dataset for research on robust indoor positioning," Data, vol. **3**, no. 1, pp. 3, 2018.

[14] Y. Shu, Y. Huang, J. Zhang, P. Coué, P. Cheng, J. Chen, and K. G. Shin, "Gradient-based fingerprinting for indoor localization and tracking," *IEEE Transactions on Industrial Electronics*, vol. **63**, no. 4, pp. 2424-2433, 2015.

[15] A. S. Al-Khaleefa, M. R. Ahmad, A. A. M. Isa, M. R. M. Esa, A. Al-Saffar, and Y. Aljeroudi, "Infinite-Term Memory Classifier for Wi-Fi Localization Based on Dynamic Wi-Fi Simulator," *IEEE Access*, vol. **6**, pp. 54769-54785, 2018.

Simulation of wear map for biodiesel interacted automotive materials

S Y Cetin[1]*, M A Maleque[2]* and H H Masjuki[1],

[1]Department of Mechanical Engineering, International Islamic University Malaysia, PO Box 10, 50728 Kuala Lumpur, Malaysia
[2]Department of Manufacturing and Material Engineering, International Islamic University Malaysia, PO Box 10, 50728 Kuala Lumpur, Malaysia

*E-mail: maleque@iium.edu.my

Abstract. Selection of material for automotive component has an important role to avoid wear problem that encountered in static and moving parts due to fuels used in the engine especially, biodiesel. Biodiesel compositions such as dissolved water, free water and emulsified water accelerate oxidation of metallic materials. During the combustion process, the biodiesel contributes to the corrosive wear of the engine components leading to the shortening of the life span. The early prediction of the wear behavior under severe conditions in presence of biodiesel would be a vital solution. Therefore, the main aim of this study is to propose a wear map simulation for examining wear phenomenon of automotive component materials in presence of biodiesel. Herewith, materials that come into contact with biofuel/biodiesel in automotive application are highlighted for better design and selection in the future. This study concluded that based on the wear map simulation proper selection of material with higher durability and wear resistant can offer a better future for economic and reliability of the components. The development of wear map with recommendation aids researchers making decisions and innovation on the metals engine components. Furthermore, a full explanation on the reaction between biodiesel and metals can increase the innovation in the automotive industry.

Keywords: wear, biodiesel, automotive materials, wear model

1. Introduction

Deformation and removal of the surfaces caused by mechanical action is called as a wear phenomenon. Wear has a direct impact to the lifespan and suitability of the material. The level of the deformation on the surface varies with the change of the material which is used. Due to these different behaviors, the selection of the material which is resistant to surface deformation, especially wear, is an important subject to considering. There is a synergistic effect between wear and corrosion processes. Although this effect expedites material loss in general, in some cases it can retard material loss. Some of wear types are;
-Adhesive wear
-Corrosive wear
-Abrasive wear
-Errosion wear

-Cavitation wear etc.

Surface coating of the materials is one of the effective method to prevent from the wear and corrosion phenomenon. Coating of the surface can increase the resistance of the material against the surface deformations. Some general methods which can easily increase the resistance of the materials are:

-Hard anodizing
-Fluoropolymer coatings
-Dry film lubricants
-Ceramic coating
-Thermal barrier coating
-Nitriding
-Plasma Electrolytic Oxidation
-Electroless Nickel

Automobiles are mainly made up of metal parts and due to high metal usage, several conditions are taken into account for selecting the metals to be used. Especially, the deformation of the parts in which the fuel comes into contact has been subject to many studies. Automakers continue to innovate on surface-resistant materials, both in the use of gasoline and diesel fuel. However, the environmental damage of traditional petroleum products has become well-known in last two decades. In the meantime, the manufacturers began to focus on clean energy. At this point, biodiesel, which can be produced from clean and renewable sources, has been able to draw attention to both manufacturers and consumers. This clean fuel is a powerful alternative to traditional petroleum fuels, with its cheap production resources opportunities and being environmentally friendly. However, with many advantages, some disadvantages of biodiesel have become difficult to use in everyday life. The most significant one is the high corrosive effect on metal parts. Biodiesel has higher hygroscopicity, higher electrical conductivity, higher oxygen content than diesel [1]. Also, some of the properties of biodiesel like, free water, dissolved water and emulsified water speed up oxidation of the metallic materials. Due to its nature, biodiesel has corrosive and wear effect on metals. Studies show that biodiesel reduces wear for short-term use. However, it has been observed to have the same effect as diesel for long-term usage [2]. Therefore, improvements to both biodiesel and metals will increase sustainability in long-term use for the automobile industry.

2. Effect of biodiesel on automotive materials

Biodiesel fuels absorb more water than conventional petroleum diesels. High water content of fuel causes corrosion and choking. This condition is caused by biodiesel and the electronegativity of oxygen due to the presence of oxygen. In brief, dissolved oxygen in biodiesel can cause corrosion [3,1]. In addition to this, the usability of biodiesel may vary depending on many parameters it includes. Undesirable substances in biodiesel negatively affect engine performance by reducing the combustion heat of the fuel, which leads to increased corrosion on metal parts [4]. There beside, along with the increase in viscosity, the combustion process in the engine and the power production of the engine are negatively affected. This causes weak atomization of the fuel. Moreover, incomplete combustion in the engine with an increase in smoke and particulate matter emission makes the fuel unsuitable for vehicle use. That's why, if the addition of antioxidant is not used in biodiesel, over long periods, the oxidation behavior of biodiesel will greater than that of diesels [5].

Friction and wear between component materials in engine applications such as, piston and piston pins, cylinder liner, crankshaft joints, valve system, etc. occurs during sliding of the mating surfaces. Oxidation of the moving motor component reacts with biodiesel, which facilitates wear and friction [6]. When the behavior of metals used in automobiles is examined with biodiesel, it is observed that stainless steel is more resistant than copper and aluminum [7]. Fazal et al. [8] investigated copper, brass, aluminum and cast iron as automotive materials under biodiesel exposure. They showed that, wear and friction increase with the temperature increases. Moreover, copper and copper based alloys that exhibit severe degradation when exposed to biodiesel are not tribologically compatible. D. Jin et al. [9] examined ASTM 1045 mild steel in palm biodiesel and they concluded that small pits have been observed to form randomly, and the size of these pits are larger than that of diesel. Herewith, studies show that the selection of the material in diesel engine is crucial. Figure 1. shows part of biodiesel/diesel engine which in contact with biodiesel.

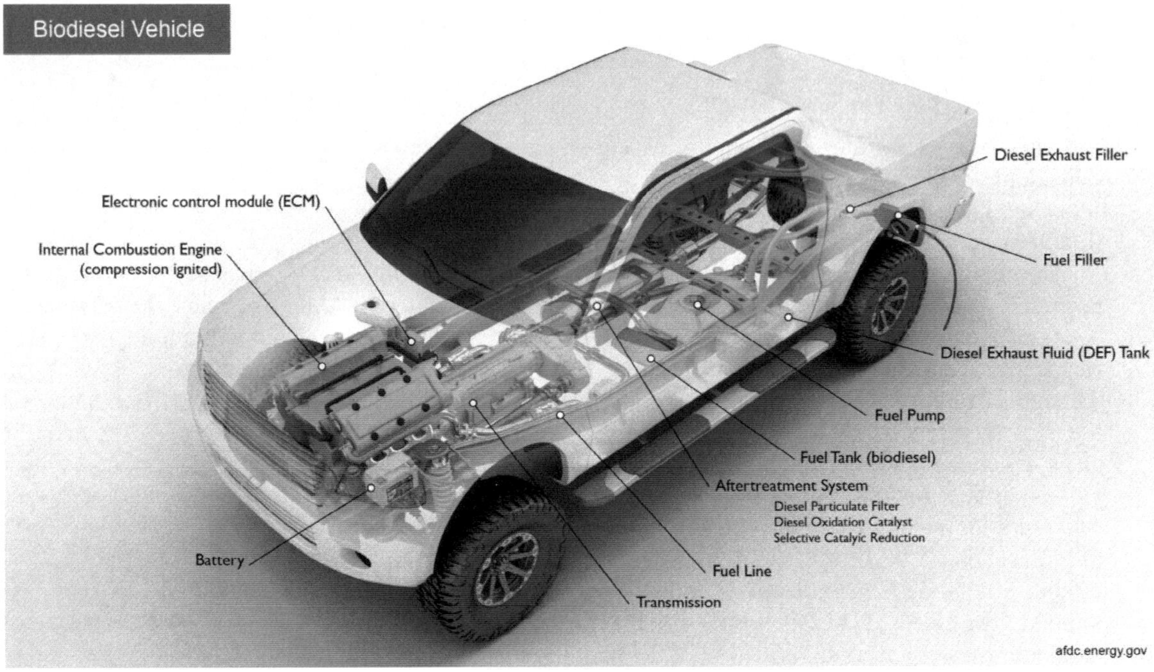

Figure 1. Biodiesel engine parts that interacted with fuel [10].

3. Wear Model

The development of the map is highly important for the wear studies. The accuracy of the path followed and the method of test directly affects the accuracy of the result. Many different test methods and laboratory setups have been suggested in previous studies [6,11]. However, in order to achieve more accurate results, the innovations to wear models can contribute to the work in this area. Research shows that the major critical point is the material selection. The good choice of the material gives better results [12,13,14]. The surface coating can increase the resistance of the material noteworthly. Thereafter, the parameters which is identified for the experiment play a decisive role in measuring the resistance of the material. The weight loss of the material is the cornerstone in evaluating the result of the study. The specific wear rate can be obtained using following equation for pin-on-disc set-up.

$$K_0 = \frac{V}{LD}\ (m^3/Nm) \tag{1}$$

Where, K0 is the specific wear rate, V is the worn material volume, L is the load in terms of Newton, D is the distance for total sliding in terms of meter. Besides, the 1worn material volume can be calculated with the density and weight of the material which is worn out. For the calculation of the sliding distance, oscillation frequency (υ), length of the stroke (A), and time period (t) should be known and the equation is;

$$-D = 2Avt \tag{2}$$

It is mentioned that the steps to be followed in wear experiments are critical to achieving the correct results. Figure 2. shows the model of wear test experiment roadmap. Before start the testing, the preparation of the materials is of great importance for the experiment. The shape, size, geometry of the material and any coating on its surface can directly affect the test result. After the materials are prepared, parameters such as load, speed, test time, frequency to be used in the experiment must be determined and set as input to the test system. After all these stages are completed, the prepared material can be examined in the experimental setup. Right after the experimental analysis, the deformations and failures on the surface of the material can be examined by surface examination tools such as scanning electron microscope. Considering the test data obtained, many parameters such as total material loss, surface deformation types, wear rate, etc. can be calculated. Finally, the results are evaluated and if the suitability of the material is acceptable, the test is completed. If the results are unacceptable, it should be returned to the first step again. With the help of this model, high-accuracy results can be obtained and also possible errors are minimized.

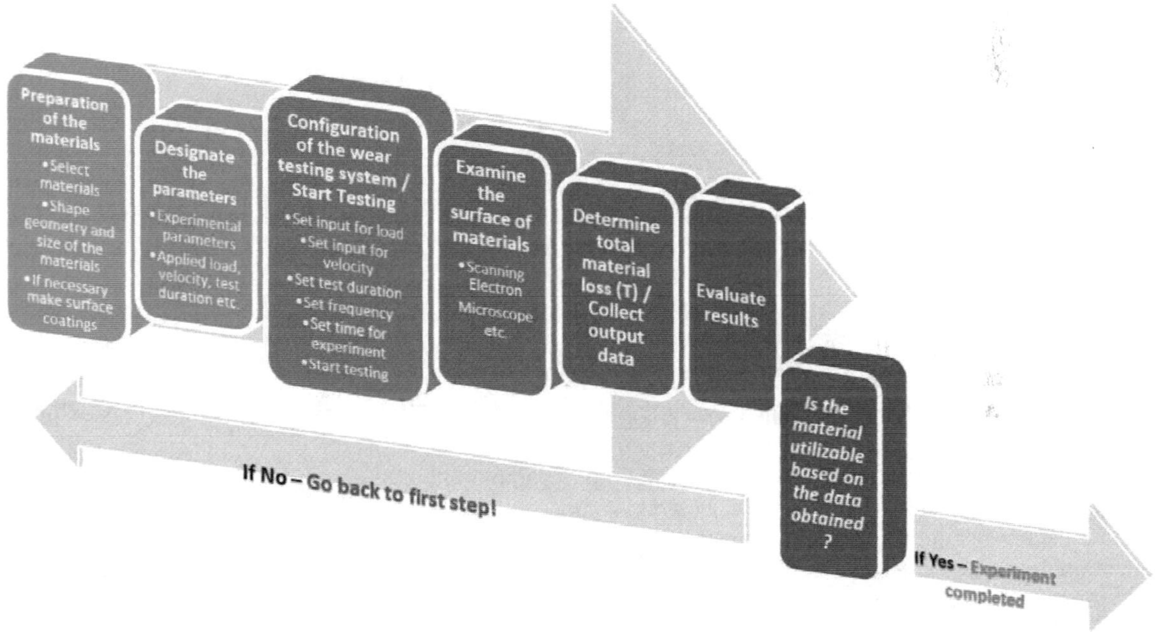

Figure 2. Roadmap model for the wear experiment

Before starting experiment, laboratory test installations have a great importance. For both electrochemical and mechanical testing, the appropriate modifications of the electrodes, the reliability of the test proper data acquisition technique and the correct positioning of the mechanical parts are very critical for getting results in high accuracy. The model for wear analyze is shown in Fig. 3. by the help of proposed model, corrosion and mechanical wear can be measured in the presence of any liquid. The model has mechanical and electrochemical setup for examine both corrosion and wear rate. The electrochemical consisting from three electrodes which are reference, working and counter electrodes.

On the other side, the mechanical part which is consisted from motor and disc-ball system shows wear rate caused by mechanical action.

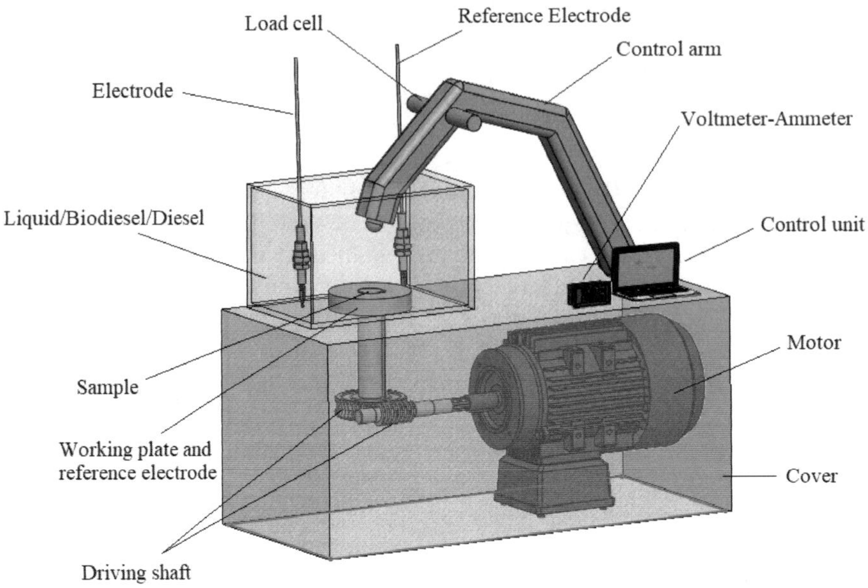

Figure 3. Model of corrosion and wear testing system

4. Wear Map

In progress of time, many wear data were recorded in the literature by using different materials in several ambient conditions. Since the obtained data are collected as a results of experiments, it is not possible to link the results with each other and achieve a standard result. In point of fact, both wear rate and wear mechanism obtained, vary according to the conditions in which the study is performed. The wear map is a graphical representation of the dependence between the wear properties of the material (wear rate, speed, time etc.) and the main test parameters (contact pressure, load etc.). Wear maps are used to observe which type of wear is dominant during wear. Furthermore, the effect of the slightest change in the rate of wear or load can be observed through the wear maps. As it mentioned, each study takes a different approach for the wear map mechanism. This study takes into account the Al-7Si alloy and the work carried out on this alloy. Wilson and Alpas (15) have developed a wear maps for the A356 alloy (Al-7Si) against SAE 52100 bearing steel. The severe and mild wear were drawn with the considering of wear rates. At the very low sliding speeds and loads, the ultra-mild wear regime observed where the magnitude decrease in wear rate occurs. It was also observed that the composites have a layer of iron oxide on the contact surfaces. This layer protects the alloy against steel from removal during sliding.

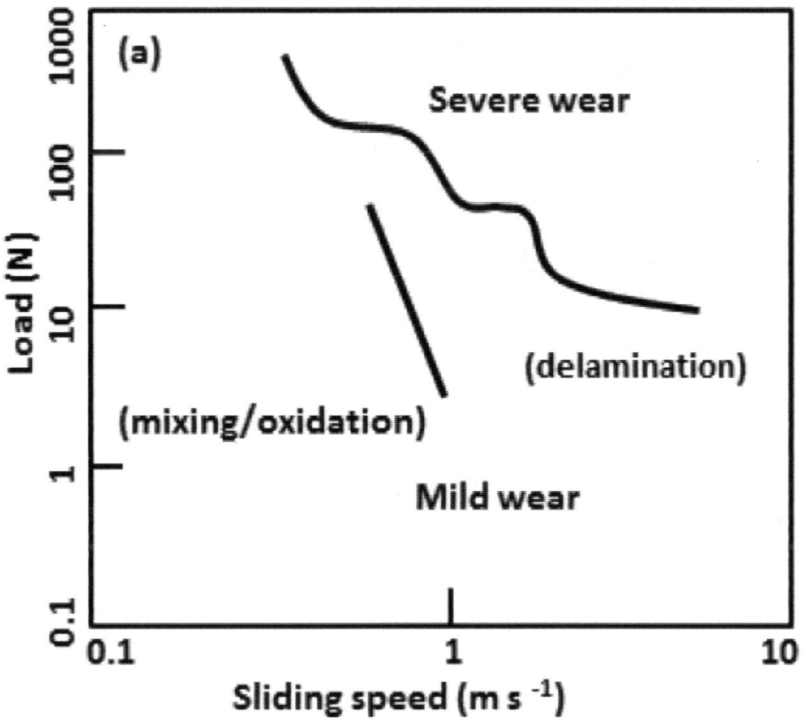

Figure 4. Wear map of Al-Si alloy with 7% silicon (A356) [16].

In this study wear map is presented by wear regime (i.e. mild, transition and severe) based on two different types of biodiesel fuels. Fig 4 shows wear maps of Al-7Si alloy to characterize main three aspects of wear in terms of operating conditions of tribosystems. In the studies considered, various loads, test speeds and wear rates of Al-7Si alloy were evaluated using different biodiesel mixtures such as mongongo oil based biodiesel (MO8, MO12) and jatropha oil based biodiesel (JBO4, JBO12, JBO20) [17, 18]. According to the results, it is observed that, mild abrasive wear exhibited at higher load, higher speed and higher blending ratio of biodiesel. In addition, with the use of higher blending ratio of biodiesel, more cracks, more grooves and more metal transfers were observed on the surface.

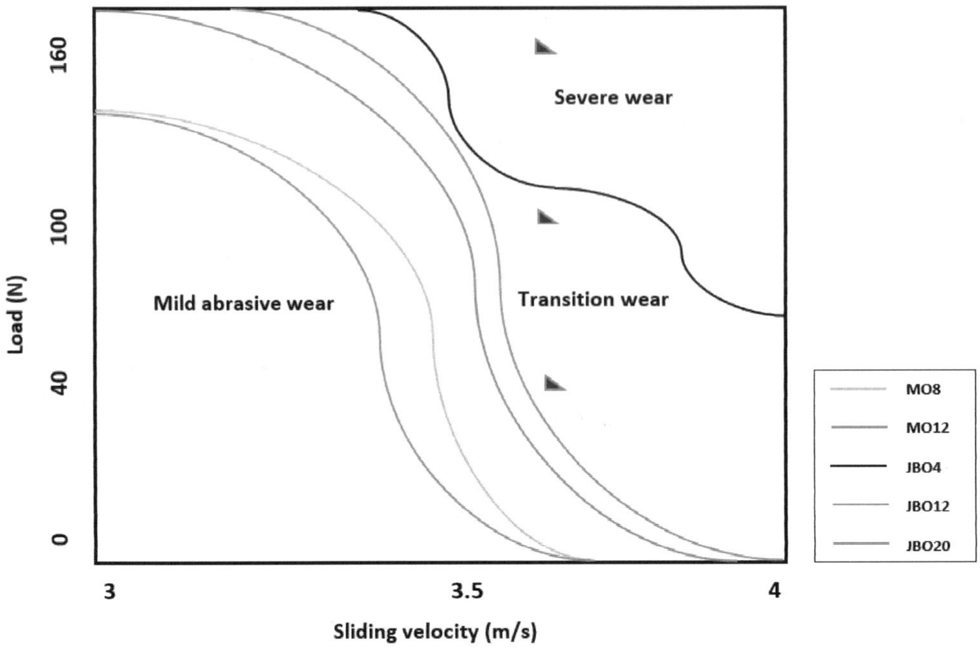

Figure 5. Wear map of Al-Si alloy with 7% silicon under different biodiesel blends exposure.

5. Conclusion

In order to minimize the negative impact of biodiesel on metals, improvements on biodiesel and the strength of metals in contact with biodiesel increasing day-by-day. As a consequence, the strength and life of materials increase against tribological reactions. In this way, it offers a better future for economic and reliability of the components. Corrosion resistance of materials with fuels or lubricant in automobile diesel engine or other relevant applications is very important for both manufacturability and safety. Biodiesel that has been used in engines is better than petroleum based diesel by 66% and hence, it would be better to use it in engine systems for reducing or controlling long term wear. For the research in this field, using the correct test equipment and methods are of great importance for the reliability of the results. This study aims to provide a way-forward model for corrosion and wear testing under biodiesel exposure. This proposed model will assist in getting the accurate result in corrosion and wear experiments. A preliminary wear map has been constructed based on the previous published data on the wear rate and mechanism of materials under biodiesel indicating many possible wear regimes, and significant differences in material loss mechanisms for different parameters such as load, speed, biodiesel concentration etc. In the created wear map, severe and mild abrasive wear boundaries were drawn for the Al-7Si alloy under the exposure of biodiesel. According to mentioned studies, mild abrasive wear observed at the high speed, load and blending ratio. Also, more cracks and grooves observed with the increasing of biodiesel blending ratio. Finally, it can be said that this map could serve as a tool which identify and display the regimes for different type of wear and the transition for biodiesel contaminant materials.

6. Acknowledgement

Authors are grateful to the Malaysian Tribology Society (MYTRIBOS) for financial support under grant No MIG2021-005 and International Islamic University Malaysia for other's that made this study possible.

References

[1] Hoang, A. T., Tabatabaei, M., & Aghbashlo, M. A review of the effect of biodiesel on the corrosion behavior of metals/alloys in diesel engines. *Energy Sources, Part A: Recovery, Utilization, and Environmental Effects*, 42(23), 2923-2943 2020

[2] Sundus, F., Fazal, M. A., & Masjuki, H. H. Tribology with biodiesel: A study on enhancing biodiesel stability and its fuel properties. *Renewable and Sustainable Energy Reviews*, 70, 399-412 2017.

[3] Zuleta, E. C., Baena, L., Rios, L. A., & Calderón, J. A. The oxidative stability of biodiesel and its impact on the deterioration of metallic and polymeric materials: a review. *Journal of the Brazilian Chemical Society*, 23(12), 2159-2175 2012.

[4] Yeşilyurt, M. K., Öner, İ. V., & YILMAZ, E. Ç. Biodiesel induced corrosion and degradation. *Pamukkale University Journal of Engineering Sciences*, 25(1) 2019.

[5] Meira, M., Santana, P. M. B., Araújo, A. S., Silva, C. L., Leal Filho, J. R., & Ferreira, H. T. Oxidative degradation and corrosiveness of biodiesel. *Corrosion Reviews*, 32(3-4), 143-161 2014.

[6] Maleque, A., & Abdulmumin, A. A. Tribocorrosion Behaviour of Biodiesel—A Review. *Tribology Online*, 9(1), 10-20 2014.

[7] Nguyen, X. P., & Vu, H. N. Corrosion of The Metal Parts of Diesel Engines In Biodiesel-Based Fuels. *International Journal of Renewable Energy Development*, 8(2) 2019.

[8] Fazal, M. A., Haseeb, A. S. M. A., & Masjuki, H. H. A critical review on the tribological compatibility of automotive materials in palm biodiesel. *Energy Conversion and Management*, 79, 180-186 2014.

[9] Jin, D., Zhou, X., Wu, P., Jiang, L., & Ge, H. Corrosion behavior of ASTM 1045 mild steel in palm biodiesel. *Renewable Energy*, 81, 457-463 2015.

[10] Alternative Fuels Data Center: How Do Diesel Vehicles Work Using Biodiesel? Retrieved 20 April 2021, from https://afdc.energy.gov/vehicles/how-do-biodiesel-cars-work 2021.

[11] .Agarwal, A. K., Bijwe, J., & Das, L. M. Wear assessment in a biodiesel fueled compression ignition engine. *J. Eng. Gas Turbines Power*, 125(3), 820-826 2003.

[12] Abdullahi, U., Maleque, M. A., & Nirmal, U. Wear mechanisms map of CNT-Al nano-composite. *Procedia Engineering*, *68*, 736-742 2013.

[13] Khan, Z. A., Saeed, A., Gregory, O., & Ghafoor, A. Biodiesel performance within internal combustion engine fuel system: a review. *Tribology in Industry*, 38(2), 197-213 2016.

[14] Sorate, K. A., & Bhale, P. V. Corrosion Behavior of Automotive Materials with Biodiesel: A Different Approach. *SAE International Journal of Fuels and Lubricants*, 11(2), 147-162 2018.

[15] Wilson, S., & Alpas, A. T. Wear mechanism maps for metal matrix composites. *Wear*, *212*(1), 41-49 1997.

[16] Rasool, G. Tribo-corrosion maps for steels, titanium and titanium carbide materials (Doctoral dissertation, University of Strathclyde) 2015 .

[17] Singh, Y., Singla, A., & Singh, A. K. Tribological characteristics of Mongongo-oil–based biodiesel blended lubricant. *Energy Sources, Part A: Recovery, Utilization, and Environmental Effects*, *39*(3), 332-38 2017.

[18] Singh, Y., Singla, A., Kumar, A., & Kumar, D. Friction and wear characteristics of jatropha oil-based biodiesel blended lubricant at different loads. *Energy Sources, Part A: Recovery, Utilization, and Environmental Effects*, *38*(18), 2749-2755 2016.

The dry sliding wear behaviour of open cell aluminum foam against mild steel

A.N. Md Idriss[1] and S. Kasolang[1,2*]
[1]School of Mechanical Engineering, College of Engineering Universiti Teknologi MARA, Cawangan Pulau Pinang, Kampus Permatang Pauh, 13500 Permatang Pauh, Pulau Pinang, Malaysia.
[2]School of Mechanical Engineering, College of Engineering Universiti Teknologi MARA, 40450 Shah Alam, Selangor, Malaysia
Email addresses; salmiahk@uitm.edu.my; ahmednazrin@gmail.com

*Corresponding author; salmiahk@uitm.edu.my

Abstract. From one perspective, wear is a phenomenon of material loss due to surface friction and their failures are characterized by classifying changes in the surface morphology of the mating component. In the case of foam materials, such as open cell aluminum foams, having interconnected porosity associated with large surface area, they acquire minimal kinetic energy for mobility during friction. Foam materials are also well known for their good qualities in thermal and noise management. This sees the foam material as an effective alternative candidate to their parents solid alloy. The objective of this work is to investigate the wear behavior of the open cell aluminum foam fabricated using matrix infiltration method on the surface of rotating mild steel under the dry sliding condition. The surface profile due to wear, microhardness and weight losses of both the open cell aluminum foam and steel disk were measured. The surface profile used the downlighting and inclined at 45° lighting techniques for the favour of distinguished appearance under the optical microscopy observations. Generally all samples showed abrasive failures due to the cutting marks without adherence. High brightness under the downlight was not able to vary the cutting marks in sizes which made among those single striations width to appear almost similar. Comparing to the 45° inclined lights, a good contrast was obtained and so ranges of cutting width were seen to be equally distributed on that open cell aluminum pin. The technique using lighting at the 45° resulted clarity on worn surface area that is less efficient by the downlight. The low hardness at 47 HV with the open cell aluminum foam showed medium wear and greater weight loss than the 155 HV mild steel sample.

Keywords: Optical microscope, wear, open cell aluminum foam, mild steel, morphology, hardness

1. Introduction

Wear is a phenomenon of surface deterioration often as a manifestation of friction arising from the relative motion of interacting mating materials. Surface deterioration may involve damage of surfaces with or without material loss [1]. For both metallic and non metallic materials, wear condition is very much influenced by the types of wear mechanism involved and many times, there are two or more mechanisms present at any time which synergistic effective. As a result, the surface morphology changes and the surface asperities often experience either elastic or plastic deformation depending upon the severity of the wear due to the effective load being applied or other factors such as temperature and dry or wet condition of the interaction. Aluminum is abundantly available and preferred due to their physical properties such as light, good corrosion resistance and thermal conductivity, cheap, low melting temperature and compatible with many alloying elements to increase the hardness hence for the reduction of wear resistance. With the solid Al 6063-TiB composite sample against the steel disk, a gradual increase of sliding speed had resulted much severe surface damage rather than increasing the load [2]. In another report, increasing the load at a given speed showed reduction of weight losses using as cast and aluminum-silicon heat treated alloy [3]. Strain that is developed at the interphase of aluminum-silicon alloy was the reason for the increased of hardness which had resulted the reduction of wear than the untreated alloy [4]. High wear damage was the resultant of strain hardening while the absence of wear losses was associated with thermal softening of the A380M aluminum alloy [5]. Lighter TiC metal matrix composite at 700 HV was 21 times better in terms of material loss than the untreated AISI 4340 low alloy steel [6]. The wear resistance of an unreinforced aluminum alloy was 4 times low than the same material added with 30 wt % alumina [7]. Composing alumina and SiC with aluminum matrix followed by heat treatment at T6 condition gave smooth surface failures compared to the same sample without that T6 treatment resulting wear severance [8]. Gradual increased of wear load reaching 3.14 Mpa resulted increased of wear rate and coefficient of friction and beyond than this load wear transition took place with decline of extreme failure that lowers those the wear rate and coefficient of friction. At the peak failure of 3.14 Mpa, abrasive and adhesive wear mechanism were the reason for severance [9]. 303 stainless steel sliding against the alumina at 3 cm/s and 4 N load showed weight loss at 44 x 10^{-7} g/m which was categorized as mild under the mild regime. When TiC is added, failure reaches very mild regime indicating increased of wear resistance [10]. It is shown that by changing the microstructural features and compositions through heat treatment, adding reinforcing material and fabrication techniques gave difference of hardness values which had resulted variety of wear failures.

Through aluminum matrix infiltration among the salt granules, architectural changes can be made instead of changing the metal composition and microstructural features formed by that matrix dissolution, segregation and solidification [11]. Salt that has the property of higher melting temperature of ~ 800°C than aluminum of about 600°C act as the space holder which helps to expand the volume of fabricated sample at less amount of solidified aluminum matrix. This explains how the aluminum can be tailored by exploiting porosity to reduce weight or relative density several times than the parent densed alloy. In general, dissolved salt granule under running water lead to the production of interconnected porosity replicated by aluminum contour that could not pass through the salt contact point during liquid infiltration. The elimination of salt granule remained the sample to have high porosity content giving low density up to 2.6 times than the solid aluminum at 2.7 g/cm^3 [12]. Infact pores are also formed in the matrix while the liquid is stirred at temperatures and speed [13]. Increasing the volume fraction of the salt against the aluminum increases the porosity content associated with reduction of density [14]. Space holder of expanded perlite with the size of 1 -1.4 mm showed higher porosity content reaching 61.88 % then perlite sizes between 4 - 5.6 mm where highest determined porosity was 59.57% both incorporated with A356 [15]. Open cell aluminum foam is known for their low in weight and melting temperature, good energy absorption capability, sound barrier and thermal control structures [16-18]. In term of wear, aluminum foam without the incorporation of SiC showed severe wear failures than those added with that hard carbide particulates forming Al-SiC metal matrix composite both slided against the steel disk [19]. Less time is taken to build the stability of friction coefficient using aluminum embedded with alumina preform when high load is employed [20]. The hard alumina network that was seen protruded within the Al-Mg matrix saved the composite against deterioration up to 10 times than sample without being reinforced [21]. The wear behaviour under the influence of oiled condition is yet to be uncovered [22-24]. Mixing soybean oil with mineral oil and additives at certain percentages was suggested suitable to lubricate in the hot and warm weather [22,23]. Increasing the load pressure from 10kN to 15 kN increases the frictional force remarkably. However at a given testing load and speed, flooding the the oil supply from 0.2 to 0.7 MPa do not significantly change the frictional force [24]. To the best knowledge of the authors, through the literature reviews conducted, there had been no work reported for the wear behaviour of the open cell aluminum foam sliding against the rotating mild steel disk by comparison. Apart from that, it came to a realization that just by inclining the projected lights at certain degree, the worn surface profile or morphology of the surface appeared different when compared to those using only downlight. With regard to these matters, an investigation was conducted to determine the influence of these lighting technique on the analysis of worn surface of open cell aluminum foam and mild steel by the dry sliding wear test. Surface failures were subjected to the DINO built in downlight microscope and externally 45° inclined light. Results pertaining to microhardness, weight loss, and surface morphology are presented in this work.

2. Experimental work

The experimental work consist of several processing steps involved from samples preparation followed by conducting the dry sliding wear test of the open cell aluminum sample with respect to their failure analysis and weight losses are shown in Fig. 1. The as received aluminum foam fabricated using the matrix infiltration method was cut to the size of the pin at the dimension of Ø 8 mm x 25 mm using CNC wire cutting technique from the bulk sample of Ø 35 mm × 50 mm as shown in Fig. 1a. The disk used sliding on the surface of the open cell aluminum foam pin was the mild steel with the dimension of 30 x 65 x 4.5 mm . The samples surface subjected to the dry sliding work were initially polished using a spinning wet pad that was filled with alumina powder until the scratches are removed in Fig. 1b . The scratch free surfaces of aluminum foam pin and mild steel disk used Mitutoyo MVK - H1 for their microhardness values at 1 kg load of 10 seconds indentation delay. DUCOM TR-20 LE wear testing machine was used to determine the wear behaviour of the aluminum pin against the steel surface at the speed of 300 rpm at 5N static load as illustrated in Fig. 1c. The track diameter was at 20 mm and testing was allowed 30 minutes before completion. The worn surfaces were carefully cleaned using alcohol and distilled water before they were subjected to morphological analysis via DINO microscope. The observed failures of both open cell aluminum foam pin and steel disk were conducted using this microscope assisted by the built-in LED lights and externally torched LED light at 45°. All micrographs were captured using DINO Capture 2.0 software attached to the laptop. Observed surface failures was torched using the downlight without the interference of the 45° inclined LED lights as shown in Fig. 1d. In comparison when using the 45° light to observe the same spot of surface failures, the downlight was swithed off (Fig. 1e). The measured weight loss was the difference of their respective weight values before and after the weighing (Fig. 1f)

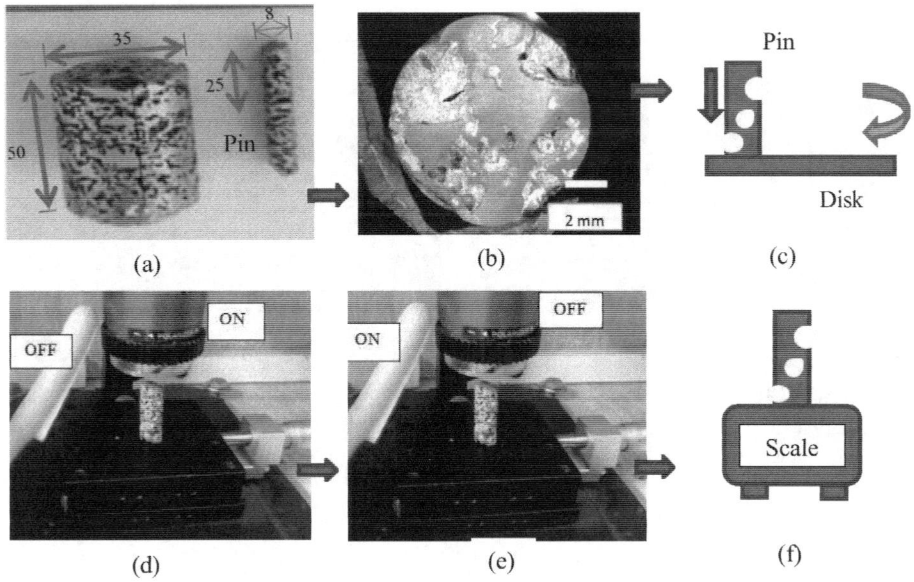

Figure 1. Processing steps of the experimental work. (a) The cut aluminum foam pin (right) from the bulk sample (left, measurements in mm), (b) surface of the prepared sample after polishing, initial weight taken, (c) the illustration showing the dry sliding wear test, (d) determining the surface failure using only the downlight. Inclined light was swithed off, (e) the downlight switched 'off' having inclined light 'on' which was to manipulate the produced shadow for a clear and distinguished observations than in (d) while (f) was weight loss measurement after wear test (between b and f).

3. Results and Discussions

The aluminum pin and disk surface failures, subjected to the dry sliding wear, were observed using 2 different lighting techniques as shown in Figs 2 to 4. Figs. 2 and 3 described the surface failures of the open cell aluminum foam pin under the built in LED downlight and 45° inclined lights respectively. The inset in Fig. 2 shows the location of the failure (by black box) on the surface of the pin sample and similar sample after polishing without wear is shown in Fig. 1b. Fig. 2 shows the poor contrast in the oval by the downlight and in Fig. 3, visible edge grooving, deep and fine cuts by the inclined are depicted. Spinned at 300 rpm. with 5 N load, Figs. 2 and 3 of the same visual area show parallel cutting marks across the sample surface as an evidence of abrasive wear. The small high porous open cell aluminum sample cut to the Ø 8 mm x 25 length was able to widthstand the speed and load and upon test completion, no surface cracking were observed on the failed surfaces. It was found that the microhardness value of the alumumium pin was at 47 HV while the couterbody spinning steel disk was at 155 HV. The third body abrasive particles and the harder steel disk at 155 HV are responsible to cause such failures on the softer aluminum surface that was at 47 HV.

Figure 2. The worn surface of the open cell aluminum foam under the downlight giving high brightness resulting poor contrast in the ovals. Inset shows the magnified area (box) for Figs. 2 and 3

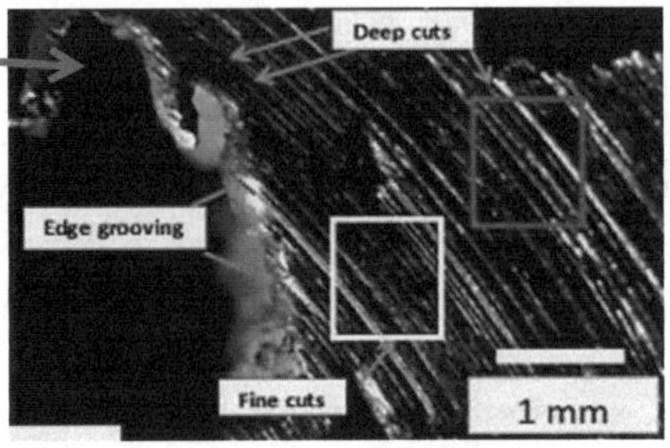

Figure 3. The worn surface of the open cell aluminum foam under the inclined 45° lights denoting good contrast among the striations. Deep and fine cuts and edge grooving appear very clear and distinct compared to that in Fig. 2 because of the poor contrast.

In this work, observation using the downlight gave almost similar range of cutting marks width distance and the reason for this is because of the high brightness that deprived the desirable contrast between the peaks and valleys among the cuts as shown in Fig. 2. Based on this result it can be said that the built in downlight seems to exemplify inferior quality in determining the shallow scratches of cut surfaces. Introduction of additional light source and manipulation to an inclination angle at 45° have resulted in an impressive contrast with distinguished features of deep and fine cuts.The edge grooving can be clearly identified (Fig. 3) which was not clearly seen at the same area before (Fig. 2). The torched and positioned at that specific inclination at 45° (Fig. 3) has potential obstruction and subsequent unwanted reflection and hence the outcome is a clear contrast that varies especially along the cutting trough, presented as shades or dark contrast. This simple technique has certainly helped to produce much appealing surface failure characterization on wear analysis. It is low cost to incorporate but with a significant difference in the results obtained

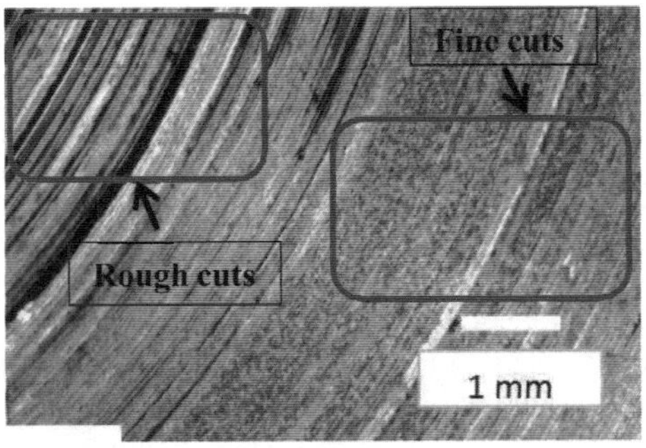

Figure 4. The worn surface of the mild steel under the 45° inclined lights depicting mild wear with rough cut and mild ones seen on the top left and right respectively

Fig. 4 shows the wear track of mild steel disk which consists of very fine cuts and rough cuts on the right and left side of this micrograph respectively. It is to be noted that the high hardness at 155 HV received milder wear profile than the 47 HV open cell aluminum foam pin. High hardness surfaces samples are less to wear damages as had been reported by many research works [6,10,25-28]. The smoother surface profile of the disk portraying lower wear losses than the aluminum corresponds to approximately at 41 x 10^{-7} gm^{-1} than at 106 x 10^{-7} gm^{-1} respectively. Classification of wear mode was referred to wear rates as reported in a few publications highlighted in Table 1 [6,10]. The very mild wear is considered for cases below than the 20 x 10^{-7} gm^{-1} while above 274 x 10^{-7} gm^{-1} the failure is considered as severe.

Table 1: Wear rates and wear regime [10]

Wear rate (gm^{-1})	Wear regime
< 20 x 10^{-7}	very mild wear
20 to 90 x 10^{-7}	mild wear
91 to 274 x 10^{-7}	medium wear
> 274 x 10^{-7}	severe wear

With the steel disk, the wear loss at 41 x 10^{-7} gm is classified as mild wear. The open cell aluminum foam at 106 x 10^{-7} gm^{-1} falls for medium wear as the weight loss was about 2.6 times higher than the steel. The previous work showed about 21 times improvement when the low steel alloy was incorporated with TiC particles [6]. Apart from the influence of the third body particles, the irregular surface profile of the aluminum pin sliding against the disk is thought to influence difference of cutting width observed and characterized by very fine and rough cuts on the steel disks. The third body particles tend to cause high stress when they are entrained between aluminum and steel surfaces and as these particles passes underneath aluminum open cell large pores, their stress development are being delayed and abrasive cutting become null. Under this circumstances, the mechanism to comminute to become smaller pieces or agglomerate become not effective as the particles are not compacted. It is to be noted that the low hardness of the open cell aluminum foam was less able to penetrate and cut along the mild steel. This also shows that the wear loss is in accordance to the archard's equation denoting that the softer aluminum would endure greater wear loss than the steel.

The severance of aluminum would require reinforcement to increase the hardness for lowering the wear despite their labyrinth known structure good for thermal management, sound and impact absorption, filtering and liquid agitations. Realizing lubrication is a way to reduce friction this dry sliding wear experimentation is an initial step to see how far the structure could tolerance wear under lubrication which is yet to be explored.

This paper is limited to investigating only the surface failure under that testing condition and no chemical analysis on either surfaces or particles were perform for composition changes. Across the softer open cell aluminum pin surfaces (Fig. 2), the difference was not remarkable and only wear due to cutting distributed seen similar sizes indicating abrasive wear. In this work, the brightness must be carefully controlled to avoid misinterpretation and the availability technique to alter brightness is very limited using the microscope. Another source of lighting that could take the advantage of the surface condition is much preferable which had set the course of this work to use inclined lights at 45°. The technique is useful not only to determine a clear variety of wear failures subjected to wear, but also helps to give special attention to characterize other organic and non organic surfaces such as stones, wood or leaves.

4. Conclusions

Experimental work to determine the dry sliding wear behaviour of the open cell aluminum foam pin against the disk and those morphologies under the built-in downlight and inclined at 45° lights were successfully conducted. The lower hardness of about 3.3 times than the steel at 155

HV had resulted in rougher surface cuts on the open cell aluminum foam. The abrasive failure of the aluminum foam and mild steel were classified as medium and mild wear from the wear rates at 106 x 10^{-7} gm^{-1} and 41 x 10^{-7} gm respectively. The severance on the aluminum foam requires reinforcement to be made for reduction of wear. The foam pin was tested under dry sliding wear without sudden rupture or breaking, indicating the metal a suitable candidate for wear under lubricative condition. The downlight exemplifies poor contrast on aluminum foam pin surface with an appearance of almost equal cutting width due to the high brightness. Good contrast using inclined lights made the surface profile standout revealing differences and distributed cutting marks at sizes which were not visible before. The irregular pin surface is the reason that had resulted in irregular cutting marks consisting rough and very fine ones on the disk. The down light images and inclination at 45° provide simplicity and easiness to affirm surface failures and the associated mechanism. This work involving the use of inclining torched lights of a degree gave changes to the visual appearance of those surface morphology. Hence, by this simple and effortless technique further explanation could be elaborated and offered precisely especially on the specific and effective mechanism for mapping and quantifying of failures established during the wear process.

Acknowledgement
The financial support for this work was provided by the UiTM Shah Alam Research Management Center under FRGS 5/3 (372/2019), FRGS 37212019 and other related MOHE grants. The authors would like to thank Associate Professor Dr M.H. Ismail for the contribution of the aluminum foam sample which was used in this work. This work is attached to RIG Tribology for Transportation Industry at the Centre for Tribology (CenTrib), School of Mechanical Engineering, College of Engineering, UiTM.

References

[1] Hutchings, I.M. (1992). Tribology: Friction and wear of engineering materials, London; Edward and Arnold.

[2] Ramesh,C.S., Ahamed, A. (2011). Friction and wear behaviour of cast Al 6063 based insitu metal matrix composites. Wear, 271, 1928-1939.

[3] Yasmin, T., Khalid, A.A., Haque, M.M.(2004). Tribological wear properties of aluminum sililicon eutectic base alloy under dry sliding condition. Journal of Material Processing Technology, 153-154, 833-838.

[4] Haque, M.M., Sharif, A., (2001), study on wear properties of aluminum-silicon pistion alloy. Journal of Materials Processing Technology 118, 2001, 69–73

[5] Islam, M.A., Farhat, Z. (2010). Wear of A380M aluminum alloy under reciprocating load. Journals of materials engineering and performance, vol. 19 (8), 1209-1213

[6] Md Idriss, A.N., Maleque, M.A., Yaacob, I.I., Nasir, R.M., and Baker, T.N. (2016). Microstructural aspects of wear behaviour of TiC coated low alloy steel. Material Science and Technology. 32(4), 303-307.

[7] Al-Qutub, A.M., Allam, I.M., Qureshi, T.W. (2006). Effect of submicron Al_20_3 concentration on dry wear properties of 6061 aluminum based composite. Journal of Material Processing Technology 172, 327-331.

[8] Rajesh, A.M., Kaleemulla, M., Doddamani, S., Bharath, K.N. (2019). Material characterization of SiC and Al_2O_3 -reinforced hybrid aluminum metal matrix composites on wear behaviour. Advanced Composites Letters, 28, 1-10.

[9] Nesarikar, A.R., Tewari, S.N. and Graham, E.E. (1991). Room temperature wear characteristics of Al2O3 - particle reinforced aluminum alloy composite. Material Science and Engineering A, 147 (2), 191-199.

[10] Rasool, G. and Stack, M.M. (2014). Wear maps for TiC composite coatings deposited on 303 stainless steel. Tribology International. 74, 93-102

[11] Metal foam - A design guide. (2000) Ashby, M.F., Evans, A.G., Fleck, N.A., Gibson, L.J.,Hutchinson, J.W., Wadley, H.N.G., Wadley. Butterworth-Heinemann.

[12] Azizan,M.A., Ismail, M.H., Mohd Salleh, N.A.,Natarajan, D.V.(2017).Sound Absorption Properties at High Sound Frequency of Open Cell Aluminium Foam. Journal of

Mechanical Engineering Vol SI 2 (1), 161-173.
[13] Kumar, M.S., Begum, S.R., Pruncu, C.I., Asl, M.S. (2021). Role of homogeneous distribution of SiC reinforcement on the characteristics of stir casted Al-SiC composites. Journal of Alloys and Compounds, 869, 159250.
[14] Aida, S.F., Hijrah, M.N., Amirah, A.H., Zuhailawati, H., Anasyida, A.S. (2016). Effect of NaCl as a space holder in producing open cell A 356 aluminum foam by gravity die casting. Procedia Chemistry, 19, 234-240.
[15] Taherishargh, M., Sulong, M.A., Belova, I.V., March, G.E., Fiedler, T. (2015). On the particle size effect in expanded perlite aluminum syntactic foam. Materials and Design 66, 294-303.
[16] Exxentis.https://www.exxentis.co.uk/Website visit on 11/2/2020
[17] Razali, R.N.M., Abdullah, B., Ismail, M.H., Ahmad, U.K., Idham, M.F., Ramli,A. (2013) Mechanical properties of aluminum foam by conventional casting combined with NaCl space holder. Applied mechanics and material, vol. 393, 156-160. DOI: https://doi.org/10.4028/www.scientific.net/AMM.393.156
[18] Goodall, R., Despois, J.F., Marmottant, A., Salvo, L., Mortensen, A. (2006). The effect of preform processing on replicated aluminum foam structure and mechanical properties, Scripta Materilia, 54, 2069-2073. DOI: https://doi.org/10.1016/j.scriptamat.2006.03.003
[19] Mondal, D.P., Das,S., Jha, N. (2009). Dry sliding wear behaviour of aluminum syntacti foam. Materials and Design, 30, 2563-2568.
[20] Sharifi, H., Ostovan, K., Tayebi, M. and Rajee, M. (2017). Dry sliding wear behaviour of open cell Al-Mg/Al_2O_3 Al-Mg/SiC-Al2O3 composite preform produced by pressureless infiltration technique. Tribology International 116, 244-255.
[21] Chang, H., Binner, J. and Higginson, R. (2010). Dry lining wear behaviour of Al(Mg)/Al_2O_3 interpenetrating produced by a pressureless infiltration technique. Wear, 268, (166-171).
[22] Ahmed, D.I., Kasolang, S., Dwyer-Joyce, R.S., Sainan, K.I. and Roselina, N.N. (2014). Formulation and physico-chemical characteristics of biolubricant. Jurnal Tribologi, 3(1), 1-10.
[23] Kasolang, S and Dwyer-Joyce, R.S. (2005). Thickness measurement, turbulence and bubbles in lubrication. Journal of Mechanical Engineering, 2(1), 63-78.
[24] Ahmad, M.A., Kasolang, S. and Dwyer-Joyce, R.S.(2013). The effect of oil supply pressure at different groove position on frictional force and torque in journal bearing lubrication. Procedia Engineering, 68, 70-76.
[25] Singh, J., Thakur, L., Angra, S. (2020). Abrasive wear behaviour of WC-10Co-4Cr cladding deposited by TIG welding process. International Journal of Refractory Metals and Hard Materials, 88, 105198
[26] Sardar, S., Pradhan, S.K., Karmakar, S.K., Das, D. (2020).Experimental analysis on tribo - performance of aluminum composites. Journal of Composite Materials. 54 (9) 2577-2598.
[27] Yan, X., Huang, C., Chen, C., Bolot, R., Dembinski, L., Huang, R., Ma, W., Liao, H. and Liu, M. (2018). Additive manufacturing of WC reinforced maraging steel 300 composites by cold spraying and selective laser melting. Surface and Coating Technology. 371, 161-171
[28] Pradhan, S., Barman, T.K., Sahoo, P., Sutradhar, G. (2017). Effect of SiC weight percentage on tribological properties of Al-SiC metal matrix composites under acid environment. Jurnal Tribologi 13, 21-35.

Modelling of Pedals Pressing Mechanisms for Low Speed Driving in Road Traffic Delay

Hazril Md Isa[1*], Salmiah Ahmad[1], Muhammad Abdullah[1], Ataur Rahman[1], NMA Ghani[2], N.M.M Noor[3]

[1]Department of Mechanical Engineering, Kulliyyah of Engineering, International Islamic University Malaysia (IIUM)
[2]Department of Electrical Engineering, College of Engineering, Universiti Malaysia Pahang
[3]Faculty of Mechanical Engineering, Universiti Teknologi MARA, Malaysia

* Corresponding author: hazrilmdisa@gmail.com

Abstract. Road traffic delay is imminent in big cities around the world including Kuala Lumpur, due to the increase in vehicle ownership, weather condition, road blocks by authorities and ongoing infrastructure projects at particular point of the road. Previous literature shows that the average time spent by drivers in driving under road traffic delay is about 30-53 minutes per day. Within the delay, regular pressings of the brake and accelerator may lead to leg and body fatigue, hence may give detrimental effect on the driver in the long run. This paper aims to mimic the pedal pressing and releasing mechanism in replacing the leg activities during road traffic delay, such that the driver will no longer need to recurrently pressing both pedals, thus reduce the fatigue. In replicating the pressing mechanism, a linear electromechanical actuator is designed for this purpose due to its high force density as well as good response time. The mathematical model of the actuator is first developed and integrated with car dynamic model in Simscape model. All parameters used for simulation purposes were taken from an actual electromechanical actuator used for the similar application. Simulation performance evaluation was conducted on a virtual environment for the integrated models and the response of the models were investigated. As a result, the actuator model is capable to produce the desired output force that is similar with human leg during pressing activity, with about 200 N force. The model was coupled with tuned PID controller, and the actuator was also able to produce an acceptable time response, where the vehicle is brought to a complete stop within less than 1 second, which is closed to the actual driver's leg activities during the pedal pressing in a road traffic delay.

1. Introduction

Road traffic delay is a common scenario in big cities around the world including Kuala Lumpur. Factors affecting the occurrence of road traffic delay include increase in vehicle ownership, weather condition, road blocks by authorities and ongoing infrastructure projects at particular point of the road. People's preferences to use private vehicles over public transports is considered as a main contributing factor to this situation, which adding the volume of vehicles at peak hours. Road traffic delay has a direct relationship with time spent in a vehicle, in which the time spent inside the vehicles increases dramatically during road traffic delay. Statistically, in ASEAN countries, the average time spent per driver in road traffic delay is about 60 minutes. Specifically, in Kuala Lumpur, the average time spent per driver in road traffic delay is around 53 minutes during normal days [1] This number can reach as high as 60 minutes and more, during festive seasons or if there is any uncontrolled event occurred along the road used [2]. This number is closed to the amount of maximum time, recommended by [3] which is 80 minutes.

Long time spent in a vehicle daily can lead to several health deteriorations such as leg, neck and back pain and stress. The effects are amplified further with external factors such as age, gender, road condition, and driver's familiarity with the vehicle [4]. The average speed of the vehicle during heavy traffic congestion is between 0 km/h to 10 km/h, depending on amount of vehicles on the road at that moment [5]. Due to imminent traffic delays in Malaysia, statistically, congestion level mark in Kuala Lumpur is put as 37%, making Kuala Lumpur is ranked 46 out of 180 major cities in the world. In other words, average time spent annually by drivers in Kuala Lumpur in the vehicle alone is 170 hours or seven days and two hours during road traffic delay, [1].

Due to extremely low distance and speed between trailing vehicles during road traffic delay, drivers tend to press brake and accelerator (and clutch pedals for manual vehicle) pedals more frequently to avoid collision with other vehicles as well as with pedestrians and motorcyclists nearby. This situation causes higher chance for fatigue to be developed in driver's leg. Incorrect sitting and leg posture while pressing the pedals amplifies further the fatigue levels during road traffic delay, especially when the delay happens towards the end of the day, [6] – [8].

Therefore, an automatic system to override the driver's leg in applying continuous force on brake and accelerator pedals is seen as a viable solution to reduce fatigue and stress level during road traffic delay. There are range of solutions in tackling this issue, but most of the existing automatic systems are generally embedded within the vehicle such as, the Adaptive Cruise Control (ACC) [9] – [10], with Stop and Go features, breake by wire [11], Active Accelerator Pedal (AAP) [12]. This embedded system is currently available only for high specifications of car, such Nissan, Audi and Tesla. Since the majority of vehicles on the road, particularly in Malaysia are from low and middle range of vehicles, therefore the embedded automatic system is not affordable and the fatigue and stress issues are not completely solved. In addition, the trending nowadays shows the number of road users are also among those who are working with delivery services, which most of them come from lower to middle income families, conforming to the inability in affording high end vehicle. Therefore, an alternative solution that is pertinent to the majority of vehicles are needed.

Therefore, this study will explore the possibility of replacing the human leg in pedal pressing activities during traffic delay using automatic pressing and releasing mechanism employing linear actuation. It is anticipated that the proposed automatic mechanism will produce sufficient insights in reducing the fatigue and stress level during driving in road traffic delay, hence improving drivers' emotion and physical condition despite of inevitable traffic congestions in big city like Kuala Lumpur.

The paper is begins with an introduction in Section 1 followed with the proposed mechanism design in Section 2. Section 3 consists of simulation results & response of the vehicle, while section 4 conclude the findings of the paper.

2. Proposed Design

The proposed design is to implement linear electromechanical actuator, in pedal pressing directly. It receives signal voltage from the controlling unit and apply the pressing force towards the pedal. The main idea is to attach the linear actuator's shaft on the surface of the pedal as shown in Figure 1.

IOP Conf. Series: Materials Science and Engineering 1244 (2022) 012010 doi:10.1088/1757-899X/1244/1/012010

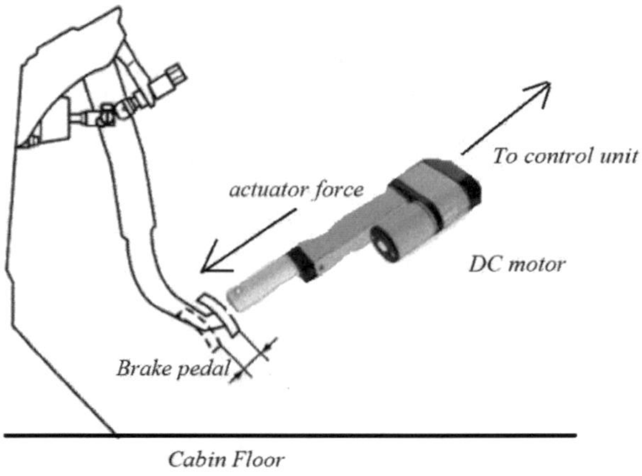

Figure 1. Linear Actuator – schematic diagram with pedal

In practice, the linear actuator can only produce linear force by converting rotational torque by means of gearbox and leadscrew. Hence, in this study, a standard model of linear actuator exist in the market will be used. On top of that, the dimension of the actuator is carefully considered so that it does not consume too much space in the driving cabin. The maximum rating voltage for this actuator is 12 V as that is the maximum voltage supplied by car battery.

In the following section, mathematical model that represents the working of the linear actuator is presented. The relationship between actuator's force and voltage signal will be extensively presented.

2.1 Mathematical Model of the Linear Actuator

The actuator converts electrical signal in form of voltage to linear force. Figure 1 shows the circuit representation of linear DC motor.

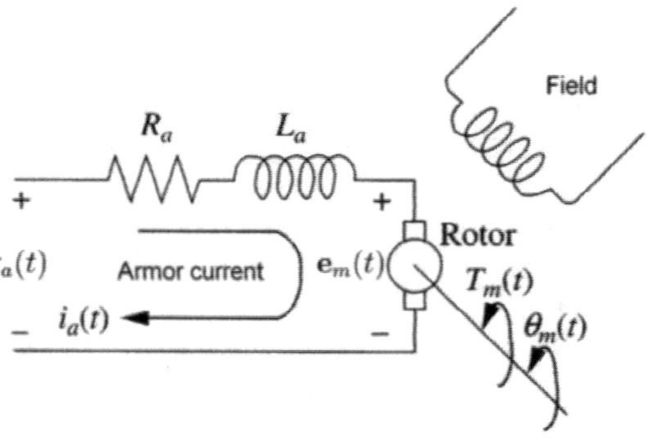

Figure 2. Circuit representation of DC motor [13]

In Figure 2, the rotational torque τ is directly proportional with current, *I*

$$r_{motor} = K_{motor}I \quad (1)$$

Where K_{motor} is the motor's constant, and it depends on physical dimension of the motor [14]

Applying Kirchoff Voltage Law (KVL) and Laplace transform on circuit in Figure 1 as well as Laplace transform yields the relationship between input voltage V_{input} and output rotational torque τ_{motor} [15], where the relationship is demonstrated by Equation (2).

$$\frac{T_{motor}(s)}{V_{input}(s)} = \frac{K_{motor}}{sL_a+R_a} \quad (2)$$

On the mechanical shaft, the same process applies in which KVL and Laplace transformation is performed and yields the relationship between the shaft's angular speed W_{shaft} and input voltage V_{input}.

$$\frac{W_{shaft}(s)}{V_{input}(s)} = \frac{1/B}{s(J/B)+1} \quad (3)$$

where B and J are shaft's rotational friction constant and moment of inertia, respectively.

2.2 Conversion to Linear Force and Linear Position

The linear force and linear speed are required to apply direct pressing on braking pedal. In actual DC motor, the gearbox and leadscrew are connected with rotational shaft to convert angular torque to linear force. By assuming the leadscrew and gear friction are negligible, the relationship between input and output torque with gear ratio is,

$$G = \frac{c_0}{c_{in}} \quad (4)$$

where G is the gear ratio, τ_o is e output angular torque and τ_{in} is the input angular torque from rotating shaft. Thus the linear force, F is produced as a result from rotating leadscrew,

$$F = G\frac{2\pi}{ly} \quad (5)$$

where η is the efficiency of the leadscrew and l is the length of one pitch on the leadscrew. As for linear position x_{linear} of the leadscrew can be expressed by the following equation, [16].

$$x_{linear} = \theta_{angular}\frac{l}{2\pi} \quad (6)$$

The linear actuator model is then built in Matlab-Simulink environment, as shown in Figure 3. The parameters used for the actuator is summarized in Table 1, and it is taken from the industry-available actuator, which suits the application.

Table 1. Linear actuator Parameters Used in Simulation

Torque constant	0.4
Winding resistance	1 Ω
Winding inductance	0.05 H
Shaft moment of inertia	0.01
Rotational friction coefficient	0.1
Screw lead pitch	0.008 m
Leadscrew efficiency	40%
Gearbox ratio	2

IOP Conf. Series: Materials Science and Engineering 1244 (2022) 012010 doi:10.1088/1757-899X/1244/1/012010

2.3 *Vehicle Dynamic Model*

The electromechanical actuator in Section 2.2 will push the braking pedal of the vehicle, causing the vehicle to experience change in speed and position. The virtual model of the vehicle model is taken from standard model in MATLAB. Table 2 shows the parameters used for the vehicle dynamic model. Figure 3 shows the dynamic vehicle model developed in the Simscape.

Table 2. Vehicle Parameters Used in Simulation

Mass of the chassis	1200 kg
Engine type	Fuel injection
Transmission	Simple gear with ratio 2
Wheel drive	Two-wheel drive
Inclination	Flat surface

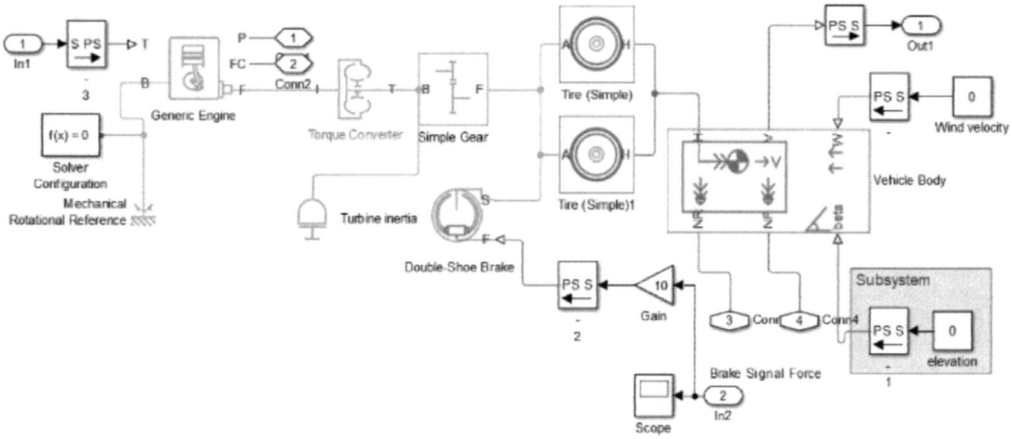

Figure 3. Vehicle dynamic model in Simscape

3. Simulation Results and Response of the Vehicle Speed

The linear actuator and vehicle model are integrated in Matlab-Simulink environment in such a way that the actuator will provide force signal towards the brake pedal. Figure 4 shows the full schematic in Matlab-Simulink to simulate the environment.

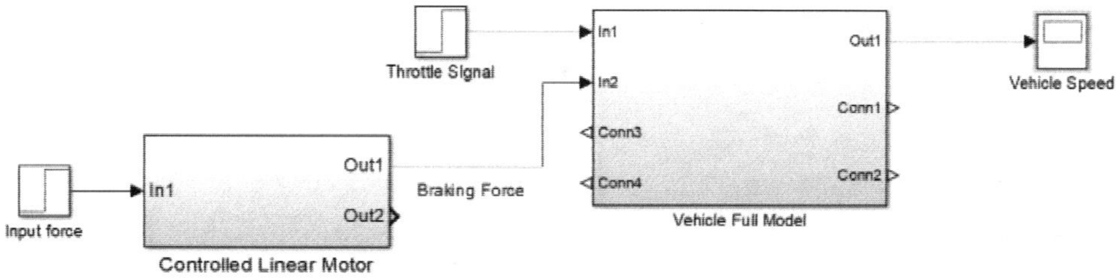

Figure 4. Linear actuator, pedals and Vehicle dynamic model assembly

In this simulation, the vehicle is assumed to have an automatic transmission, in which when at extremely low speed, the vehicle will move with idle speed. The throttle signal in the simulation indicates that the fuel valve opening at the engine is around 0.1 %, causing the vehicle to move at around 2 km/hr; a standard speed in a traffic jam.

3.1 Actuator Response

The complete model was simulated to examine its output linear force and the linear position. A 1 s step voltage signal of 12 V is injected into the actuator, and Figure 5 shows the force response of the actuator.

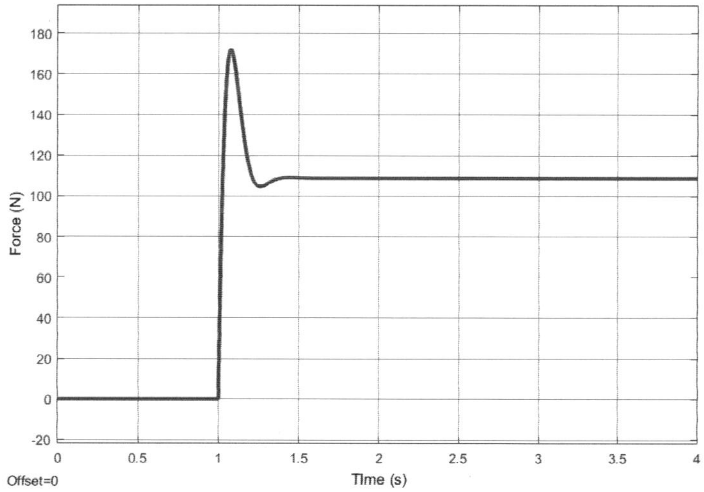

Figure 5. Linear Actuator for open loop response

It is observed that the actuator is able to produce a steady force of about 100 N when stimulated with 12 V step voltage, which is enough to mimic the typical human leg force on brake pedal. In addition, the response time of the actuator is less than 0.2 s, quick enough to press the pedal to quickly control the vehicle speed during traffic jams.

A Proportional-Integral-Derivative (PID) controller is designed and optimized to improve further the force response in terms of time. The equation for PID controller can be written as

$$u(t) = K_p e(t) + K_d \frac{de(t)}{dt} + K_i \int_0^i e(t)dt \tag{7}$$

where $u(t)$ is the input voltage to the pedal and to the vehicle dynamic, $e(t)$ is the feedback error between the desired input force and the measured of the input force. The K_p, K_i and K_d are the proportional integral and derivative gains respectively and are automatically tuned in suitability with the plant implemented, as shown in Figure 6.

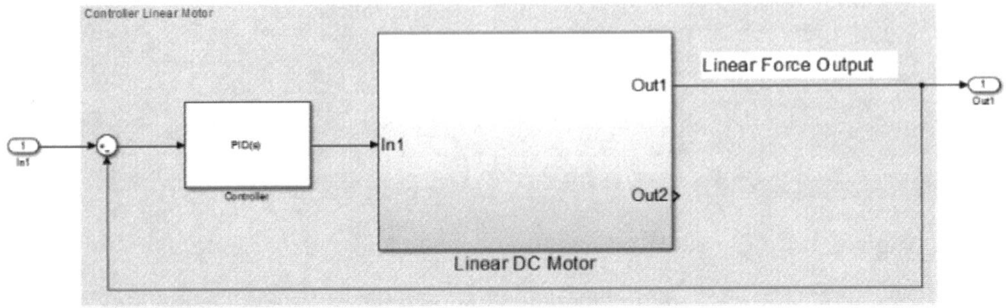

Figure 6. Linear Actuator force response with PID controller

From Figure 7, it is shown that the settling time is improved as well as overshoot magnitude. In the next section, this controlled actuator will be implemented on the vehicle brake to observe the speed response.

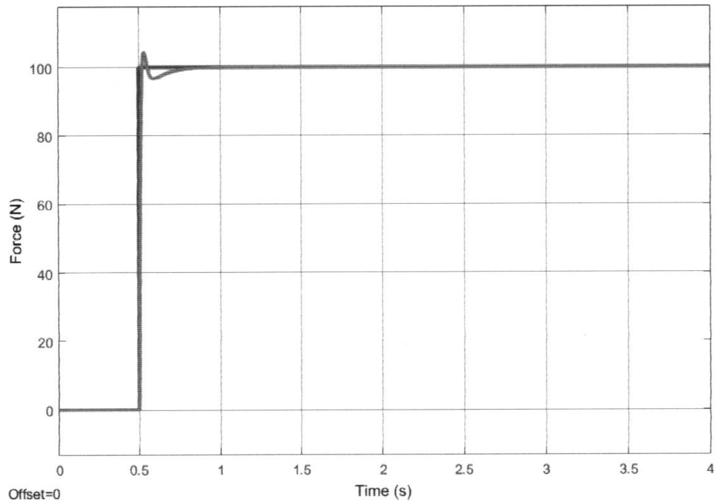

Figure 7. Linear Actuator force response with PID controller

3.2 *Vehicle Speed Response*

Figure 8 shows the vehicle speed response when controlled linear actuator is fed with force to the brake pedal. In this simulation, the force output required to be produced by the actuator to brake pedal is 100 N, acting within less than 0.5 seconds.

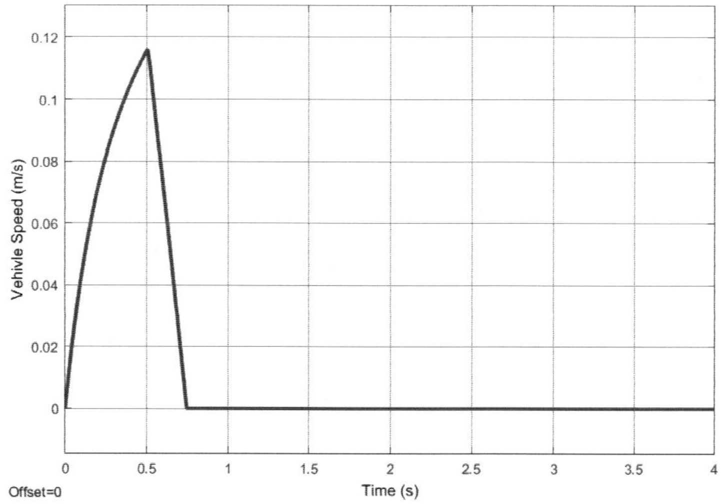

Figure 8. Vehicle speed response when brake pedal is pressed with 100 N from the controlled linear actuator

It is shown that when the linear actuator is pressing the pedal, the vehicle is brought to a complete stop within less than 1 second, as compared to a typical human brake reaction time which is maximum 2.5 seconds [18]. Figure 9 shows the actuator response when the random inputs are injected into the actuator. The minimum and maximum force is 0 N and 180 N respectively. Figure 10 shows the vehicle speed response when pushed with random signals with no throttle was set. But if the throttle was set to have about 0.1 %, which in this case was taken to be the inverse of the brake force, then the speed responses can be seen in Figure 11.

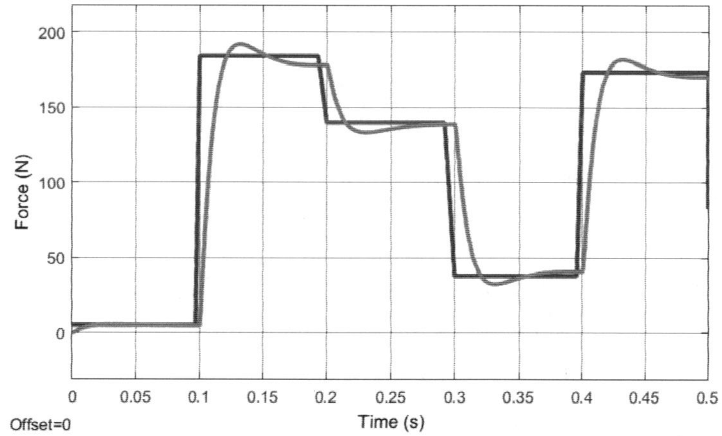

Figure 9. Random input signal with controlled output force of the actuator

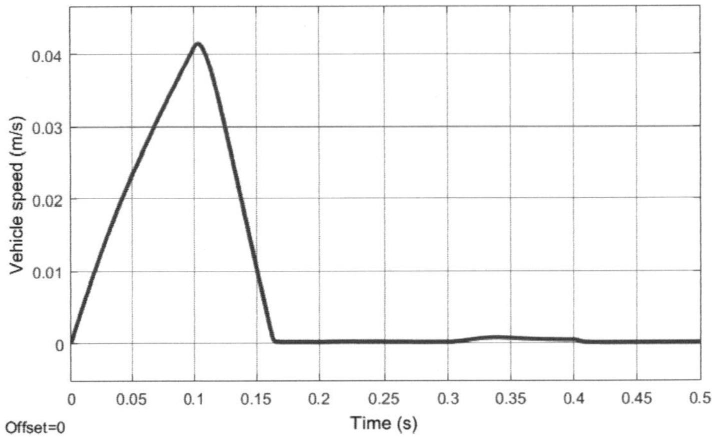

Figure 10. Vehicle speed response when brake pedal is pressed with random output from actuator and zero throttle.

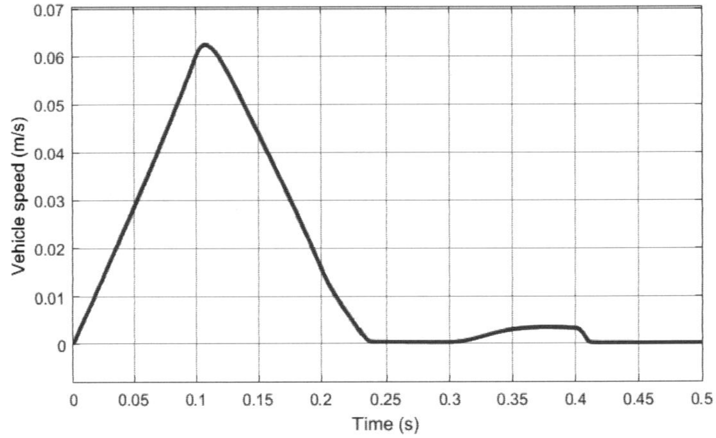

Figure 11. Vehicle speed response when brake pedal is pressed with random output from actuator and 0.1 % throttle opening.

It has shown that the linear actuator works well under random signals, making it potentially able to replace the human legs in performing repeating pedal pressing during traffic delay.

4. Conclusion & Future Works

In this study, it is proven that the actuator is able to produce acceptable force output and response time towards brake pedal within supplied voltage range. Potentially, the actuator is able to replace the human leg in performing the pedal pressing. Fail-safe mechanism will be implemented in the near future in the case of the presser failure with an alarm. For future works, the intelligent controller will be implemented in the system to improve further the response time as well as force output considering variation of constraints of road, vehicle, driver as well as the environment.

Acknowledgement
The authors would like to acknowledge the sponsors of this research, which is under the IIUM-UMP-UITM Sustainable Research Collaboration Grant 2020 (SRCG) - SRCG20-015-0015.

References

[1] NST, "NST Leader: Traffic-Choked in KL," *NST*, Mar. 11, 2020.
[2] A. Ashaari, "Malaysians Could Spend Well Over An Hour On The Road Daily This Raya," *Malaymail*, Mar. 2019.
[3] P.-H. Ting, J.-R. Hwang, J.-L. Doong, and M.-C. Jeng, "Driver Fatigue And Highway Driving: A Simulator Study," *Physiol. Behav.*, vol. 94, no. 3, pp. 448–453, 2008.
[4] C. M. Martinez, M. Heucke, F.-Y. Wang, B. Gao, and D. Cao, "Driving Style Recognition for Intelligent Vehicle Control and Advanced Driver Assistance: A Survey," *IEEE Trans. Intell. Transp. Syst.*, vol. 19, no. 3, pp. 666–676, 2018, doi: 10.1109/TITS.2017.2706978.
[5] Verhoef, E., & Rouwendal, J. (2004). A behavioural model of traffic congestion: Endogenizing speed choice, traffic safety and time losses. *Journal Of Urban Economics*, *56*(3), 408–434.
[6] Macuzic, S., Lukic, J., Glisovic, J., & Miloradovic, D. (2017). Pedal Force Determination Respect To Ride Comfort. *Materials And Science Engineering*. https://doi.org/10.1088/1757-899X/252/1/012010
[7] Gyulyev, N., Voronko, V., Ostashevskiy, S., Ponkratov, D., & Psol, S. (2020). Development of Models for Assessing a Driver's Failure•free Operation in a Transportation System Under Conditions of Traffic Congestion. *Eastern-European Journal of Enterprise Technologies*, *1*(3).
[8] Meng, F., & Yang, L. (2019). Temporal Patterns Of Driving Fatigue And Driving Performance Among Male Taxi Drivers In Hong Kong : A Driving Simulator Approach. *Accident Analysis & Prevention*, *125*, 7–13.
[9] Lin, Y., McPhee, J., & Azad, N. (2020). Comparison of Deep Reinforcement Learning and Model Predictive Control for Adaptive Cruise Control. *IEEE Transactions On Intelligent Transportation Systems*, 1–1.
[10] Munyaneza, O., Munyazikwiye, B. B., & Karimi, H. R. (2015). Speed control design for a vehicle system using fuzzy logic and PID controller. *International Conference on Fuzzy Theory and Its Applications (IFUZZY)*.
[11] Long, M., & Jing, H. (2020). Active Deceleration Control Of Brake-By-Wire System. In IEEE (Ed.), *Chinese Control Conference*. IEEE.
[12] Rosier, C., & Cole, D. (2017). Model Predictive Control To Represent Interaction Of Driver And Active Accelerator Pedal. In P. & P. P, Edelmann (Ed.), *Advanced Vehicle Control* (pp. 100–105). Taylor & Francis.
[13] R. Dindorf, J. Takosoglu, and P. Wos, "Using Of The Brain Activity Of A Disabled Driver To Exert A Pneumatic Actuator Pressure On An Secondary Foot Brake Pedal During Emergency Braking," in *2020 XII International Science-Technical Conference Automotive Safety*, 2020, pp. 1–6.
[14] M. Ariyanto, G. D. Haryadi, M. Munadi, R. Ismail, and Z. Hendra, "Development of Low-Cost Autonomous Emergency Braking System (AEBS) for an Electric Car," in *5th International Conference on Electric Vehicular Technology (ICEVT)*, 2015, pp. 167–171, [Online]. Available: https://ieeexplore.ieee.org/abstract/document/8628442.
[15] H. M. Isa, W. N. Liza, R. Ramli, and M. A. Z. Abidin, "A Review On Electromagnetic Suspension Systems For Passenger Vehicle," in *International Conference On Electrical, Control And Computer Engineering 2011*, 2011, p. 5, [Online]. Available: http://ieeexplore.ieee.org/xpl/login.jsp?tp=&arnumber=5953914&url=http://ieeexplore.ieee.org/iel5/5945018/5953830/05953914.pdf?arnumber=5953914.
[16] Z. G and Zhang S, "Novel Electromechanical Brake Actuator Adopting the Two Way Ball Screw," *SAE Tech. Pap.*, 2015, doi: https://doi.org/10.4271/2015-01-2698.
[17] J. Schwartz, "Design Of An Automobile Accelerator/Brake Pedal Robot For Advanced Driver Assistance Systems," Purdue University, 2017.
[18] D. Clover, "Brake Reaction Time," *Geometric Design: Theory & Concepts*, 2021. https://www.webpages.uidaho.edu/niatt_labmanual/chapters/geometricdesign/theoryandconc

epts/BrakeReactionTime.htm#:~:text=Extensive research has shown that,by the brake reaction time.

IOP Conf. Series: Materials Science and Engineering 1244 (2022) 012011 doi:10.1088/1757-899X/1244/1/012011

Frictional Power Evaluation of Additive-Mixture in Trimethylolpropane Trioleate Oil using Single Cylinder Diesel Engine

A. Z. Syahir[1,*], H. H. Masjuki[2], M. N. A. M. Yusoff[1], T. M. Ibrahim[1]

[1] Centre for Energy Sciences, Department of Mechanical Engineering, Faculty of Engineering, University of Malaya, 50603 Kuala Lumpur, MALAYSIA
[2] Department of Mechanical Engineering, International Islamic University Malaysia, 53100 Kuala Lumpur, MALAYSIA

*syahiramzar@gmail.com

Abstract. Rapid depletion of petroleum sources and increase of demand for environmentally friendly products has urged the need for biodegradable lubricants to be utilized in various applications, especially for internal combustion (IC) engine lubrication. To perform on par with the commercial engine oils, bio-based lubricants need to be further improved through the addition of appropriate additives. This study investigates the synergistic performance of additive-mixture comprising of ionic liquids (IL), glycerol monooleate (GMO), molybdenum dithiocarbamate (MoDTC) and titanium dioxide (TiO_2) nanoparticle in trimethylolpropane trioleate (TMPTO) ester base oil. Lubricity performances of the prepared sample were evaluated in the context of tribological improvement using laboratory tribotester as well as friction power reduction in actual single-cylinder diesel engine. From a prior study, it was found that additive mixture containing 0.93 wt.% IL, 1.49 wt.% GMO, 0.52 wt.% MoDTC, and 1.25wt.% TiO_2 was able to provide substantial friction and wear as well as extreme pressure improvements. The present study was conducted to evaluate whether the formulation can offer the same performance on the actual engine in conjunction to the laboratory bench test. To do so, two different types of engine tests were conducted which are that standard engine performance test and also the Willans line test to evaluate the engine frictional power. The performance of the additives-TMPTO mixture was compared directly to those of commercial SAE 0W-30 engine lubricant. In the standard engine performance test, it was found that the additives-TMPTO mixture performed better than commercial SAE 0W-30 engine lubricant in terms of 2.5% higher maximum torque value, 3% higher overall brake power output, 10% lower brake specific fuel consumption and 12% higher brake thermal efficiency throughout the conducted test speed of 800 to 2300 rpm. Through the Willans line test, it was found that the additives-TMPTO mixture showed better results than the commercial SAE 0W-30 engine lubricant in terms of 20-25% lower frictional power at 900, 1200 and 1500 rpm. Overall, these results further justify the enhanced lubrication properties of the formulated bio-based oil in addition to the prior conducted tribological analyses.

1. Introduction

Globally, the production of oil has been experiencing high demand lately especially with the growing energy demand from all over the world. However, there are issues regarding fulfilling the energy demand because the supply of fossil fuel has been consistently depreciating. Since lubricant oil is among the biggest share in the energy demand, there will be problems of producing engine lubricant and improving engine efficiency if this problem goes unsolved. For instance, piston ring-cylinder is one of the components in internal combustion engine that will be affected if it is not lubricated properly. Without proper lubrication, it will produce more frictional loss from the reciprocating motion between the piston ring and the cylinder.

Simultaneously, researchers have also started to investigate the utilization of bio- based lubricants for variety of applications including internal combustion engine. Bio-based lubricants are oil that is derived from renewable sources. The common source of bio-based lubricant is vegetable oil. Unlike conventional lubricants, bio-based lubricant has minimal pollution risk and inexpensive. Hence, bio-based lubricant has always been claimed to be an excellent candidate to be a lubricant because of the environmental and economic concerns it fulfills. Bio-based lubricants are known to have excellent lubricity, high flash point, high viscosity index and good resistance to shear compared with mineral oils, which make them highly potential for IC engine application [1-3]. Nevertheless, to achieve maximum performance, incorporation of additives is still required. The challenges to date include the suitability of existing additives with bio-based oils, as opposed to the conventional mineral and/or synthetic hydrocarbon base oils.

For better lubrication performance, specific additives are added into the lubricant to enhance the lubricant ability, mainly to resist friction and wear. Many research have been done regarding this matter to innovate lubricant formulation that possesses excellent tribological properties. There are currently various types of additive groups which responsible to tackle different aspects of lubrication performance. However, only a few from those groups have a direct influence on the tribological characteristics, which are friction modifiers, anti-wear and extreme pressure. It is known that other additive groups also have indirect effects towards tribological behavior of the lubricant mixture due to their effect on the properties and rheology. Based on an unpublished, earlier studies conducted by the Authors, the incorporation of trihexyltetradecylphosphonium bis(2,4,4-trimethylpentyl)phosphinate ionic liquid ([$P_{14,6,6,6}$][TMPP]), TiO_2 nanoparticles, glycerol monooleate (GMO) and molybdenum dithiocarbamate (MoDTC) into trimethylolpropane trioleate (TMPTO) bio-based oil produced in overall a satisfactory tribological improvement in terms of 30% better extreme pressure characteristics, 45% lower coefficient of friction (COF) as well as lower wear and smoother surface finishes that those observed by using neat TMPTO. The test was conducted using high frequency reciprocating rig (HFRR) tribotester.

Thus, the present study aims to test and further justify the good performance of the current formulation using an actual engine test bed. The idea is that since the additive-mixture produced satisfactory results in laboratory tribological bench test, it would produce a similar improvement when used in an actual engine, thus, translates to better engine performance and lower fuel consumption.

2. Methodology

In the present study, all material were obtained commercially and used directly without modification. TMPTO ester was selected as the bio-based base stock in this study as it has been widely reported to possess good lubricity and oxidation stability. Also, the viscosity of TMPTO falls within the range of SAE 30 viscosity grade, thus making it a promising base stock for the formulation of engine oil. Two types of friction modifiers (FM), GMO and MoDTC have been widely used as FM in today's engine lubricant to enhance the tribological performance at boundary lubrication regime. Nanoparticles on the other hand, was added as an extreme pressure agent. Nano-titanium dioxide (TiO_2) was selected owing to its nearly-spherical shape has been reported to provide better load carrying capacity during surface interaction.

The additive-mixture was added into TMPTO base oil at concentration of 0.93 wt.% IL, 1.49 wt.% GMO, 0.52 wt.% MoDTC, and 1.25wt.% TiO_2. The composition of additives is based on the Author's previous analysis via a series of optimization processes. Magnetic stirrer and sonicator was used for the additive dispersion process. The performance of the optimized bio-based lubricant (codenamed as OPT) was evaluated using YANMAR TF 120-M single-cylinder diesel engine attached to DYNOmite liquid-cooled eddy-current dynamometer. The engine specifications and the overall set-up are as presented in Figure 1 respectively. Data monitoring, control and analysis was done using the DYNO-MAX™ software which is linked to the dynamometer's data acquisition system. Using the same set-up, two different types of performance tests were conducted which are: (1) the determination of the standard engine performance parameter and (2) the determination of engine frictional power at a constant speed using the Willans line method.

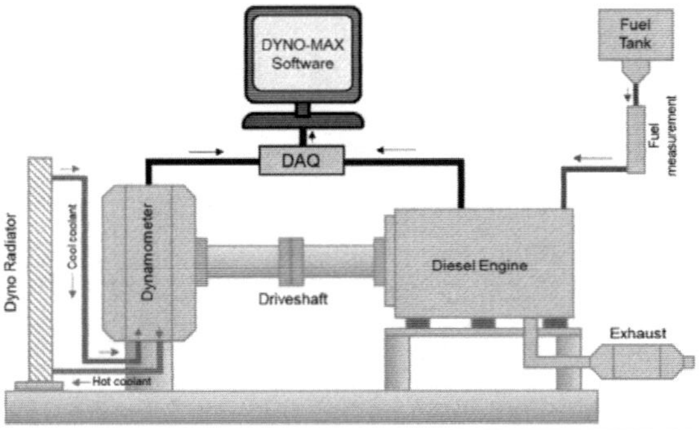

Parameter	Specification
Model	TF 120-M (YANMAR, Japan)
Configuration	Naturally aspirated, single cylinder, manual throttle body control
Maximum speed	2400 rpm
Rated continuous output	7.8kW at 2400 rpm
Injection type and timing	Direct injection, bTDC 17°
Displacement	0.638 L
Lubrication system	Forced lubrication with trochoid pump & hydraulic regulator valve
Lubricant capacity	2.8 L

Figure 1: Engine test bed set-up and specifications

In both tests, the performance of the OPT sample was compared directly to those of fully formulated synthetic engine lubricant (FFL), PETRONAS Syntium 7000 0W-30, which has the same viscosity grade. In terms of fuel, PETRONAS Dynamic Diesel Euro 5 (B7) was used. In the first engine performance test, typical engine parameters such as engine torque, brake power (BP), brake specific fuel consumption (BSFC), and brake thermal efficiency (BTE) were obtained. The engine was set to run at the full-throttle condition with varied engine speeds starting from 2300 to 800 rpm with 300 rpm decrement. In the second test which is the Willans line test, a line of gross fuel consumption against BP at a constant speed condition was plotted using the same diesel engine. The line was extrapolated to zero fuel consumption and such point where the line cuts the BP axis (horizontal axis) is an indication of the frictional power of the engine at that particular speed. In this study, the Willans line technique was conducted at three different engine speeds which are 900, 1200 and 1500 rpm

3. Results and Discussion

3.1. Engine Performance Parameter

Engine performance of an internal combustion engine is usually characterized in terms of engine torque (T), brake power (BP), brake specific fuel consumption (BSFC) and also brake thermal efficiency (BTE) under the specified engine operating condition. While the engine performance test is mainly carried out to determine the performance of fuel used, the same test procedure is also applicable to test the performance of lubricant as it indirectly gives indications of how well the engine system operates [4-6].

As seen in Figure 2, for both tested oil samples, the engine torque increases from 800 rpm and peaks at 1400 rpm, before declining afterward. The maximum torque at 1400 rpm was obtained when using optimized sample (OPT) with a torque value of 34.29 Nm, whereas FFL produced the torque value of 33.57 Nm at the same speed. As the engine speed increases beyond 1400 rpm, the torque gradually decreases due to several reasons. At high speed, lower torque is produced due to insufficient air being supplied due to shorter air intake time, or also known as fuel choking [7]. In addition, mechanical losses are more significant at such high speed [8].

In terms of brake power, it can be seen from Figure 2 that the brake power increases in a pretty much linear trend as the engine speed is increased from 800 to 2300 rpm. At 2300 rpm, the higher brake power value of 7.62 kW was recorded when OPT sample was used. Such value is roughly 3.5%

higher when compared to that of produced by the FFL sample at the same speed. Similar to as observed for torque, the OPT sample produced a relatively higher power output throughout all tested engine speed than FFL. Such improvement is believed due to better lubricity of the OPT sample than FFL.

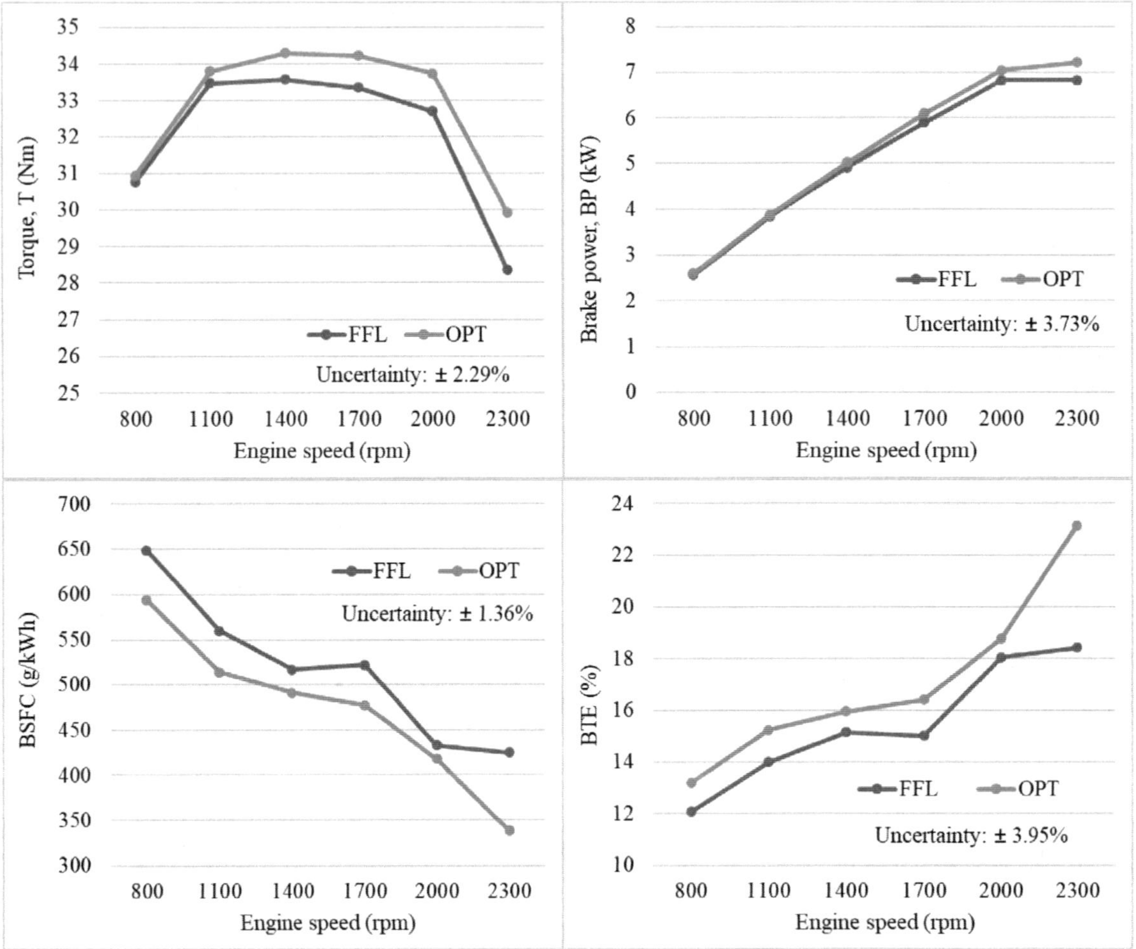

Figure 2: Variation of torque (top-left), brake power (top-right), BSFC (bottom-left) and BTE (bottom-right) against engine speed for FFL and OPT samples

From Figure 2, At 2300 rpm where the highest brake power is recorded, OPT sample showed better fuel efficiency through the approximately 14% lower BSFC recorded compared to that of FFL. Needless to say, lower engine fuel consumption translates to better economic benefits in the long run. In terms of BTE, it is shown in Figure 2 that the BTE is lower at low BP and engine speed and eventually increases with higher BP and engine speed. This is due to the time available for the heat to be transferred to the cylinder walls is relatively longer per cycle at low engine speed condition, thereby allowing more heat loss to occur [9]. Such occurrence results in lower combustion efficiency. Comparing the two tested samples, OPT produced roughly 9 and 16 % higher BTE at 800 and 2300 rpm respectively when compared to those of FFL.

3.2. Willans Line Test

The engine's Willans line represents the relationship between fuel energy input and engine output, while the engine speed is held constant [10]. Figure 3 represent the Willans line obtained at 900, 1200 and 1500 rpm respectively when run using OPT and FFL samples. In each plot, the line is extrapolated and the brake power values when the fuel consumption is zero (x-intercept values) are labelled. Such values correspond to the frictional power of the engine at that particular speed [11]. It can be seen that the trend of the plotted Willans line varies with each operating engine speed. For the line of 900 rpm, the line increases linearly until a point where it starts to grow exponentially. A similar trend is observed for the line of 1200 rpm, however, the exponential increase is less significant when compared to that observed for 900 rpm. At 1500 rpm, the line was observed to increase only in a linear manner.

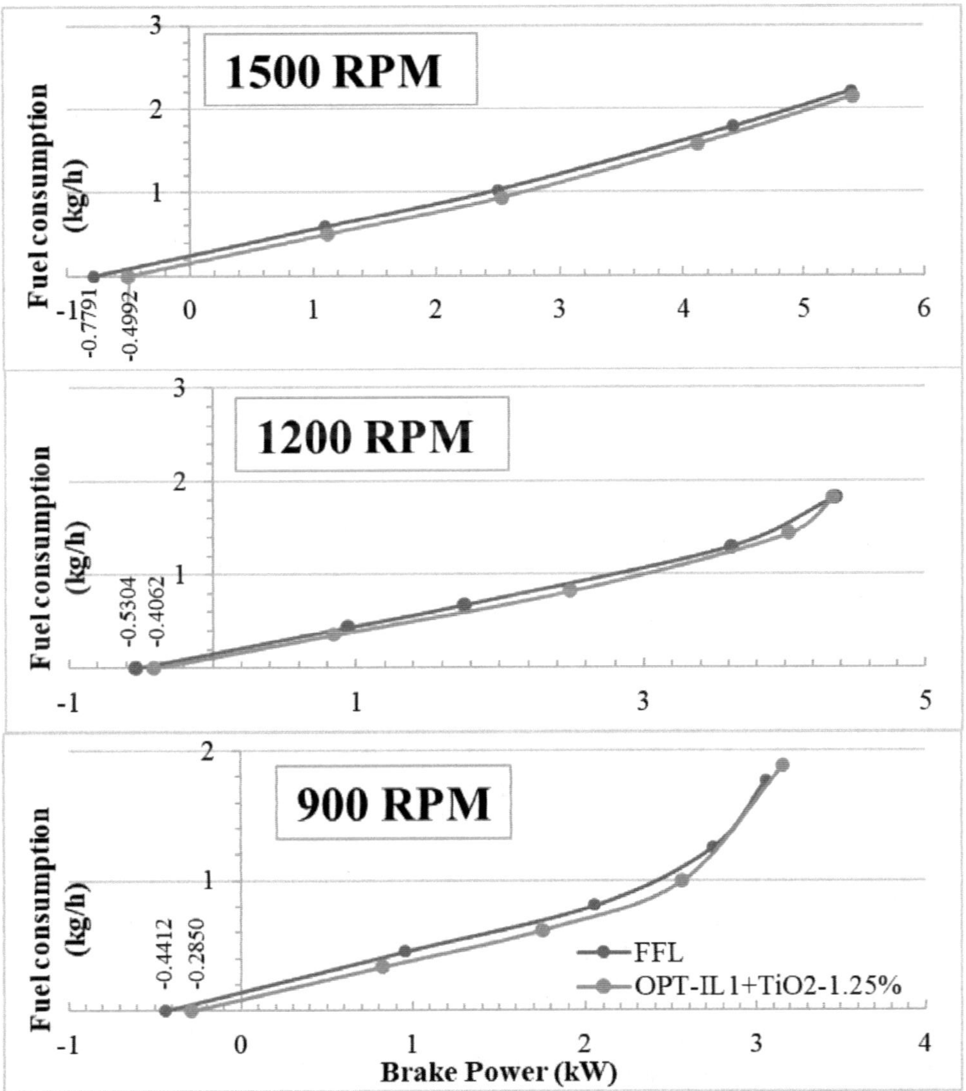

Figure 3: Extrapolated Willans line at the RPM of 1500 (top), 1200 (middle), 900 (bottom) for FFL and OPT samples

It was found that the usage of OPT lubricant resulted in roughly 20% lower frictional power at 900 rpm and 25% lower frictional power at both 1200 and 1500 rpm. Such improvement can be

IOP Conf. Series: Materials Science and Engineering 1244 (2022) 012011 doi:10.1088/1757-899X/1244/1/012011

attributed to the effectiveness of the OPT sample in lubricating the moving engine components. This is in coherent with the results obtained through the previously carried out tribological bench test, where it was deduced that optimized OPT sample (IL1+1.25%-TiO_2) performed better than FFL. The superior friction reducing capability of OPT sample is achieved through various tribological mechanisms which include tribolayer formation via surface reactions as well as mechanical roller bearing actions from the combination of TMPTO base oil, IL1, MoDTC, GMO and TiO_2 nanoparticles. Therefore, it can be said that in addition to the tribological bench test, the synergistic tribological improvement of the base oil and additives combination is further justified through the engine performance and Willans line test.

4. Conclusion

Through the Willans line test using a single cylinder diesel engine, it was found that the formulated bio-based oil comprising of 0.93 wt.% IL1, 1.49 wt.% GMO, 0.52 wt.% MoDTC and 1.25 wt.% TiO2 nanoparticle showed better results than the commercial SAE 0W-30 engine lubricant in terms of 20-25% lower frictional power at 900, 1200 and 1500 rpm. In the standard engine performance test, the formulated bio-based oil performed better than commercial SAE 0W-30 engine lubricant in terms of relatively higher maximum torque value, higher overall BP output, lower BSFC and higher BTE throughout the conducted test speed of 800 to 2300 rpm. Overall, these results further justify the enhanced lubrication properties of the formulated bio-based oil in addition to the prior conducted tribological analyses

It is worth mentioning that apart from all the conducted analysis prior to this engine test, the formulated bio-based lubricant in this project is still far from the standard of those commercially available FFLs in terms of other lubricant properties which are not related to tribology i.e. deposit control, low temperature properties, etc. Nevertheless, the results obtained from these tests can still provide an early indication regarding the performance of the additive-TMPTO mixture in IC engine.

References

[1] Salimon, J., N. Salih, and E. Yousif, *Biolubricants: Raw materials, chemical modifications and environmental benefits.* European Journal of Lipid Science and Technology, 2010. **112**(5): p. 519-530.

[2] Fernando, S., M. Hanna, and S. Adhikari, *Lubricity characteristics of selected vegetable oils, animal fats, and their derivatives.* Applied Engineering in Agriculture, 2007. **23**(1): p. 5.

[3] Soni, S. and M. Agarwal, *Lubricants from renewable energy sources – a review.* Green Chemistry Letters and Reviews, 2014. **7**(4): p. 359-382.

[4] Ali, M.K.A., et al., *Fuel economy in gasoline engines using Al2O3/TiO2 nanomaterials as nanolubricant additives.* Applied Energy, 2018. **211**: p. 461-478.

[5] Chua, M.I.H., et al., *The hBN Nanoparticles as an Effective Engine Oil Additive to Enhance the Durability and Performance of a Small Diesel Engine.* Journal of Mechanical Engineering, 2017. **SI**: p. 103-112.

[6] Öğüt, H., et al., *The effects of the use of vegetable oil based as engine lubrication oil on engine performance and emissions in diesel engines.* Energy Sources, Part A: Recovery, Utilization, and Environmental Effects, 2019: p. 1-16.

[7] Ralbovsky, E., *An Introduction to Compact and Automotive Diesels.* 1997: Delmar Publishers.

[8] CarBike Tech. *What Is Engine Torque? It's Characteristics & Formula.* Technical Anatomy 2019 January 2020]; Available from: https://carbiketech.com/engine-torque/.

[9] Alahmer, A., *Influence of using emulsified diesel fuel on the performance and pollutants emitted from diesel engines.* Energy Conversion and Management, 2013. **73**: p. 361-369.

[10] Pachernegg, S., *A closer look at the Willans-line.* 1969, SAE Technical Paper.

[11] Mathur, M.L. and R.P. Sharma, *Internal combustion engines.* 2005: Dhanpat Rai Publ.

IOP Conf. Series: Materials Science and Engineering 1244 (2022) 012012 doi:10.1088/1757-899X/1244/1/012012

Synthetization of TiC surface hardening using TIG melting technique - The effect of working distance

A. N. Md Idriss[1], M. A. Maleque[1]* and A Afiq[2]

[1]Department of Manufacturing and Material Engineering, International Islamic University Malaysia, P.O. Box 10, 50728 Kuala Lumpur, Malaysia
[2]Materials Engineering and Testing Unit, TNB Research Bhd, Selangor

*Corresponding author's email address: maleque@iium.edu.my

Abstract. Surface hardening with and without hard reinforcing material can be achieved by melting technique on the top substrate area via high energy input fusion. The aim is to apply TIG torch heat input of 1344 J/mm at two different working distances such as 0.5 mm and 1.5 mm to incorporate the preplaced TiC particulates on the surface of AISI 4340 low alloy steel. Results pertaining to the microstructural features, melt pool sizes, defects, topographies and microhardness were investigated using the optical microscope, scanning electron microscope and Vickers microhardness tester. The results showed that the heat loss through the arc column at 1.5 mm of working distance resulted small melt pool geometry and poor dissolution of TiC particulates. Lower working distance increases heating value leading more precipitation of TiC whilst minimizing those the undissolved particulates. The surface submerged at working distance of 0.5 mm produced microhardness ranging from 800 HV to 1500 HV. The work suggested that heat fusion can be enhanced by marginal working distance whilst saving the energy to melt instead of increasing the used voltage, current, decreasing scanning speed, or adding fluxes or shielding with special gasses.

1. Introduction

Many works have shown that greater melting depth from extensive reinforcement dissolution with substrate were achieved by gradual increase of energy inputs using lasers or TIG [1-5]. Others have reported that melt pool size expansions coupled with high dissolution are achieved by manipulating types of used shielding gasses to aid the expansion of melt pool under the TIG works [5,6]. Used of fluxes served the same purpose for diluting the molten metal [2]. Increasing heat inputs, using special shielding gasses, or even adding fluxes to expand the melt volume may incur some amount of additional cost or complexity processing setup and techniques for experimental works. Very little information is available through publication to explain the beauty of manipulating working distances hence endeavoring bigger resolidified melt [7,8]. Altering working distances in favour to have less or more melt involved no additional cost. Our experimental work showed that fusing the TIG arc at certain heat input of 1 mm working distance is limited to give dissolution of the TiC on the AISI 4340 low alloy steel substrate [9]. This work delivers information of melt pool sizes with regards to their HAZ, defects, microstructural features, and microhardness from used working distances at 0.5 mm and 1.5 mm. Current results compares with previous published work at 1 mm working distance under the same processing condition as mentioned elsewhere [1].

2. Experimental works

In this work, the TiC particulates was used as the reinforcing material. The size of the particulate's specification given by the manufacturer ranges from 45 μm to 100 μm. The substrate uses machined low alloy steel sample to 40 × 100 × 14 mm followed by thorough cleaning to remove debris or oil on the surfaces. Liquid PVA binder was mixed with weighed TiC particulates to form slurry before they were manually preplaced on the surface of low alloy steel substrate at 1 mg/mm^2 content. Subsequently, the sample was heated to 80°C to for 1 hour to remove the moisture from the particulates paste. The Telwin TIG 165 machine was used to produce melt pool at the speed of 1 mm/s. Keeping the voltage at 35V and arc current of 80 amp, the single layer was fabricated at 1.5 mm, and 0.5 mm of working distances. In this work, the heat input at 1344 J/mm calculated from a simple mathematical equation [10] is kept constant. Only working distances are varied to distinguish difference of processing conditions. Argon gas at 20 lit/min was used to shield the melt pool against excessive oxidation while melting the surfaces. The illustration describing the TIG process melting the powder followed by the substrate are shown in Fig 1. The sectioned samples were subjected to grinding and polishing using SiC abrasive papers and padded alumina powders that spinned on the rotating disk. Etching was conducted using nital solution for a few second serving the purpose to reveal the microstructural features before being held under the JEOL-5600 scanning electron microscope for further observations. Nikon Epiphot 200 optical microscope was used to classify the topography of all samples while Wilson Wolpert microhardness testing machine was used to determine the samples hardenability of 500 gf indentation load at 10 seconds delay.

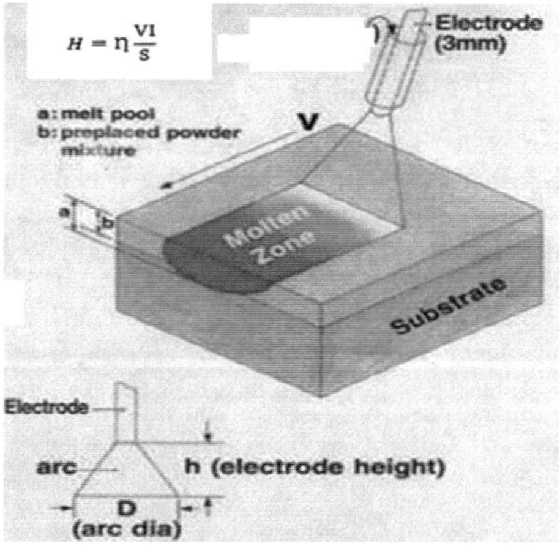

Figure 1. Illustration describing the preplaced powder on the substrate followed by TIG scanning forming the molten zone. The effect of varying the electrode height corresponding to diverse arc diameter (melt width) is also shown in the figure.

3. Results and discussions

The melt geometrical features associated with respective HAZ fused from 1.5 mm to 0.5 mm of working distances are shown in Table 1. The hemispherical melt shapes exclusively contain carbides and passing over than the interphase produced martensitic structure because of the rapid cooling from the austenitic condition [1,9]. The results in Table 1 shows that the decreased of working distance increases the melt pool sizes and the HAZ. During the melting process, increased of arc length to 1.5 mm caused the reduction of heat intensity because of heat losses through the arc column. Arc would not bridge to the metal surface when very large working distances are employed. Arc possesses conical shape and increasing the working distance increases the spot size resulting fusion energy to be much more distributed described in Fig 1.

When this happens, the arc would lose its intensity to melt. Therefore, having higher working distances reduces melt pool sizes to 0.834 mm and 3.15 mm of depth and width respectively in the presence of energy losses through the column. Adding more voltage is one technique to instantly stabilize the arcing problem seeing this method potential to capture more preplaced powder and substrate to melt. Others had increased of hardness is associated with increased of working distance because of the melt that rapidly solidifies [7] As the gap is narrowed from 1.5 mm to 0.5 mm, fusion energy were high enough to cause extensive TiC and substrate melting followed by the enlargement of those HAZ. Under this condition the Gaussian energy distribution grows making increased of heat intensity which lead to the high melt depth and width of 1 mm and 3.47 mm respectively. This work was limited to 1.5 mm of working distances. Increasing over than 1.5 mm may cause uncertainty for arcing to initiate and prone for oxidation. So many produced tracks are kept at 1 mm in the previous investigation. A work showed that gradually increasing the working distances from 1 to 7 mm increases the oxygen content in the melt pool [8].

Table 1. Measured melt geometrical figures and HAZ at working distances. All dimensions in mm.

Working Distance	Melt depth	Melt width	HAZ
0.5	1	3.47	1
1.0	0.88	3.2	1.17
1.5	0.834	3.15	1.1

The surface topographies, microstructures and the EDX results are shown from Fig.2 (a - f) of the melted samples from 1.5 mm to 0.5 mm working distances. All micrographs exhibited crack free topographies and across cross sectioned surfaces. This shows that the coexistance of TiC and the low alloy steel forming the melt pool are compatible and tolerates developed stresses while heating and cooling fused under the arcing TIG method. In fact, this flaw was absent using other processing condition [1,9]. To have surfaces consisting of ripples the melt must be high and are achieved by fusing high energy input at low percentage of TiC amount. This is to prevent the sudden solidification. In surface melting, sudden solidification prevents vigorous melt turbulence, and their solidification may appear as smooth surface as shown in Fig. 2(a) at 1.5 mm working distance followed by bulging at the edges potential enough to increase the surface roughness [11]. At 1 mm and 0.5 mm rippling marks were seen as shown in Fig. 2(b) and (c) and these happened because the slow solidification of melt waves from the beyond the liquidus melting temperature. One may suggest that the appearance of rippling marks is a clear identification that high dissolution of matrix and particulates took place underneath the surface. Greater length or sizes of the ripples exert extensive dilution in the melt pool. Rippling marks from the two working distances of 1 mm and 0.5 mm Fig. 2 (b, c) affirmed the presence of the high fusing temperatures that were responsible to give large melt pool sizes in Table 1.

Micrograph in Fig. 2(d, e) exhibited the cross sectioned surface of the samples fused at 1.5 mm and0.5 mm respectively near the arc source while the EDX result of a reprecipitated carbide from Fig. 2 (e) is shown in Fig. 2f. Pores are seen especially at the melt matrix interfaces due to the entrapped gasses from the burnt binder as in the inset measuring about 100 µm (2d). At the melt matrix interphase, the sluggish melt restrained the gas from being escaped thus many seen along this region. Formation of entrapped pores of gasses from the burnt PVA binder was also being reported in one latest research of preplaced NiTi powder on titanium surface overlapped at certain percentages [12]. Another work showed that carboxymethylcellulose (CMC) concentration of 2 wt% added with water produces 0.5 % of porosity compared to the CMC concentration at 5 wt% showing released of entrapped gasses at 3.0% both exemplified by the Fe/TiC composite layers under the reactive casting coatings [13]. Under this circumstance, it is postulated that the increase of powder content through powder preplacement method may increase the number of gasses released into the melt pool because the presence of more binder. The micrograph in Fig. 2d shows undissolved and reprecipitated TiC microstructures at an overwhelmingly large amount of steel matrix. The poor heat fusion at 1.5 mm of working distance did not successfully melt most particulates even they are the closest to the arc source. The remaining of many large particulate is an evident of reprecipitation and dispersion of those carbide are not encouraged. As TiC is lighter than the steel matrix, they tend to float as solidification proceed upon high dissolution of those particulates. Reprecipitation produces the TiC in the form of cubic, globular, or dendritic microstructures which can be seen in Fig. 2e (inset). Depending on the amount of heat input and amount of preplaced TiC, successfully dissolved carbide would lead to the homogeneous dispersion of reprecipitated ones that narrows the area of the matrix. In this work the TiC microstructure is confirmed by EDX analysis depicting association of Ti and C as shown in Fig. 2(f).

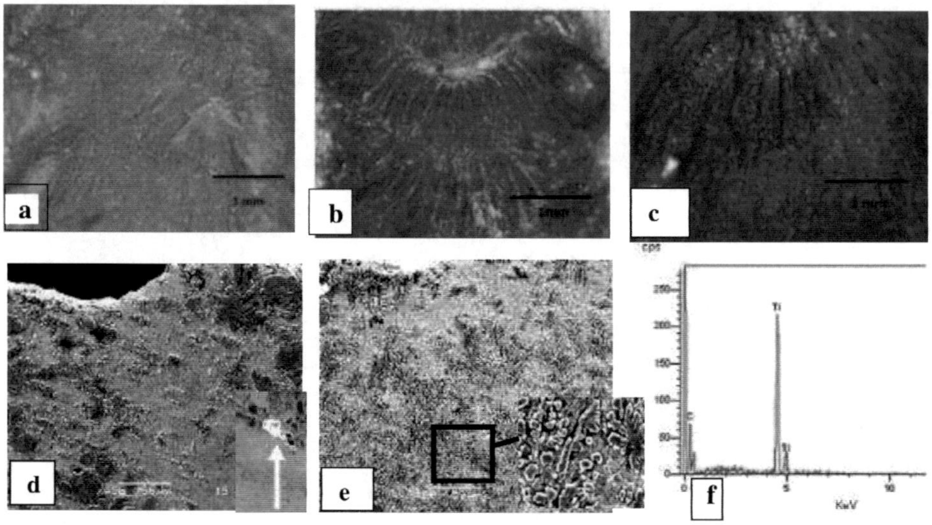

Figure 2. Pictures of surface topographies from melted tracks at 1.5 mm, 1.0 [1] and 0.5 mm of working distances from (a)-(c). The cross sectioned surfaces showing more undissolved with an inset of a pore(d) at 1.5 mm working distance. Many reprecipitated TiC (e) at 0.5 mm near the arc source magnified shown in an inset and the EDX result in (f) of a carbide from (e).

4. Microhardness

The microhardness profile of the produced track at the 0.5 mm and 1.5 mm working distances are shown in Fig. 3. Each sample contains 3 hardness clubs and ranges of high hardness from about 700 HV to 1500 HV in the melt pools followed by martensitic HAZ leveraging at 600 HV and the lowest of about 300 HV across the undisturbed substrate. In comparison hardness with the 0.5 mm working distance was found higher reaching 1500 HV than the 1.5 mm one of only approximately 1030 HV. The average microhardness of the melted surface at 0.5 mm working distance from 9 indentations from the melt pool top to the bottom showed 1120 HV. In comparison the 1.5 mm working distance gave 890 HV. Developed microhardness showed that the coatings can protect the substrate against abrasions. The molten pool enlarges causing more steel to be dissolved under the 0.5 mm working distance and is sufficient to produce the high amount of carbide near the last region to solidify which is near the arc source. Because they populated in this region microhardness is high seeing over than 1000 HV up to 600 μm of depth. At 500 μm from the surface, the hardness was 1500 HV which is about twice higher that the overlapped TiC metal matrix composite sample used for the room temperature sliding wear work [14]. Below than 500 μm the hardness gradually declines reaching 1 mm with a similar behaviour reported elsewhere [15].

The implication of increasing the working distance narrowed the melt pool size to the smaller one (Table 1). Reprecipitation is scarce even though at the highest heating region underneath the arc source, more matrix was seen rather than TiC microstructures. Despite less amount of carbide, the undissolved TIC interphase bonding is not created and is considered as a factor that also reduces hardness. The sluggish melt far away from the arc source is less efficient to turbulence the liquid which had led mostly undissolved particulate to sitting just above the melt matrix interphase (Fig. 3, inset). The microhardness was seen to abruptly reduce from the surface at a little over than 1000 HV to 720 HV of 500 μm depth as carbide decreases. Below than this depth hardness is brought higher due to the mixed population of undissolved and reprecipitated carbide reaching 1000 HV above the interphase (inset). The 1 mm working distance microhardness was at the middle between those 0.5 mm and 1.5 mm while the microstructure was a blend of homogenously distributed undissolved and re-precipitated TiC [1].

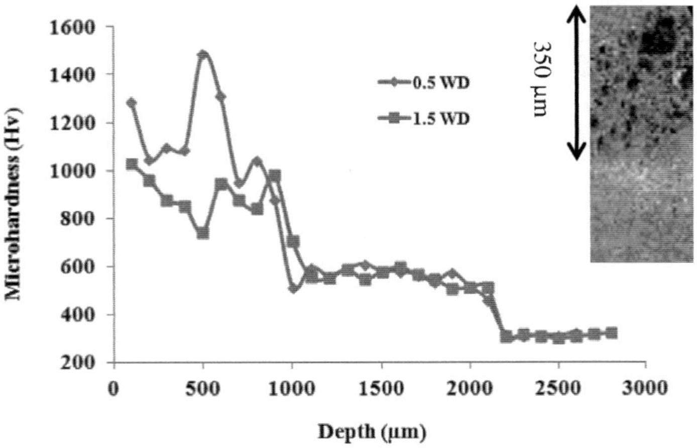

Figure 3. Microhardness profile of the samples produced at 0.5 mm and 1.5 mm working distances. Inset shows the populated TiC microstructures above melt matrix interphase by the 1.5 mm sample.

During this investigation, one disadvantage of bringing the gap closer for melting is that the arcing electrode may bump on the coincided TiC powder preplaced on that AISI 4340 low alloy steel making the process to stop. It needs a proper configuration to restart the process if this was likely to happen. Nevertheless, reducing gaps helps to overcome the limitation of the amount of heat input that the TIG machine could give hence securing the machine from oversupply of energy produced. When the pools are excessively melted, it would require overlapping distances to be reduced to maintain equal depth on the weak substrate. The number of passes needed to span for a coverage is also reduced when overlapping percentages are reduced [16-18].

5. Conclusion

Coated tracks fused at 1344 J/mm of 0.5 mm and 1.5 mm working distance were successfully produced to incorporate the TiC particle on the surface of AISI 4340 low alloy steel. The samples were crack free associated with some trapped pores which was due to the burnt binder forming gasses that were not being able to escape. Energy losses via extending the arc column resulted less heat input as the arc width broadened as so retained more of undissolved TiC particulates, hence reducing the hardness. Reprecipitation masses of carbide encouraged hardness seeing majority depth over than 1000 HV. This work exemplifies that reducing the working distance helps to increase TiC dissolution on the surface of the substrate at no cost and is an alternative way to increase dissolution instead of increasing current, voltage, adding fluxes, mixing variety of gasses to enhance heat absorption or by preheating.

References

[1] Mridha, S., Md Idriss, A.N., Maleque, M.A., Suryanto and Souad, A. (2012). Effect of voltage on the consolidation of TiC particulates on steel substrate fused by TIG welding arc. International Journal of Mechanical and Materials Engineering, 7, 48-

53.
[2] Paraye, N.K., Neog, S.P., Ghosh, P.K., Das, S. (2020). Surface modification of AISI 8620 steel by in situ grown TiC particle using TIG arc. Surface and Coating Technology, Article in press, https://doi.org/10.1016/j.surfcoat.2020.126533.
[3] Maleque, M.A., Bello, K.A.., Md Idriss, A.N., Mridha, S. (2013). Processing of TiC-CNT hybrid composite coating on low alloy steel using TIG torch technique. Applied Mechanics and Materials (378), 259-264.
[4] Sekhar, B.R., Nayak, R.K., Rout, S.R., Masanta, M. (2020). Wear characteristic of TiC coated AISI mild steel fabricated by TIG cladding method. Materials Today: Proceeding 26, 3288- 3291.
[5] Muñoz-Escalona, P., Mridha, S., Baker, T.N. (2014). Effect of shielding gas on the properties and microstructure of melted steel surface using a TIG torch. Advances in Material and Processing Technologies, 1 (3-4), 435-443.
[6] Muñoz- De Escalona, P. Mridha, S., Baker, T.N. (2016) Effect of silicon carbide particles size on microstructure and properties of a coating layer on steel produced by TIG technique. Advances in Material and Processing Technology. 2(4), 451-460.
[7] Singh, J., Thakur, L. and Angra, S. (2019). Effect of Argon flow rate and standoff distance on the microstructure and wear behaviour of WC-CoCr TIG cladding. Journal of Physics: Conference series 1240, 012162
[8] Li, D., Lu, S, Dong, W., Li, D., Li, Y. (2012). Study of the law between the weld pool shape variations with the welding parameters under two TIG processes. Journal of Materials Processing Technology, vol. 212, 128-136.
[9] Md Idriss, A.N. and Mridha. (2012). Microstructure of TIG melted composite coating on steel produced using 1.0 and 1.5 mg/mm^2 TiC at an energy input of 2640 J/mm. Advance Material Research, 576, 467-470.
[10] Easterling, K.E. (1992). Introduction to Physical Metallurgy of Welding, Butterworth-Heinemann, London.

[11] Mridha S and Baker, T.N. (2007). Incorporation of 3 μm SiCp into titanium surfaces using a 2.8 Kw laser beam of 186 and 373 MJm^{-2} energy density in a nitrogen environment. Journal of Materials Processing Technology, 185 (1-3), 38-45.
[12] Prasad, R., Tanaji, Waghmare, D.T., Kumar, K., Manoj, M. (2020). Effect of overlapping condition on large area NiTi layer deposited on Ti-6Al-4V alloy by TIG cladding techique. Surface and Coating Technology, 385, 125417.
[13] Szymański, L., Olejnik, E., Tokarski, T., Kurtyka, P., Drożyński, D., Żymankowska-Kumon, S. (2018). Reactive casting coatings for obtaining in situ composite layers based on Fe alloys. Surface & Coatings Technology, 350, 346-358.
[14] Md Idriss, A.N., Maleque, M.A., Yaacob, I.I., Nasir, R.M., and Baker, T.N. (2016). Microstructural aspects of wear behaviour of TiC coated low alloy steel. Material Science and Technology. 32(4), 303-307
[15] Lailatul, P.H., Maleque, M.A. (2017). Surface modification of duplex stainless steel with SiC preplacement using TIG torch cladding. Procedia Engineering 184, 737-742.
[16] Arbogast, A., Roy, S., Nycz, A., Noakes, M.W., Masuo, C., Babu, S.S. (2020). Investigating the linear thermal expansion of additively manufactured multi material joining between Invar and Steel. Materials, 13, 5683.
[17] Mridha, S., Md Idriss, A.N., Maleque, M.A.,Yaacoh, I.I. and Baker, T.N. (2015). Melting of multipass surface tracks in steel incorporating titanium carbide powders. Materials Science and Technology, vol. 31 (11), 1362-1369.
[18] Sahu, A., Raheem, A., Masanta, M., Sahoo, C.K. (2020). On the constancy in wear characteristic of large area TiC-Ni coating developed by overlapping of TIG arc scanning. Tribology International, 151, 106501.

Green Vehicles, Incentives and Policies: A View from Logistics Companies in Malaysia

Nur Fadiah Mohd Zawawi[1,*] , Mohd Rafi Yaacob[1], Sazali Abd Wahab[2], Khairil Wahidin Awang[3] , Samihah Ahmed[4] , Ruslee Nuh[5]

[1]Faculty of Entrepreneurship and Business, Universiti Malaysia Kelantan, City Campus, Pengkalan Chepa, 16100 Kota Bharu, Kelantan, Malaysia.
[2]Putra Business School, Office Building of the Deputy Vice Chancellor (Research and Innovation), 43400 UPM Serdang, Selangor, Malaysia.
[3]Faculty of Hospitality, Tourism and Wellness, Universiti Malaysia Kelantan, City Campus, Pengkalan Chepa, 16100 Kota Bharu, Kelantan, Malaysia.
[4]Malaysian Technology Development Corporation, Ground Floor, Menara Yayasan Tun Razak, Jalan Bukit Bintang, 55100 Kuala Lumpur, Malaysia.
[5]Faculty of Islamic Sciences, Prince of Songkla University Pattani Campus Rusamilae, Meang Pattani, 9400, Thailand.

*Corresponding author: rafi@umk.edu.my

Abstract. There are limited green solutions have been offered to reduce pollutions caused by logistics transportation as compared to passenger vehicles. Thus, this paper aims to reveal insights from logistics companies on the perspective of policies of sustainability including financial and non-financial incentives from the government of Malaysia in assisting them to obtain socio-environment sustainability and consequently to contribute to the effectiveness of the country's Sustainable Development Goals. To achieve this objective, data were gathered from 134 logistics companies in Malaysia, listed in the Malaysian Logistics Directory 2018/2019. They were analysed using IBM SPSS Statistics 25 software. Utilizing descriptive and correlation analyses, the findings indicated that logistics companies inherently need government support in terms of incentives and supportive policy in order to attain the country's socio-environmental sustainability. Based on the findings, this paper, firstly suggests that the government needs to find a better solution to reduce the environmental impact from logistics transportation from its root, which is achieved by integrating green motor vehicle into logistics transportation. Secondly, the government should develop relevant sustainable policies which guide logistics companies to comply accordingly. This study contributes a new insight into socio-environmental sustainability of logistics industry.

1. Introduction

Environmental pollution due to vehicles' emission has been discussed widely over the past two decades. Countries with high traffic volumes of transportation activities face big problem of environmental pollution. Indonesia for example, one of the biggest populations in the world recorded a 50% increased of carbon emissions level from 2008 till 2018 and an increase of 313% of greenhouse gas emission (GHG) from 1990 level [1]. Jakarta alone, the capital city of Indonesia had more than 8 million of population [2], making it the 10th on the list of most congested cities in the world [3]. Population and transports are therefore contributing greatly towards the GHG emission level. In case of Malaysia, the carbon dioxide (CO2) emission increased dramatically from 126.5 million tons in year 2000 to 250.1 million tons in year 2019. These conditions intrude and impose challenges on the sustainable development of the world as promoted by the United Nations through Sustainable Development Goals (SDGs) or also called as the Global Goals. The main objective of the SDGs is to make a universal call to put an end to the poverty, protect the earth and to ensure people live in peace and prosperity by 2030 [40]. Thus, protect the environment is one of major focuses of SDGs.

Unnaturally, CO2 emission is caused by the combustion of fuel inside the internal combustion engine (ICE), both from spark ignition gasoline engine and compression ignition diesel engine. Although research and development has been done over 30 years ago on reducing the emission including reducing ICE emissions such as nitrogen oxides (NOx) and particulate matter by more than 99% to abide by EPA emissions standards [5], the residue emissions still accumulating day by day, therefore reaching the high level of GHG index. Hence, at international level, various efforts have been done to reduce this environmental pollution, including using Alternative Fuel Vehicle (AFV) [6] and the development of electric vehicle (EV) [7]. They are called as green vehicles (GV). According to United States Environmental Protection Agency (EPA), GV are vehicles which are more efficient and produce less pollution to the air [8]. The AVF and EV functions without gasoline or diesel, they operated based on electricity, methanol, biodiesel, propane, natural gas, hydrogen or ethanol [6]. Though some of these energy sources may produce heat-trapping gases and some other pollution, it still produces less compared to fuel-engine vehicles [9].

Though, till now, most efforts are concern on personal vehicles, without giving due attention to logistics transportation. Logistics transportation becomes a backbone to the world today especially in Covid-19 pandemic, as people under lockdown almost totally changed from physical buying to online purchasing either from locals or internationals. Hence, the significant surge of demand makes logistics companies increased the number of their vehicles especially for road transportation in order to ensure the delivery to end consumers reach them on the estimated time. This happened around the world. During the early hit of Covid-19 in March 2020, logistics services were highly demanded to cater medical and food supplies throughout the globe by using air freight and trucks. Simultaneously, logistics services were critically requested by the people who do online shopping due to lockdown [10]. In this time around, logistics service providers (LSPs) actively rearranged and added more vehicles in order to manage the drastically increased of delivery service especially road transportation like trucks and vans which were needed to complete the supply chain, from initial point like suppliers and ports, to the end point, customers [11]. In this sense, logistics vehicle should also not be left behind in the effort of reducing the environmental pollution, which then leads to socio-environmental sustainability.

Yet, the GV is price sensitive. Most people and companies need huge encouragement to opt for GV as they considered about cost of buying, efficiency, cost saving and reliability, especially for EV. Studies showed that they expected for government supports including policies and incentives to help them to adopt and use GV. Government of China, India and Europe for instance has developed their own policies in influencing their folks to involve in GV for the sake of protecting the socio-environment sustainability [12]. Therefore, against this background, the objective of this study is to view the insight of LSPs in Malaysia especially on the government roles in encouraging the LSPs to involve closely in sustainability efforts. This paper is divided into several sections. The first section discusses literature review, followed by research methodology, findings and discussion, research implication and last but not least is conclusion.

2. Literature Review

2.1 Financial and Non-Financial Incentives and Policies for Sustainability

Literature studies reviewed those matters regarding GV or hybrid vehicle always needed government support to influence people to choose them [12,13]. According to Gong, Ardeshiri, and Rashidi [14], this happened because there were some major obstacles in widespread adoption of the EV including cost of purchasing, limitations of battery technology, lack of charging infrastructure and high cost of battery [15]. Additionally, consumers also less confident on the performance ability of the EV technology [16,17]. As a result, many countries like Australia, China, Bhutan and India have developed policies regarding this issue and using incentives mechanism to help improving the adoption of EV and other GV, as well as to overcome the available barriers [18,19]. Some countries offer investment supports for technology implementation, some give incentives to buyers and some provide infrastructure-related investment [14].

The government of India has taken this issue very seriously. Various efforts were done to encourage the usage of GV in the country for the purpose of reducing the GHG effects including the launching of National Electric Mobility Mission Plan 2020 (NEMMP) in year 2013 to promote EV. In its policy, the Indian government provides several incentives such as encourage research and development in enhancing technology of battery, testing infrastructure, system integration and promote charging infrastructure. Other than these incentives, the government also works on special scheme which an EV can be purchased zero of cost; zero down payments and monthly payments out of saving of fuel cost. Based on a study done on 220 respondents from manufacturing and service backgrounds in India, the data showed respondents are all set to accept the EV, but they expect the government roles to support such for the said cause [12].

In Bhutan, its government has a big goal to become the world leader of EV by introducing an enormous proportion of the government fleet. The government aims to change police cars, service cars, public transport, school busses, tourists' vans and the government cabinet transport into EV to help reducing environmental pollution [20]. Moreover, in order to encourage their folks to use EV, the Bhutan government also suspended import duty and planned to reduce the import of fossil fuel by 70% in order to inspire their people to change to EV [21].

Currently in Australia, where it counters a very small EV market, some small incentives provided by the government such as discounts on energy bill and parking fees were well-received and helped them to slowly increase the demand of EV. Looking at the positive feedback, a survey was done in Australia upon 1076 respondents to study the type of incentives the people intended to have in order to own the EV. The finding shows that residents intended to have both financial and non-financial support from the government. As for financial support, the findings resulted into two categories which are one-off incentives and on-going incentives (operating incentives). One-off incentives cover rebates on the upfront cost and reduce the price as EV is price sensitive. The price of EV is quite high for middle-income categories thus these incentives were favorable among the respondents. On-going incentives included rebates on parking fees until 2025 and discounts on energy bills until 2025. Between these two, discount for energy bills was the most favorable among the respondents. Meanwhile, in terms of non-financial support, respondents only requested the government to provide access for bus lane for EV since traditional vehicles were not allowed to travel on the bus lane. This would save their travel time [14].

In Malaysia, the usage of GV is still relatively very low and in its infancy compared to traditional ICE vehicle. According to a study regarding the influencing factors of GV adoption in Malaysia among 107 respondents found that government tax incentives, sales incentives and fuel subsidy policy were significant factors to the adoption of hybrid vehicles. For example, the government tax exemption reduced the price of hybrid vehicle significantly compared to price without tax exemption. The price of Honda Civic hybrid was MYR162,800 without tax exemption, but the price only at MYR120,000 with tax exemption. This leads to increase the adoption of hybrid vehicle in Malaysia [22]. On the other side of coin, the Green Technology Policy 2009 and the National Automotive Policy launched by the government must be referred to allow this sector to accelerate and transform the automotive industry and to be one of important contributors for the national economy. However, some parties think that investment in public charging is still considered as expensive approach to save petroleum fuel [23].

Hence, from the literature review, it is observed that many countries put their struggles to conserve environment for the sake of socio-economy sustainability. They encouraged the adoption of GV among their people and government fleet inclusive of police cars, government's cabinet transport, and vans for tourists through incentives and policies, but they seem forgetting to overlook at logistics vehicles into account while their movements on road are increasing day by day due to tremendous demand of logistics services nowadays. This is especially true in pandemic Covid-19 that takes the world by storm and Malaysia not spared. In this sense, this study fills the gap by viewing the insight of logistics service providers (LSPs) in Malaysia on how government policies and incentives could work together to significantly contribute to sustainability of socio-environment. In logistics sector, the Malaysian government recently highlights its focus on sustainability issues, especially related to environmental protection [24]. The government acts to protect environment through sustainability and environmental policies, and encourage logistics players to take part in sustainability efforts by providing support through tax incentives and funds [25,26].

2.2 Socio-environment Sustainability

Socio-environment sustainability is a combination of social and environmental sustainability. In many cases environment and social sustainability are related with each other. Healthy environment leads to healthy social performance. Similarly, this paper discusses how government incentives and policies can help LSPs to provide services that do not contribute to the environmental pollution and GHG effects, which then lead to unhealthy social performance. Hence nowadays, LSPs are aimed for sustainability performance approach, concerning on the Triple Bottom Line elements; the environmental, social and economic performance [27], as a result of growing awareness of the unsustainability of logistics services which threaten the environment directly [28], social and economic performance indirectly [29]. However, only environmental and social sustainable performance is being concerned in this study.

LSPs which involve and concern in environmental activities will bring sustainable performance for both LSPs and Earth simultaneously; a balance win-win situation in terms of (i) saving Earth from climate change effect, (ii) healthy and harmony Earth for living creatures to live in, (iii) increase economic performance of LSPs and (iv) social welfare of employees are taken care of [30]. As an additional advantage, they will be much favorable among stakeholders and shareholders [31-33]. This becomes an extra motivation for LSPs to seriously take action regarding the adoption of GV in their delivery services.

According to literature, environmental sustainability means the effects of environmental footprint are relatively lesser than the capacity of the environment can accommodate. One of major contributors to the environmental sickness is transportation [34]. The transportation mainly contributes to negative effects through their ICE which produces (i) GHG emissions which contains CO2 and NO2 [34,35], (ii) other hydrocarbon particles [34] and (iii) noise, dust, water and ocean pollutions [34,36]. Meanwhile, social sustainability in the context of logistics and supply chains are (i) process or products which affect the safety and welfare of people [37], (ii) "the management of social issues such as equity, safety and health, product responsibility, human rights, and philanthropy throughout the supply chain" [38] and (iii) "the health and wellbeing of people in the supply chain and impact on society" [39]. Thus, it can be literally said that social sustainability is focusing on human right, safety and welfare.

In this sense, environmental sustainability comes from strong economic and monetary policies, together with environmental policy across transport, energy, trade and finance policy sectors [40] and social sustainability is facilitated by strong formal institutions which provide strong social policies [41]. Thus, Welter [42] concluded that social sustainability comprises of human, environment and economic well-being resulted from government policies, rules and regulations. Previously, most companies work for sustainability in order to obey law and regulations, as well as for the purpose of competitive advantage. But nowadays they concern on sustainability due to their consciousness of healthy environment, well-being of society and not forgotten to increase their value in global market [43], thus the companies develop sustainable business model which creates long term environmental, social and economic values [44].

3. Hypothesis Development and Conceptual Framework

Based on the literature review, four hypotheses were developed:

H1: LSPs strongly agreed they need financial and non-financial incentives in order to involve in adoption of green technology.
H2: LSPs strongly agreed that they need policies for sustainability in order to involve in adoption of green technology.
H3: Financial and non-financial incentives have a significant positive effect on the socio-environment sustainability of LSPs.
H4: Policies for sustainability have a significant positive effect on the socio-environment sustainability of LSPs.

Consequently, a conceptual framework was established. Figure 1 shows the conceptual framework.

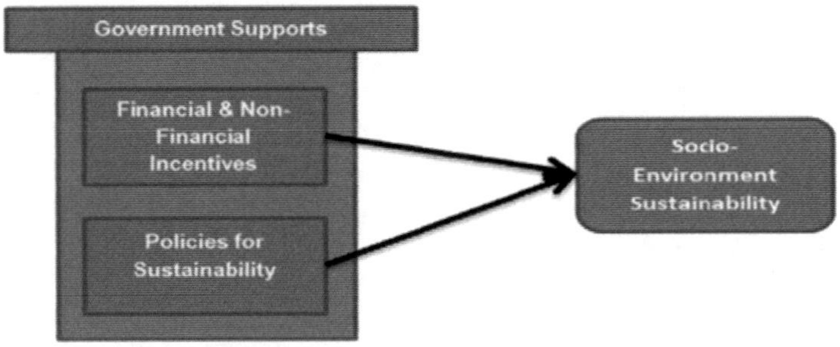

Figure1: Conceptual Framework

4. Research Methodology

A total of 600 questionnaires were sent through e-mail to the 600 LSPs in Selangor, Johor and Penang which were registered under the Federation of Malaysia Freight Forwarders, listed in the Malaysia Logistics Directory 2018/2019 [45]. The questionnaires were directed to top management of companies, ranging from chief executive officer (CEO) to senior executive. A total of 134 returned questionnaires were usable, which is equal to 22.3% of response rate.

Section A of the questionnaire contains three items measuring policies of sustainability, Section B comprises four items which measure financial and non-financial incentives, Section C covers nine items about socio-environment sustainability of LSPs and last but not least Section D asked about demographic of company and the personnel who answered the questionnaire. Respondents were requested to answer on a 7-point Likert scale ranges from 1 (strongly disagree) to 7 (strongly agree). Table 1 below shows the sources of items adopted in this study.

Subsequently, the data were analyzed using IBM SPSS 25.0 statistical software. Further analysis comprised of descriptive and correlation analyses. According to Pallant [46], if the value of Pearson correlation coefficient, r is in the range of 0.1 - 0.29, the correlation is small; when $r = 0.3 - 0.49$, it is assumed that the relationship possesses a medium strength and when $r = 0.50 – 1.0$, the correlation is considered large, suggesting that the relationship of both observed variables is strong.

Table 1: Sources of Measurement

Variable	No. of Item	Source
Policies for sustainability	3	Wu, Ding, and Chen (2012), Ye, Zhao, Prahinski, and Li (2013) and Chu, Yang, Lee, and Park (2017).
Financial and non-financial incentives	4	J. Hong, Feng, Wu, and Wang (2016), Crago and Chernyakhovskiy (2017) and Ethe Raj (2019).
Socio-environment sustainability of LSPs	9	Paulraj (2011), Huo (2012) and Huo, Gu, and Wang (2019).

5. Results and Discussion

5.1 Demographic of Respondents

This section is divided into two subsections - profile of company and profile of respondent. The demographic information is important in order to recognize the characteristics of sample for this study. From 134 of respondents, Table 2 presents the findings of descriptive analysis on the personnel's demographic who answered the questionnaire on behalf of each company.

From the descriptive analysis, Table 2 shows that majority personnel who answered the questionnaire have Bachelor degree as their highest education (41.8%), followed by Diploma (26.9%), Master degree (17.2%), SPM (9.7%), PhD/DBA (3.0%) and STPM (1.5%). In case of position in company, 49 out of 134 personnel were senior executive, representing 36.6% of the sample, 25 were managers (18.7%), 23 were senior managers (17.2%), 18 were managing directors (13.4%), 17 were general managers (12.7%) and two of them were CEOs (1.5%). In terms of personnel experience for working in current job and current organization, both show that almost 50% of them experienced more than 10 years of experience in logistics companies, covering the largest percentage of sample size, 43.3% and 37.3% respectively. Only small portion worked 1 year or less in the logistics companies; 9% worked 1 year or less in current job and 14.2% worked in current organization.

Table 2 presents the findings of demographic analysis for LSPs'. As for age of companies, majority of companies involved in this study (72.4%) established more than 10 years in the market. Only 6% established 2 years or less. 14.2% of respondent companies sustained in the market for three to six years and 7.5% already established since seven to ten years ago. In terms of company's size, more than 50% were large companies (54.5%), 36.9% were small and medium size and only 8.2% of companies were categorized as micro.

In general, majority of the LSPs which involved in this study operated for more than 10 years, large and local companies. Almost all of them provided transportation service as their core logistics service (112 companies), followed by freight forwarding (98 companies), warehousing (93 companies), packaging (64 companies), inventory management (57 companies) and cross-docking (55 companies). The number of total services offered (based on types of services) reflects that those available services appeared manageable to cater the demand of logistics services in Malaysia. However, in the pandemic Covid-19 as a new norm, the LSPs increased their availability for transportation service due to sudden increment in the demand, as people mainly use digital platform to shop.

Table 2: Descriptive Statistics for Demographic of Personnel

Background	Information	N (t=134)	Percentage (%)
	Personnel		
Highest Educational Qualification	SPM	13	9.7
	STPM	2	1.5
	Diploma	36	26.9
	Bachelor Degree	56	41.8
	Master Degree	23	17.2
	Ph.D./DBA	4	3.0
Current position in your organization	Senior Executive	49	36.6
	Manager	25	18.7
	Senior Manager	23	17.2
	General Manager	17	12.7
	Managing Director	18	13.4
	Chief Executive Officer (CEO)	2	1.5
Years in current job	1 year or less	12	9.0
	2 - 5 years	42	31.3
	6 - 10 years	22	16.4
	More than 10 years	58	43.3
Years of working at the organization	1 year or less	19	14.2
	2 - 5 years	42	31.3
	6 - 10 years	23	17.2
	More than 10 years	50	37.3

5.2 Financial and Non-financial Incentives

Table 3 shows the results for descriptive analysis on the financial and non-financial incentives provided by the government. There were four items in this variable, asking the respondents on the importance of financial and non-financial incentives to their companies. The first item 'Tax incentives pertaining to sustainability are important financial based incentives' obtained mean of 6.045. The second item 'Price subsidy for sustainable-related equipment is important financial based incentives' attained mean value of 6.022. The mean value for the third item 'Grants is important financial based incentives' was 5.828 and the last item 'Non-financial incentives' are equally important' had mean of 5.918.

Table 3: Descriptive Analysis for Financial and Non-financial Incentives

Item	N	Minimum	Maximum	Mean	Standard Deviation
Tax incentives pertaining to sustainability are important financial based incentives	134	2.00	7.00	6.045	1.054
Price subsidy for sustainable-related equipment is important financial based incentives	134	2.00	7.00	6.022	1.015
Grants are important financial based incentives	134	1.00	7.00	5.828	1.211
Non-financial incentives are equally important	134	2.00	7.00	5.918	1.069

With average mean of the four items were 5.953 (close to 6.000), these findings indicate that the LSPs mostly agreed on tax and price subsidies were important as their companies' financial incentives. Furthermore, the LSPs also found that grants were important for financial incentives and non-financial incentives were equally important for them in order to adopt sustainable technology including GV. Green technology especially for vehicles are price sensitive, thus they cannot stand alone without supports from the government. These findings supported by studies elsewhere by Malik et al. [12], Axsen et al. [18] and Holtsmark and Skonhoft [19] who concluded that people need government supports like tax incentives, price subsidy, energy subsidy, parking subsidy and related infrastructure to help them with the implementation of sustainability technology, focusing of green vehicles. These supports can be categorized into financial and non-financial incentives from the government. Hence, H1 is supported by this study.

5.3 Policies for Sustainability

Table 4 demonstrates the results for descriptive analysis pertaining to the second variable - policies for sustainability. This variable was established in this study purposely to investigate the view of LSPs on the importance of government policies on the sustainability upon their companies. Mean values for items 'Environmental protection policies and society's rights protection force our company to exercise environmental protection' was 6.127, 'Encourages our environmental management in our company' obtained 6.030 and 'Potential conflicts between our services and policies affect company's environmental management' attained 5.900.

Similar to financial and non-financial incentives, these findings explained that the LSPs mostly agreed (average mean of 6.019) that environmental policies are important for their companies to practice environmental protection. This is because environmental policies seem to motivate their companies to apply sustainability efforts. Opposing the government policies on sustainability will cause conflict thus affecting their environmental management. After all, the findings are supported by aforementioned studies by Malik et al. [12] and Tshering [20]. They found that the government of India and Bhutan respectively developed an important policy regarding this issue in the effort to reduce GHG emission. Furthermore, Adnan et al. [23] also found that the Malaysian government has slowly developed several policies to reduce the price of GV to encourage people to buy GV in that country. Thus, H2 is supported.

Table 4: Descriptive Analysis for Policies for Sustainability

Item	N	Minimum	Maximum	Mean	Standard Deviation
Environmental protection policies and society's rights protection force our company to exercise environmental protection	134	1.00	7.00	6.127	1.086
Encourages our environmental management in our company	134	1.00	7.00	6.030	1.047
Potential conflicts between our services and policies affect company's environmental management	134	1.00	7.00	5.900	1.146

5.4 The Effects of Financial and Non-financial Incentives, and Policies of Sustainability on Socio-economic Sustainability of LSPs

This study examines the relationship between financial and non-financial incentives and socio-economic sustainability of LSPs by adopting Pearson correlation analysis. Table 5 displays the result for this relationship. From the Table 5, the result shows that the Pearson correlation coefficient, $r = 0.631$ and significant at $p = 0.01$, indicating that financial and non-financial incentives have a strong significant positive effect on the socio-economic sustainability of LSPs. Therefore, H3 is supported.

Table 5: Correlation Analysis for Relationship between Financial and Non-financial Incentives and Socio-economic Sustainability of LSPs

		FNI	SES
FNI	Pearson Correlation	1	.631**
	Sig. (1-tailed)		.000
	N	134	134
SES	Pearson Correlation	.631**	1
	Sig. (1-tailed)	.000	
	N	134	134

***. Correlation is significant at the 0.01 level (1-tailed).
FNI–Financial and non-financial incentives, SES–Socio-environment sustainability

Table 6 presents the results of correlation analysis for the relationship between policies for sustainability and socio-environment sustainability. Based on the results, the Pearson correlation coefficient, $r = 0.763$ with significant $p = 0.01$, demonstrating that financial and non-financial incentives have a strong significant positive effect on the socio-economic sustainability of LSPs. Then, H4 is found supported by this study.

Table 6: Correlation Analysis for Relationship between Policies for Sustainability Incentives and Socio-economic Sustainability of LSPs

		PS	SES
PS	Pearson Correlation	1	.763**
	Sig. (1-tailed)		.000
	N	134	134
SES	Pearson Correlation	.763**	1
	Sig. (1-tailed)	.000	
	N	134	134

***. Correlation is significant at the 0.01 level (1-tailed).
PS – Policies for sustainability, SES – Socio-environment sustainability

The findings of correlation analysis for both relationships suggest that financial and non-financial incentives, as well as policies for sustainability strongly affect the socio-economic sustainability of LSPs in Malaysia. According to respondents, with the supports from the government in terms of tax incentives, price subsidy, grants and non-financial incentives for instance were all important for them to involve in sustainability efforts. Among these four items, the respondents seem to favour tax incentives and price subsidy a little more compared to grants and non-financial incentives. They might assume that tax incentives and price subsidy give extra freedom for them to make choices. Subsidize of grants normally tie with certain contract that they in return should fulfil. This seems to give a slight pressure for them. Meanwhile, non-financial incentives were equally important for them, especially when they come to related equipment that the LSPs could not afford to provide or service subsidy that can help reducing their operational cost such as routine maintenance. These kinds of incentives spark robust motivation for LSPs to invest in the change of their ICE type vehicles to the environment-friendly type of vehicles. This is so called 'cure from the root'.

Furthermore, the findings also reveal that LSPs need 'a soft force' from the government, which are policies for sustainability. From their views, the policies regarding sustainability like environmental and social protection policies forced and pushed their companies to pay attention and take appropriate actions to exercise sustainable logistics operations as well as encourage the companies to have a proper environmental management. With the availability of these policies, the LSPs are enforced to follow in order to prevent potential unwanted conflicts happened between their companies and authorities. Conflicts will affect their business operations and damage their reputations in the market.

Henceforth, when the two influential factors are fulfilled, the socio-environment sustainability is obtained for both LSPs and society. LSPs which use GV in their delivery services will significantly cut the volume of CO2 and NO2 discharges to the air, thus reducing the GHG emissions. In other words, environment-friendly vehicles actually decrease the consumption of harmful and toxic materials, gasoline and diesel. Additionally, LSPs use many trucks for the delivery process and the trucks operated based on diesel fuel. Unfortunately, diesel produces 13% more CO2 in 1 litre diesel compared to gasoline [47] and 60% of NO2, compared to 0-30% in gasoline [48]. Therefore, it is very much significant for the government to develop sustainability policies and provide suitable incentives to encourage LSPs transform their vehicles from ICE to the GV. Consequently, it will reduce environmental impacts and lessen risks on the society. It goes without saying that societies who live in clean environment is healthier and safer, especially for the most vulnerable to environmental pollution - old folks and kids. Particles in exhaust emissions might critically harm respiratory system. Therefore, their welfare is more assured.

The positive effects due to the transformation of logistics vehicles' engines towards the environment and society contribute significantly to the SDGs. In order to achieve it, green logistics sector fulfils five out 17 pillars which are Goal 3 – "Ensure healthy lives and promote well-being at all ages", Goal 8 – "Promote sustainable economic growth, full and productive employment, as well as decent work for all", Goal 9 – "Build resilient infrastructure, promote sustainable industrialization and foster innovation", Goal 11 – "Make cities and human settlement safe, resilient and sustainable" and Goal 17 – "Strengthen the means of implementation and revitalize the global partnership for sustainable development" [49].

6. Research Implication

Based on the findings, this study implies that LSPs need government supports to become environmentally friendly and consequently supporting SDGs. The government should take immediate action on this issue because transforming vehicles from ICE to GV is an effective way to help reducing the level of environmental pollution and GHG emission. For example, a research from Stanford University, East Anglia University, Exeter University and the Global Carbon Project found that the movement restriction during Covid-19 pandemic has caused CO2 emissions due to combustion of fossil in ICE vehicle declined by approximately 2.4 billion tonnes in 2020. It was a world record since the past century. Year 1981 and 2009 recorded 0.5 million tons decreased of CO2 (GtCO2), 1992 with decrement 0.7 of GtCO2 and 0.9 GtCO2 at the end of the World War II in 1945. The reduction of 2.4 billion tons emissions seems to equivalent of taking 500 million cars off from the world's roads in that year. Inferring the emissions from transports are the greatest contributions to the global environmental quality degradation. [50].

LSPs could not afford to stand alone, hence, without supports from government, this mission is seen as implicit, and hence they must work hand in hand with the government. Their cooperation in this matter promises a healthy environment and society. Moreover, it becomes a competitive advantage in the market by means of stakeholders around the world also now aware about the sustainability of the world as a whole. They tend to choose companies who care about environment and society. This helps LSPs in terms of sustainable economic performance as well.

7. Conclusion

In conclusion, this study reveals the insights of LSPs pertaining the importance of policies for sustainability and financial and non-financial incentives from the government to help them adopting green type of vehicles as a substitution to vehicles with ICE in order to achieve socio-environment sustainability. The gathered data from 134 LSPs in Klang, Johor and Penang were analysed in IBM SPSS 25.0 statistical software. Based on the results of descriptive analysis, LSPs were strongly agreed that government supports like policies for sustainability and financial and non-financial incentives were vital for them to implement sustainable technology in their vehicles as the GV are price sensitive. Moreover, the correlation analysis exposed those policies for sustainability and financial and non-financial incentives had significant positive relationships with socio-environment sustainability of LSPs. Therefore, all four developed hypotheses were supported.

However, this study is not without limitation. The data collection process happened during the total lockdown due to Covid-19 pandemic in Malaysia, thus only 134 samples were successfully collected. It is suggested that future research could consider a bigger sample size and to take a deep-dive into specific incentives required by LSPs in Malaysia to contribute to social-environmental sustainability.

Acknowledgement

This research is funded by Ministry Higher Education (Malaysia), under Fundamental Research Grant Scheme (FRGS), grant number R/FRGS/A1100/01670A/001/2018/00547.

References

[1] Orim, Ruby. 2020. "Indonesia Has Seen a Whopping 313% Increase in Greenhouse Gas Emissions Since 1990." Climate Scorecard Accessed 14th April 2021 https://www.climatescorecard.org/2020/12/indonesia-has-seen-a-whopping-313-increase-in-greenhouse-gas-emissions-since-1990/.
[2] Worldometer. 2020. "Indonesia Population (LIVE)." Worldometer Accessed 14th of April, 2021. https://www.worldometers.info/world-population/indonesia-population/.
[3] Tempo.co. 2020. "Jakarta Ranks 10th on World's Most Congested Cities List ". Tempo.co Accessed 14th of April, 2021. https://en.tempo.co/read/1303052/jakarta-ranks-10th-on-worlds-most-congested-cities-list.
[4] UNDP. 2021. "Sustainable Development Goals." United Nations Accessed 21th of April, 2021. https://www.undp.org/content/undp/en/home/sustainable-development-goals.html.
[5] Energy.gov. 2013. "Internal Combustion Engine Basics ". U.S Department of Energy Accessed 14th of April, 2021. https://www.energy.gov/eere/vehicles/articles/internal-combustion-engine-basics.
[6] Erdoğan, Sevgi, and Elise Miller-Hooks. 2012. "A green vehicle routing problem." Transportation research part E: logistics and transportation review 48 (1):100-114.
[7] Mom, Gijs. 2013. The electric vehicle: Technology and expectations in the automobile age: JHU Press.
[8] United States Environmental Protection Agency. 2019. "Learn About Green Vehicles - What Makes a Vehicle Green?". United States Environmental Protection Agency Accessed 10 June 2021 https://www.epa.gov/greenvehicles/learn-about-green-vehicles.
[9] Union of Concerned Scientists. 2015. "How Do Battery Electric Cars Work?". Union of Concerned Scientists Accessed 15th of April. https://www.ucsusa.org/resources/how-do-battery-electric-cars-work.
[10] World Bank Group. 2020. The Impact of COVID-19 on Logistics. International Finance Corporation.
[11] Ames, Ben. 2020. "Covid-19 shortages: brokers shift trucking capacity to “essential” sectors." ePublishing Accessed 11 March 2021. https://www.dcvelocity.com/articles/45546-covid-19-shortages-brokers-shift-trucking-capacity-to-essential-sectors.

[12] Malik, Yashpal, Nirupama Prakash, and Ajay Kapoor. 2018. "Green Transport: A Way Forward for Environmental Sustainability." In Environment, Politics, and Society. Emerald Publishing Limited.

[13] Gallagher, Kelly Sims, and Erich Muehlegger. 2011. "Giving green to get green? Incentives and consumer adoption of hybrid vehicle technology." Journal of Environmental Economics and management 61 (1):1-15.

[14] Gong, Shuangqing, Ali Ardeshiri, and Taha Hossein Rashidi. 2020. "Impact of government incentives on the market penetration of electric vehicles in Australia." Transportation Research Part D: Transport and Environment 83:102353.

[15] Egbue, Ona, and Suzanna Long. 2012. "Barriers to widespread adoption of electric vehicles: An analysis of consumer attitudes and perceptions." Energy policy 48:717-729.

[16] Lévay, Petra Zsuzsa, Yannis Drossinos, and Christian Thiel. 2017. "The effect of fiscal incentives on market penetration of electric vehicles: A pairwise comparison of total cost of ownership." Energy Policy 105:524-533.

[17] Matteson, Schuyler, and Eric Williams. 2015. "Learning dependent subsidies for lithium-ion electric vehicle batteries." Technological Forecasting and Social Change 92:322-331.

[18] Axsen, Jonn, Joseph Bailey, and Marisol Andrea Castro. 2015. "Preference and lifestyle heterogeneity among potential plug-in electric vehicle buyers." Energy Economics 50:190-201.

[19] Holtsmark, Bjart, and Anders Skonhoft. 2014. "The Norwegian support and subsidy policy of electric cars. Should it be adopted by other countries?" Environmental science & policy 42:160-168.

[20] Tshering, L. 2014. "Bhutan electric vehicle initiative: A step towards zero emission.". United Nations Accessed 15th of April. http://www.uncrd.or.jp/content/documents/22548EST-P2_Bhutan.pdf.

[21] Sundas, S. . 2015. "On the road to zero emissions, Bhutan hits a few electric car bumps." Accessed 15th of April 2021. http://www.reuters.com/article/2015/06/09/us-bhutan-climate-change-autosales-idUSKBN0OP0 W720150609.

[22] Hong, Yong Hoe, Nasreen Khan, and Muhammad Madi Abdullah. 2013. "The determinants of hybrid vehicle adoption: Malaysia perspective." Australian Journal of Basic and Applied Sciences 7 (8):347-454.

[23] Adnan, Nadia, Shahrina Md Nordin, and Imran Rahman. 2017. "Adoption of PHEV/EV in Malaysia: A critical review on predicting consumer behaviour." Renewable and Sustainable Energy Reviews 72:849-862.

[24] Maheswari, Hesti, Gatot Yudoko, and Akbar Adhiutama. 2018. "Theory building of quattro bottom line approach for sustainable reverse logistics from government perspective: The Indonesia evidence." Advance in Science, Technology and Engineering Systems Journal 3 (3):83-98.

[25] Cohen, Steven. 2014. "The Role of Government in the Transition to a Sustainable Economy." Huffpost Accessed 30 May 2021. https://www.huffpost.com/entry/the-role-of-government-in_b_4759621.

[26] MIDA. 2019. "Incentives in Services Sector." Accessed 16th of April, 2021. http://www.mida.gov.my/home/incentives-in-services-sector/posts/.

[27] Elkington, John. 2013. "Enter the triple bottom line." In The triple bottom line, 23-38. Routledge.

[28] Peterlin, Judita, Noel J Pearse, and Vlado Dimovski. 2015. "Strategic decision making for organizational sustainability: The implications of servant leadership and sustainable leadership approaches." Economic & Business Review 17 (3).

[29] Markman, Gideon D, and Daniel Krause. 2016. "Theory building surrounding sustainable supply chain management: Assessing what we know, exploring where to go." Journal of supply chain management 52 (2):3-10.

[30] Macharis, Cathy, Sandra Melo, Johan Woxenius, and Tom Van Lier. 2014. Sustainable logistics: Emerald Group Publishing.

[31] Byrne, PJ, Paul Ryan, and Cathal Heavey. 2013. "Sustainable logistics: a literature review and exploratory study of Irish based manufacturing organizations." International Journal of Engineering and Technology Innovation 3 (3):200-213.
[32] Werbach, Adam. 2009. Strategy for sustainability: A business manifesto: Harvard Business Press.
[33] Wolf, Christina, and Stefan Seuring. 2010. "Environmental impacts as buying criteria for third party logistical services." International Journal of Physical Distribution & Logistics Management 40 (1/2):84-102.
[34] Mosaberpanah, M.A., and S. Darban Khales. 2012. "The Role of Transportation in Sustainable Development." International Conference on Sustainable Design, Engineering, and Construction 2012, Fort Worth, Texas.
[35] United States Environmental Protection Agency. 2017. "Overview of Greenhouse Gasses." Accessed February, 5th, 2021. https://www.epa.gov/ghgemissions/overview-greenhouse-gases.
[36] McKinnon, Alan, Arni Halldorsson, and Christophe Rizet. 2014. Theme issue on sustainable freight transport. Elsevier.
[37] Wood, D.J. 1991. "Corporate social performance revisited." Academy of Management Review 16 (4):691–718.
[38] Mani, Venkatesh, Angappa Gunasekaran, and Catarina. International Journal of Production Economics Delgado. 2018. "Enhancing supply chain performance through supplier social sustainability: An emerging economy perspective." 195:259-272.
[39] Pulis, A., and S. Dovers. 2006. "Environment and Sustainability Policy — Creation, Implementation, Evaluation." Journal of Environmental Assessment Policy and Management 8 (3):403-406.
[40] McCall, L., and C. Percheski. 2010. "Income inequality: New trends and research directions." Annual Review of Sociology 36:329-347.
[41] McCall, L., and C. Percheski. 2010. "Income inequality: New trends and research directions." Annual Review of Sociology 36:329-347.
[42] Welter, Friederike. 2011. "Contextualizing entrepreneurship—conceptual challenges and ways forward." Entrepreneurship theory and Practice 35 (1):165-184.
[43] Agrawal, Saurabh, Rajesh Kr Singh, and Qasim Murtaza. 2016. "Triple bottom line performance evaluation of reverse logistics." Competitiveness Review 26 (3):289-310.
[44] Wales, Terry. 2013. "Organizational sustainability: what is it, and why does it matter?" Review of Enterprise and Management Studies 1 (1):38-49.
[45] Malaysia Logistics Directory. 2018/2019. Malaysia Logistics Directory. Malaysia: Marshall Cavendish Business Information.
[46] Pallant, Julie. 2020. SPSS survival manual: A step by step guide to data analysis using IBM SPSS: Routledge.
[47] The International Council on Clean Transportation. 2019. Gasoline vs diesel. Comparing CO2 emission level of a modern medium size car model under laboratory and on-road testing conditions. Accessed 18th of April 2021.
[48] The Transport and Environment. 2015. Five facts about diesel the car industry would rather not tell you. Accessed 18th of April 2021.
[49] United Nations. 2015. "Sustainable Development Goals." Accessed February, 6th 2021. http://www.un.org/sustainabledevelopment/sustainable-development-goals/.
[50] Stanford Earth. 2020. "COVID lockdown causes record drop in carbon emissions for 2020 ". Stanford University Accessed 21th of April 2021. https://earth.stanford.edu/news/covid-lockdown-causes-record-drop-carbon-emissions-2020#gs.zc2rat.

Structural Analysis of Continuous I Beam of Steel with Crack Using Finite Element Simulation

Mumtahina Rahnuma[1], Md. Mehadi Hassan Chowdhury[2], Nihal Ahmed[3*], Ashfaq Ahmed[4]

[1]Department of Civil Engineering, Rajshahi University of Engineering & Technology, Rajshahi, Bangladesh
[2]Department of Civil Engineering, Sylhet Engineering College, Sylhet, Bangladesh
[3]Researcher, National Nanotechnology Research Center, Bilkent University, Ankara, Turkey
[4]Graduate Student Researcher, Department of Thermal and Fluid Sciences, Brown University, USA

Corresponding author: nihal.ahmed@bilkent.edu.tr

Abstract. The statically indeterminate beam that is supported by three or more supports and at least one support that develops a reaction along the beam axis is known as a continuous beam. This type of beam is used in a large variety of structures as they show better resistance to seismic loads and lower deformity. However, due to the decay of steel or the increase of stresses, sometimes surface or sub-surface fissures develop in the material which is known as cracks. Because of the cracks in a structure, it may be vulnerable and start to age rapidly, which will eventually lead the structure to collapse. In this study, Abaqus CAE, a widely operated commercial software is used to simulate the cracks in a continuous I beam by the Finite Element Method. The surrounding crack zone is meshed with a smaller area due to its sensitiveness. Initially, suitable steel is selected among the major types of steel by evaluating their stress-strain curve. Then a continuous I beam of that steel with crack is simulated and the effect of the crack on the stresses is analyzed. The crack is safer when it forms on the part of the structure that is over-support. The risk of failure of a structure increases with the increasing distance of the crack from support. The greater crack depth poses a greater risk to the structure on the other hand increasing crack width has less effect on the stability of the structure

Keywords: Continuous I Beam, Abaqus CAE, Finite Element Method, Crack

1. Introduction

The precision with which a structure's forces and responses are calculated is crucial to its protection and serviceability. In structural engineering, beam analysis and design continue to play a significant role. It requires precise care and precision as a major structural component.

Continuous beams are ones with many spans that extend through intermediate spans. Continuous beams are most commonly used in bridges and building structures. Again, span lengths may be uneven, and the cross-moment section's of inertia may differ from one span to the next. Beams are made continuous over the supports to improve structural stability. A continuous beam provides an alternative load direction if a section collapses. In seismically vulnerable portions of a superstructure, continuous beams and frames are recommended. A continuous beam is a structure that is statically indeterminate. Continuous beams have a higher vertical load capacity than simply supported beams for the same span and section. Continuous beams have a less deflected mid-span section and may have a reduced section depth than a simply supported beam of the same span, enabling them more material-efficient. In contrast, continuous beams require fewer anchorages and tendons..

Continuous beams, on the other hand, require complex analysis and design techniques. Continuous beams offer construction challenges, particularly with precast parts. Increased frictional loss is seen in continuous beams due to changes in cable profile curvature and increased beam shortening, resulting in lateral load on supporting columns.. Secondary stresses develop in Continuous beams due to time-dependent effects like creep and shrinkage, settlement of support, and variation of temperature. A structural member with a crack introduces local flexibility that affects its dynamic response. Furthermore, depending on the loading conditions and vibration amplitude, the crack can open and close over time. So it is necessary to identify how vulnerable or sustainable the structure becomes according to crack position and opening. This is why crack simulation in structural analysis is essential.

Several types of research have been made on Structural analysis of I beams. Ahmed et al [1] analyzed the change in the natural frequency of such I beam for various parameters. Runsheng et al [2] worked on the stress distribution of such beams. Works related to shear stress has been done by other researchers in different fields as well [3]. Most of the researchers used the finite element method and commercial software for solving such problems[4–8].

Though a potential number of researches has been made under various categories in continuous beams there is no significant research to determine the sustainability of the structure according to position, depth, and opening of the crack along with a continuous I beam.

In this study, five representative types of steel (mild, carbon, alloy, AISI 4140, Grade 304, S 420) are selected, and among the variants structural steel (S420) showed a better stress-strain graph in the elastic zone and thus chosen for further analysis. Then cracks were analyzed in continuous I beam and the impact and performance of the beam are observed. The principal stress increases when the position of the crack is in the middle of the greater span. The crack depth and opening also affect the stress. An increase in stress can cause the propagation of the crack ultimately leading to structural failure.

2. Numerical methodology

The Euler-Bernoulli equation for a static beam is

$$\frac{d^2}{dx^2}\left(EI\frac{dw^2}{dx^2}\right) = q \tag{1}$$

Here, *w(x)* is the deflection of the beam along the z-axis, q is the distributed load, E is the modulus of elasticity and I is the second moment of inertia of beam section. The general equation of the shear and bending moment of a beam can be calculated by the following equation.

$$M = EI\frac{d^2y}{dx^2} \tag{2}$$

$$V = EI\frac{d^3y}{dx^3} \tag{3}$$

Here, y is the deflection along Y-axis. This equation is solved in the finite element method where the stiffness matrix of the continuous beam used in this study is

$$\begin{bmatrix} M_1 \\ M_2 \\ M_3 \end{bmatrix} = \frac{EI}{L} \begin{bmatrix} 6 & 3 & 0 \\ 3 & 18 & 6 \\ 0 & 6 & 12 \end{bmatrix} \begin{bmatrix} \theta_1 \\ \theta_2 \\ \theta_3 \end{bmatrix} \tag{4}$$

The boundary conditions for the continuous beam are,

$$w(0) = 0, \quad w\left(\frac{2}{3}L\right) = 0, \quad w(L) = 0, \quad \frac{d^2}{dx^2} w\,(0) = 0, \quad \frac{d^2}{dx^2} w\,(L) = 0$$

3. Beam modeling

I beam, also known as H beam is popular in the structural industry due to its high functionality. The cross-section of I beams is excellent for unidirectional bending parallel to the web. The flanges of the I beam resists the bending moment while the web resists the shear force. I beams are also cost-effective. The detailed specifications of the I beam used in this study are given in table 1.

Table 1. Beam Specifications

Beam type	Beam length	Height	Width	Web thickness	Flange thickness	Process fillet
I beam	1000 mm	100 mm	100 mm	6 mm	8 mm	10 mm

3.1. Support condition

In this study, the beam is continuous with two unequal spans divided by 3 hinged support. Hinged support can resist deflection but cannot resist rotation. The arrangement of the beam with support is shown in figure 1.

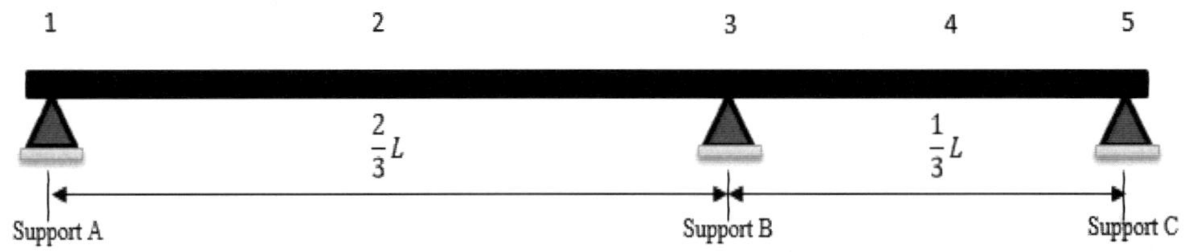

Figure 1. Beam Arrangement

3.2. Crack design

Here wedged-shaped breathing cracks are designed. The position of the cracks is shown in figure 1 and summarized in table 2.

Table 2. Position of the Cracks

Name	Location of the crack on the beam
Position 1	On support A
Position 2	In the middle of support A and B
Position 3	On support B
Position 4	In the middle of support B and C
Position 5	On support C

Three different crack depths of 3mm, 4mm, and 5mm are assumed while the opening is kept fixed (0.6mm) for determining optimum crack depth. Again to determine optimum crack opening, the depth of the crack is kept fixed (5mm) and the opening is changed to 0. mm, 0.6mm and 0.8mm consecutively.

3.3. Mesh design
For mesh sensitivity check, the crack is assumed to be on position 2 of the beam and their maximum principal stress (Von Mises) is taken into account to compare among seven different mesh sizes.

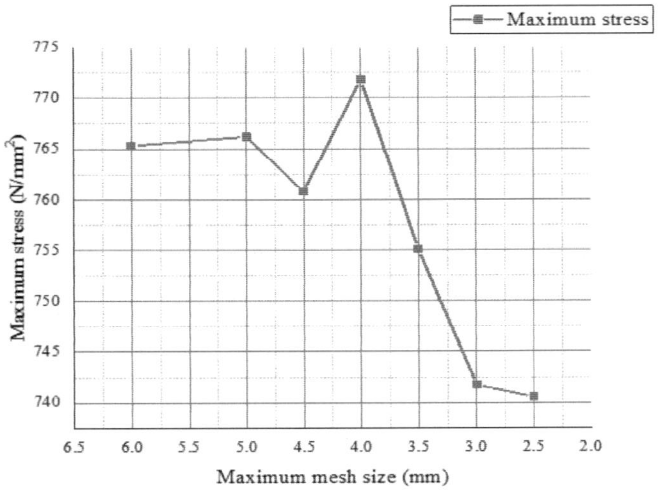

Figure 2. Mesh sensitivity

The crack of the beam is assumed to be 4 mm in depth and 0.6mm in the opening. It can be observed that the change in maximum stress follows a shallow curve from 2.5mm mesh size to 3mm then the curve becomes steep. So the simulation of the beam is done with a maximum mesh size of 3mm and a minimum mesh size of 1mm. The zone around the crack is meshed in a smaller wedged shape considering its sensitiveness while the rest part of the beam has meshed in relatively bigger hexagonal sizes.

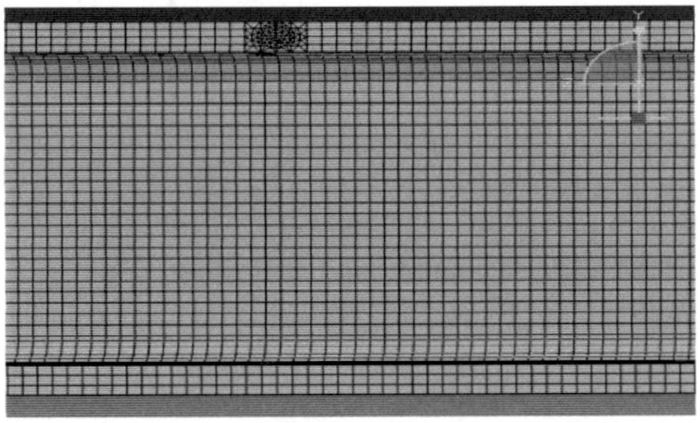

Figure 3. Meshing on the I beam

3.4. Validation
In Wen Runsheng and Zhang Yang [2], the same I beam without crack is simulated with a fixed support on the left end of the beam, and its energy vs. time graph is drawn. Here, the energy vs. time graph is generated following the same criteria, and the two graphs are compared for validation.

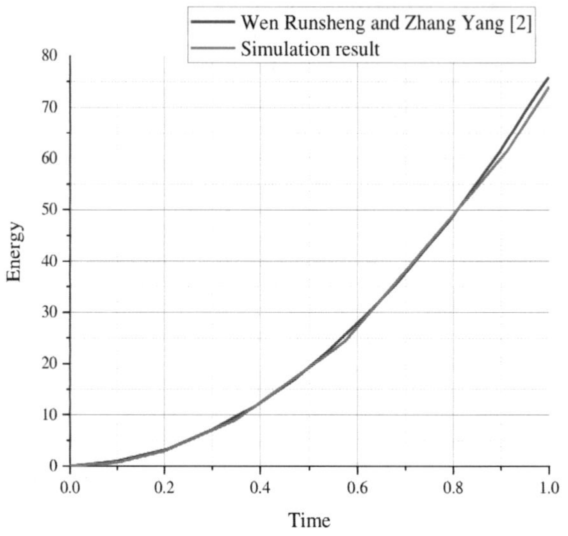

Figure 4. Energy vs. Time graph

The graph shows that, the simulation result and result from Wen Runsheng and Zhang Yang [2] almost co-inside with 2% error. So the simulation result of this study can be accepted.

4. Result and Discussion

In this section, the results are compared and the possible reason behind these results are discussed.

4.1. Material selection

To determine the optimum material for the simulation, five representative types of steel such as high carbon steel, mild steel, alloy steel (AISI 4140), stainless steel (grade 304), and structural steel (S 420) are selected among the huge varieties of steel. Their properties are given in table 3.

Table 3. Properties of the materials

Types of steel	Density (g/cm^3)	Young's modulus (Gpa)	Yield strength (Mpa)	Poisson's ratio
High carbon steel (0.45% to 0.75% carbon content with 0.3 to 0.9% manganese content)	8.26	210	525	0.3
Mild steel (0.05 to 0.25% carbon content)	7.85	370	210	0.29
Alloy steel (AISI 4140)	7.85	415	210	0.29
Stainless steel (Grade 304, contains at least 10.5% chromium and less than 1.2% carbon	7.85	200	215	0.265
Structural steel (S420)	7.85	210	420	0.3

The stress-strain curve of these materials within the elastic limit is drawn from simulation data. From figure 5, it is evident that for a specific strain, Structural steel (S420) can endure more stress than the other types of steel. Thus the further study has proceeded with S420.

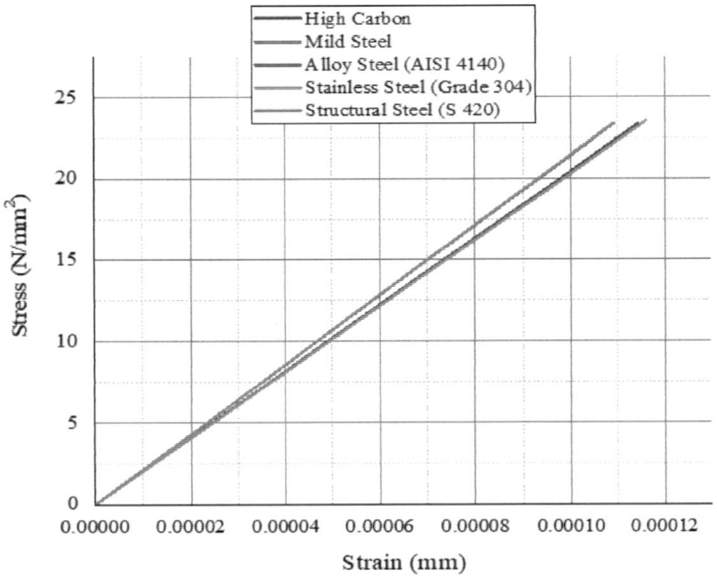

Figure 5. Stress-strain curve (in elastic zone) for various types of steels simulated in Abaqus CAE

4.2. Effect of crack position

The path along which the stresses will be analyzed in figure 6.

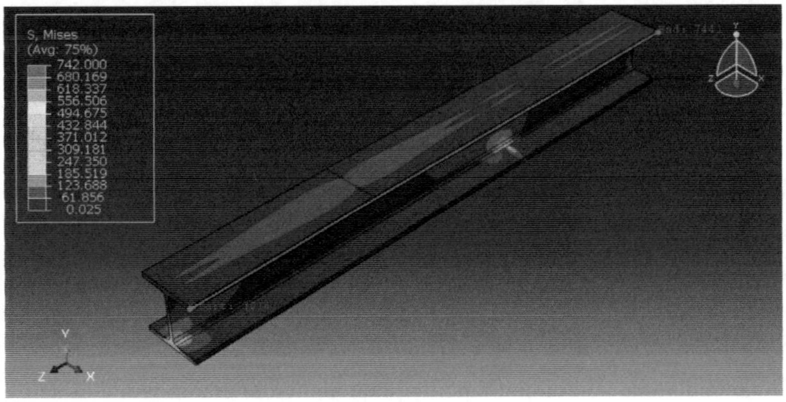

Figure 6. The path along Z axis

The crack is placed on 5 positions of the beam (table 2) one by one to calculate its effect on the principal stress. The opening and depth of the crack are assumed to be 0.6mm and 4mm.

From figure 7, it can be observed that the stress is higher when the crack is in the middle of support A and B. The span length between support A and B is greater than the other span length. So here the deflection of the beam is comparatively higher. It can be said that if the crack forms in the middle of a greater span the structure are more likely to fail. Again the crack has less effect on the stress of the beam when it is above support.

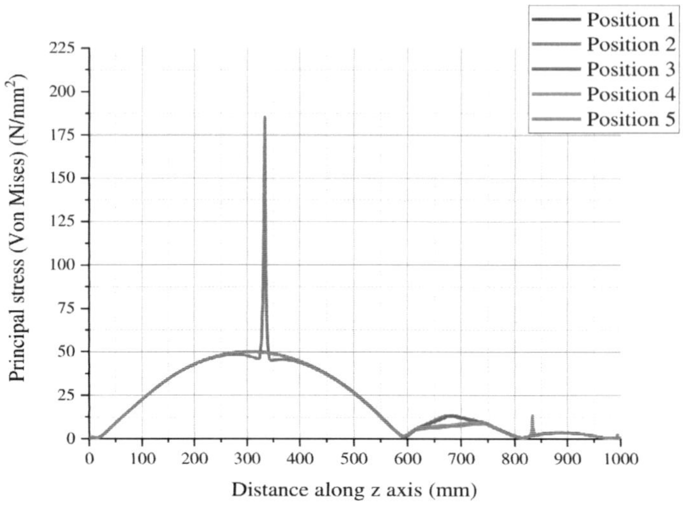

Figure 7. Principal stress distribution at various crack location

4.3. Effect of crack depth

The path along which the stresses will be analyzed in Figures 9 and 10.

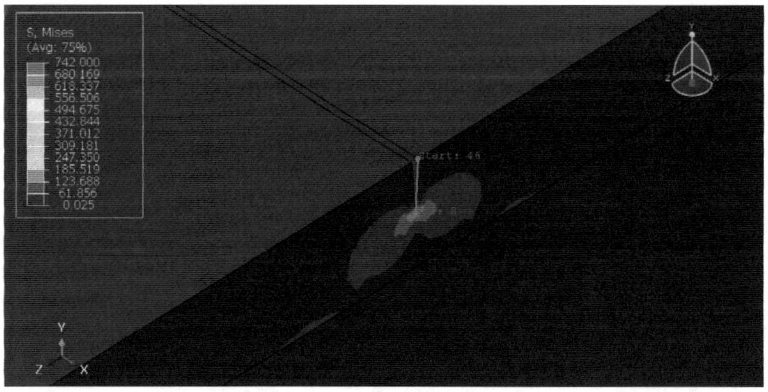

Figure 8. Path along the crack

As the crack in position 2 is the most critical state, crack of different depth such as 3mm, 4mm, 5mm is placed there. Their effect on principal stress is shown in figure 9.

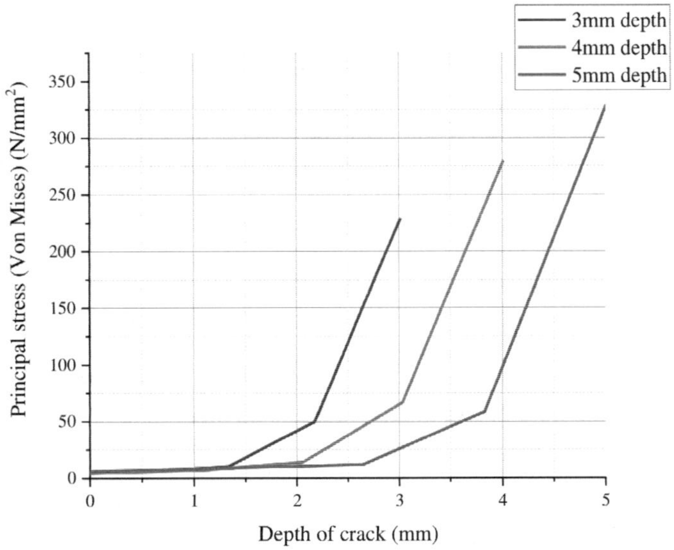

Figure 9. Principal stress for various crack depth

The stress increases with the depth of the crack. As here the crack is wedged shaped and the opening is kept the same, the tip of the crack becomes narrower. It makes the propagation chances of the crack higher which eventually leads to structural failure.

4.4. *Effect of crack opening*

Here three crack openings of 0.4mm, 0.6mm, and 0.8mm are placed in position 2 to visualize their effect on principal stress. All the crack depth is kept at 5mm as it shows maximum stress for different crack depths in Figure 10.

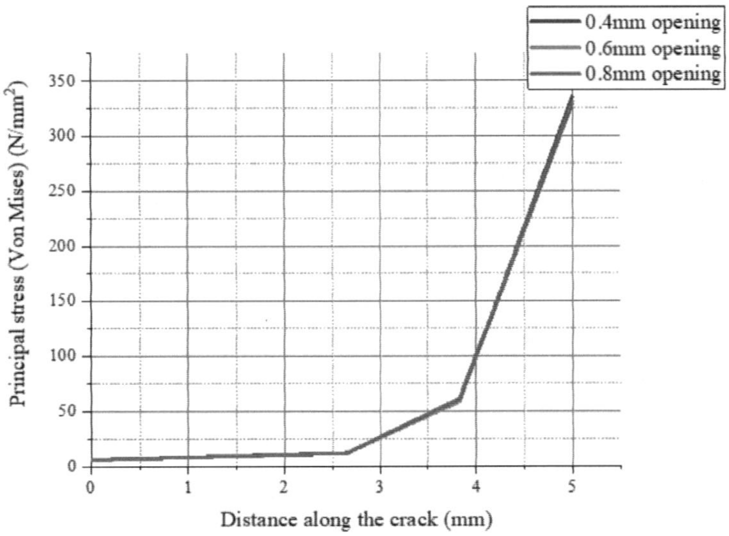

Figure 10. Principal stress for various crack opening

Figure 10 indicates that the crack openings have a very small effect on the principal stress of the beam. A smaller crack opening shows a slight increase of stress than the larger opening. The small crack opening indicates a narrower crack tip which may make the structure more vulnerable.

5. Conclusion

The beam shows different behavior when it is exposed to a crack. Among the different varieties of steel, Structural Steel (S420) is one of the suitable materials for I beam in terms of stress-bearing capacity. If the crack is located in the middle of the beam with a greater span length, then the structure becomes more unstable. On the other hand, the beam is relatively safer when the crack forms in the portion of the beam that is above support. The stress of the beam along the crack increases with the depth of the crack. The opening of the crack does not have a significant effect on the stress of the beam. However, a smaller crack opening shows a slight increase in stress.

From this study, it is evident that, if the crack has greater depth and located in the middle of the span of the beam, then the beam is at more threat than other cases.

6. References

[1] N. Ahmed, M. Mamun, and A. Ahmed, "Vibration Response Analysis of a Structural Metal Mild Steel under the Effect of Crack," *IOP Conf. Ser. Mater. Sci. Eng.*, vol. 1092, no. 1, p. 012037, 2021, doi: 10.1088/1757-899x/1092/1/012037.

[2] W. Runsheng and Z. Yang, "Simulation of Stress Distribution and Deformation Analysis for Cantilever I-beams Using ABAQUS Software," in *IOP Conference Series: Materials Science and Engineering*, 2019, vol. 626, no. 1, p. 12010.

[3] Ahmed, N., Ahmed, A., & Imran, S. B. (2020, November). Analysis of the Effect of Atherosclerosis with the Changes of Hematocrit: A Computational Study on the Hemodynamics of Carotid Artery. In 2020 2nd International Conference on Advanced Information and Communication Technology (ICAICT) (pp. 336-341). IEEE.

[4] N. A. Meshram and V. S. Pawar, "Analysis of crack detection of a cantilever beam using finite element analysis," *Int. J. Eng. Res. Technol.*, vol. 4, no. 04, 2015.

[5] K. Cui, W. H. Yang, and H. Y. Gou, "Experimental research and finite element analysis on the dynamic characteristics of concrete steel bridges with multi-cracks," *J. Vibroengineering*, vol. 19, no. 6, pp. 4198–4209, 2017, doi: 10.21595/jve.2017.18084.

[6] B. L. Wahalathantri, T. H. T. Chan, and & Fawzia, "a Material Model for Flexural Crack Simulation in Reinforced Concrete Elements Using Abaqus Wahalathantri," pp. 260–264, 2008.

[7] S. Khuda and A. Anwar, "Design AID for Continuous Beams," *MIST J. GALAXY*, vol. 3, no. 1, 1970, doi: 10.3329/mist.v3i0.8051.

[8] T. G. Chondros, "The continuous crack flexibility model for crack identification," *Fatigue Fract. Eng. Mater. Struct.*, vol. 24, no. 10, pp. 643–650, 2001.

IOP Conf. Series: Materials Science and Engineering 1244 (2022) 012015 doi:10.1088/1757-899X/1244/1/012015

Numerical simulation of axially impacted elastic bar

A. Hamdani[1,*], P. Pankaj[2]

[1]Department of Mechanical Engineering, International Islamic University Malaysia, Malaysia
[2]Institute for Bioengineering, The University of Edinburgh, Edinburgh, United Kingdom

*Corresponding author: hadib@iium.edu.my

Abstract. This numerical analysis considers a model of an elastic cylindrical bar impacted by rigid mass, where the accuracy of the model is benchmarked by a closed-form solution developed for this impact process. The analytical solution provides theoretical background and understanding, however, practical problem is not always restricted to a simple one-dimensional geometry and boundary conditions of a purely elastic material behaviour. Hence, the current work numerically simulates the impacted elastic bar, then explores parameters which are not considered in its theoretical model. A two-dimensional (2D) solid axisymmetric system is considered, in which rigid impactor is assigned with initial velocity corresponding to certain drop height and a 2D contact type is defined between interacting bodies. Both elastic and time-dependent viscoelastic material models are considered in this study. The simulation results reveal time intervals gradually increase in every sequential intervals, while relatively small discrepancies are recorded for peak load and pulse width outputs. The parameters of impactor's mass, drop height and structural stiffness are varied; showing how these parameters individually affect the resulting force response at end struck. The models then evaluate time-dependent viscoelastic material model; showing stiffer response of the resulting force in comparison to models assigned with long-term elastic moduli. Derived elastic modulus-output variable relations show comparable mathematical forms to corresponding equations formulated analytically.

1. INTRODUCTION

This numerical work simulates an impact problem based on Timoshenko and Goodier [1], which involves a restrained elastic bar subjected to rigid mass as shown in figure 1. The simulation is undertaken using a finite element (FE) commercial package, LS-DYNA (smp d R7.1.1, LSTC, Livermore, California). The available analytical model is considered as the bar is treated as purely elastic, then it is extended to more complex time-dependent elastic behaviour. The study also evaluates relations between several parameters and corresponding output variables.

The dynamic properties of trabecular bone were characterised by Shim et al. [2]: the time-dependent viscoelastic properties using one Maxwell element in parallel with a nonlinear Newtonian dashpot constructing a nonlinear viscoelasticity model. The one-dimensional (1D) model was extended to a three-dimensional (3D) model by incorporating 3D strain tensor into 1D strain variable in the non-linear strain rate dependency constitutive model.

Time-dependent properties of bovine trabecular bone samples were characterised in Manda et al. [3] using uniaxial creep tests. The research utilised the creep compliance function, which is easily convertible to the relaxation modulus function, E(t) by using numerical interconversion methods [4].

As a result, three-term Prony series of compliance-relaxation functions were derived describing linear viscoelastic response of the bone specimens subjected to compressive creep load, which are related to their bone volume fraction (BV/TV).

2. PROBLEM DEFINITION

Consider a stationary elastic cylinder with diameter, $2r$, and length, l, subjected to impact due to an incoming rigid impactor at one end (called the end struck), while the other end is restrained as shown in figure 1. Numerical solutions are important as not all problems are elastic, have a simple one-dimensional geometry and boundary conditions as in theoretical model by Timoshenko and Goodier [1] and Rosli [5]; consequently, closed-form solutions are not possible. The closed-form solutions, however, help in evaluating and benchmarking the accuracy of numerical methods, such as this attempted work.

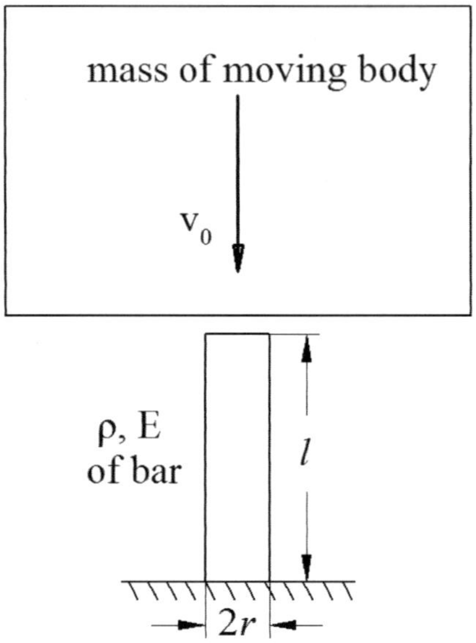

Figure 1. Axial impact problem [1] to be numerically modelled.

Explicit code of LS-DYNA is utilised for this work, which uses explicit time integration based on central difference method (CDM) to obtain nodal displacements from their velocities and accelerations in every time step. The time step is calculated by the programme at every stage in simulation by dividing the smallest characteristic length with the continuum wave speed [6]; indicating the time required for wave speed to traverse the smallest element. The current study is limited to elastic and viscoelastic material models without modelling their post-elastic behaviour.

3. NUMERICAL MODEL

FE analysis requires four basic main inputs: the geometry with corresponding FE mesh; boundary conditions defined for the models; load application and the material behaviour of the models. This section discusses the definitions for each of these required inputs in the current study.

3.1. Discretisation of axisymmetric system

Impact mechanics involves more than one body in the collision [7], which implies the necessity of defining a minimum of two geometrical models. Planar symmetry of the cylindrical shape in the defined problem as shown in Fig. 2 is exploited. Two-dimensional (2D) solid models in LS-DYNA are based on integral difference scheme, which defines the geometries in global XY plane. In axisymmetric case, radial direction corresponds to X-axis, while the axis of symmetry lies on the Y-axis. LS-DYNA offers two options of axisymmetric elements, namely Petrov-Galerkin and Galerkin finite element approaches, generally referred as 'area-weighted' and 'volume-weighted' methods respectively. The former is designed suitably for situations where the developed pressures in the model constitute a large fraction of the elastic modulus, i.e. a hydrodynamic case [6].

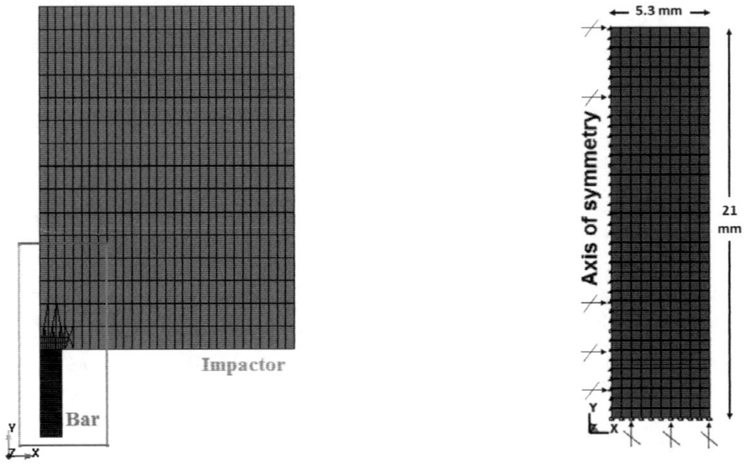

(a) Axisymmetric element (b) Dimensions and constraints

Figure 2. Finite element (FE) models. The box in (a) focuses on the region of interest.

The volume-weighted Galerkin method is chosen in this work since there is no possibility of a hydrodynamic scenario arising in the problems being considered. The bar is discretised into 4-noded axisymmetric 10×40 elements to provide almost square 2D elements, with similar 1:2 diameter-to-length ratio (figure 2b). The rest of this primitive geometry are theoretically revolved 2π radians around the axis of symmetry to create a virtual cylindrical model of the bar as shown in figure 2.

Accordingly, the impactor is also assumed to have axisymmetric shape with arbitrary radius-length dimension, while ensuring the impactor's mass is 2.5 kg simulating the experimental drop test or around 26.6 g for a light-weight dropped hammer [5, 6]. Figure 2(a) shows the impactor as coarsely meshed in order to reduce the computational cost, while refinement is applied at the bottom elements which involve with contacting the end struck of the bar. This mesh refinement is required to ensure each slave node (on the impactor) are properly projected to its master segment (the bar struck). Node-collapsing method is applied to some three-noded elements in this region of refined mesh.

3.2 Boundary conditions

By the definition of axisymmetry in LS-DYNA, the global Y-axis represents theaxis of symmetry, which for the problem considered automatically constrains the nodes on this axis in the global X-translation (radial). In addition, all bottom nodes are restrained in vertical direction as shown in figure 2(b) which is sufficient and consistent with the closed-form model in figure 1.

3.3 Loading conditions: Contact and initial velocity

Contact definition is required to establish interaction between the contacting surfaces of impactor and bar. The contact definition allows the incoming slave nodes of the impactor to recognise the top of the bar (which contains master segments). The mesh refinement size between the two contacting models is comparable. Standard penalty-based contact is employed for this purpose, which uses numerical springs to prevent penetration of each slave nodes through master segments as well as to transfer loads between contacting parts [6].

The *2D Automatic Surface to Surface* contact type is employed which uses penalty forces to prevent penetration between external faces of 2D continuum elements. In addition, *Force Transducer* option is used in conjunction with the previous contact card to measure forces generated by that contact definition. In the current axisymmetric model, the force output is per unit radian. Normal direction of both slave and master surfaces are determined automatically by the programme, in which the initial model set up already has orthogonal contact surfaces prior to the impact [6].

The hammer is assigned with corresponding initial velocity from drop height with only small gap is introduced between the models in order to avoid initial penetration leading to instability in the calculation of the contact forces [6]. The gap size is generally balanced between minimising initial computational cost (caused by a large gap) and avoiding initial penetration (due to gap being too small) between contacting solids. The best decision depends on the pre-calculation of the defined time step of the smallest element, the initial distance between the first expected contacting elements and the assigned nodal velocity.

3.4 Material models

The bar's material is defined as elastic, which is similar to the closed-form solution in [1, 5], or viscoelastic. In both cases, the bar has homogeneous and isotropic mechanical properties. The impactor is defined as rigid material which matches the problem definition and justifies the irregular mesh of the impactor near the impacted region. In addition, this material modelis a cost-effective option because its model does not store history variables due to its undeformed body restriction.

Even though the rigid definition itself implies infinite stiffness, values of both Young's modulus, and Poisson's ratio need to be assigned to determine the bulk modulus for calculating the contact stiffness. In addition, the explicit time integration requires the mass density of the impactor in order to construct mass matrix in the CDM solution for the nodal displacements, even though the wave propagation is not calculated for the rigid material.

4. COMPARISON WITH THE CLOSED-FORM SOLUTION

The force response of a cylindrical FE model (10.6 mm diameter and 12 mm length) impacted by a light-weight 26.6-g hammer is first compared with the closed-form (CF) solution in terms of the resulting peak load and pulse width. The solid axisymmetric bar is converted to a one-dimensional model by assuming the Poisson's ratio to be zero, i.e. there is no transverse response in the 2D solid element. The light hammer is chosen to increase the mass ratio, which in turn would result in only seven stress intervals in the pulse response, allowing easy comparison to be made.

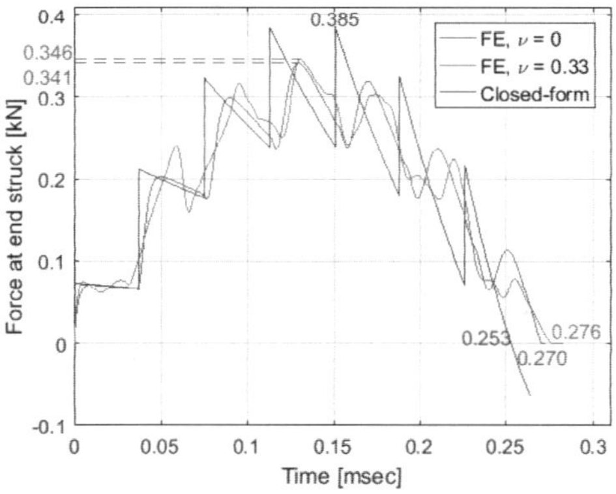

Figure 3. Verifying force response from FE of zero Poisson's ratio ($\nu = 0$) with the closed-form solution [5]. The effect of a non-zero ν FE model is also shown.

Figure 3 compares the FE response (zero Poisson's ratio) to that from the closed-form solution. It can be seen that there are seven force intervals in both FE and closed-form responses, which they are similar in the form of bell-like shape. The sharp peak-and-trough of vertical shape in every interval due to the instantaneous rise of force by CF solution [1, 5] is not recreated by the numerical simulation, nonetheless it is still fairly visible.

It is apparent that the closed-form solution returns consistent time interval, following strictly the governing equation of interval time [1, 5], whereas this time from FE solution gradually increases in every sequential interval. This discrepancy is explainable by understanding the ideal scenario in the closed-form solution in which the supposed deformation of bar is neglected in the analysis. On the other hand, the FE model experiences numerical damping which is a default setting for all contact options in the LS-DYNA [6] causing longer time for the travelling stress wave. Table 1 shows comparison of the two main outputs under investigation, namely peak load ($F_{\max}$) and pulse width (t_{pulse}) between the two solutions. The FE method records 11.43% lower peak load, while its pulse width is 6.72% larger in comparison to its benchmark results by the CF solution. The slightly high discrepancy recorded by this FE method is explicable by re-evaluating the element size of elastic bar which has 1:2 ratio of rectangular shape. It will be shown in the subsequent sections when the elements have 1:1 ratio, the outputs improve significantly as square-like 2D elements give more accurate results. It is worth mentioning that in both groups of bar models, the total number of elements are similar viz. 400 which are considered as sufficiently refined.

Table 1. Comparing one-dimensional FE model ($\nu = 0$) with the closed-form solution

Output variables	Peak load [kN]			Pulse width [msec]		
Models	**FE**	**CF**	**% diff.**	**FE**	**CF**	**% diff.**
	0.341	0.385	-11.43	0.27	0.253	6.72

The effect of Poisson's ratio ($\nu = 0.33$) is shown in figure 3, in which higher frequency distortion in its overall resulting pulse pattern is observed causing slightly unclear pattern of seven intervals. This happens perhaps due to the effect of wave bouncing from the sides. Nonetheless, the following peak load and pulse width values are not significantly affected, such that only 1.5% and 2.2% differences with the one-dimensional FE model. It can be concluded that assigning a reasonable value of Poisson's ratio does not drastically alter the response, in particular the output variables under interest, which are $F_{\max}$ and t_{pulse}.

5. EFFECT OF DIFFERENT PARAMETERS ON THE RESULTING ELASTIC PULSE

The effect of varying several parameters: mass of impactor;drop height; and structural stiffness; towards the resulting force at end struck by using numerical simulation is considered. This elastic study of axial impact on trabecular bone samples is based on report by Rosli [5]; some of the parameters are not varied i.e. density: $\rho = 1.31 \times 10^{-6}$ kg/mm^3; bar's length, $l = 21$ mm; and Poisson's ratio, $\nu = 0.33$. The axisymmetric model is discretised using 10×40 elements in radial and axial directions respectively, which is almost 1:1 of bar's diameter-to-length ratio or even slightly less for models with smaller radius. The analysed output is the summation of eleven (11) nodal forces histories of the nodes at the top of axisymmetric bar model, multiplied by 2π to obtain the total force acting on the surface of end struck.

5.1 Impactor's mass and drop height

Two values of impactor's mass are chosen, which are 1.807 and 2.5 kg, while the two drop heights are 100 and 200 mm, making up four (4) FE models for this analysis. The models have geometry of 10.6 mm of diameter and 21 mm of length, with elastic modulus is assigned to be 531 MPa as reported by Xie et al. [8] and falls within ranges reported by many studies including [9-14] for trabecular bone. The drop height affects only the peak load, while pulse width is independent of this input parameter as shown in figure 4; this finding by numerical method is consistent with theoretical parametric analysis in Rosli [5]; it was also shown in the study the peak load has direct proportionality to the square-root of drop height, h. As seen in figure 4, this is satisfied by the numerical solution. The pulse width is also almost identical for impactors of the same mass dropped from different heights. Figure 4 also shows that a higher drop height results in a fractionally smaller pulse width.

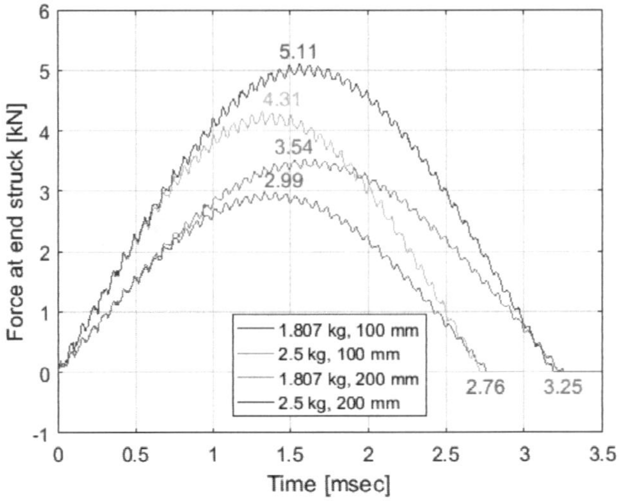

Figure 4. Effect of drop height, h and impactor's mass, m_{impactor} on the resulting force at end struck.

A larger mass of the impactor leads to larger peak loads and also pulse width, which is consistent to the analysis of the closed-form solution in Rosli [5]. In addition, it is observed in figure 4 that with the increase in m_{impactor}, the force responses appear to be almost similar initially, then the load for higher mass increases at a substantially higher rate and as expected comprises of larger number of stress intervals.

5.2 Structural stiffness

The structural stiffness of a simple column under compressive load is a function of its material (Young's modulus) and dimensions (cross-sectional area and length). The pulse exhibits stiffer response when the combination of higher peak load and lower pulse width is obtained [5].

In this parametric study, 2.5 kg impactor is dropped from 100 mm height. The parametric study with respect to length was examined in the closed-form solution [5], therefore only elastic modulus and diameter (representing area) are varied in this analysis. Similar elastic modulus of 531 MPa is chosen, with addition of arbitrary materially-stiffer models of 1 GPa. The diameters are either 10.6 mm [5] or 8.6 mm, in which a smaller area is expected to exhibit a less stiff behaviour.

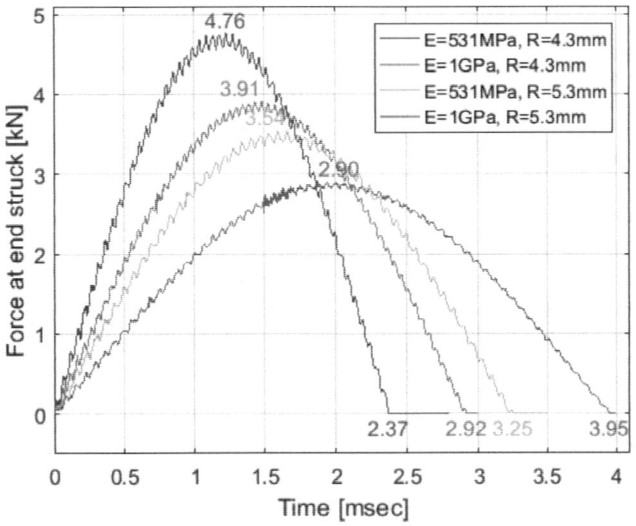

Figure 5. Effect of structural stiffness (elastic modulus, E and radius, R representing material and geometry components respectively) on the resulting force at end struck.

The resulting force responses at the end struck are shown in figure 5 for four varying-stiffness models. The stiffest response is shown by the model with highest elastic modulus and cross-sectional area, i.e E = 1 GPa, radius, R = 5.3 mm; followed by models with lower structural stiffness contributors. These demonstrate the influence of elastic modulus and cross-sectional diameter on the peak load and the pulse width. The peak load is seen to be directly proportionalto the elastic modulus as would be expected [5]. The peak load is also directly proportional to the cross-sectional area. The effect on pulse width is simply the opposite to that for the peak load i.e. pulse width decreases with increase in elastic modulus and radius.

6. IMPACT RESPONSE OF MODELS WITH TIME-DEPENDENT MATERIAL BEHAVIOUR

The effect of employing a time-dependent material model, viz. viscoelasticity (VE) is analysed in comparison to simple time-independent linear elastic model (abbreviated as TI-E) assigned to the impacted axisymmetric bar. In VE material model, the relaxation modulus can be defined using the Prony series [4] as

$$E(t) = E_{\text{eqm}} + \textstyle\sum_{i=1}^{n} E_i e^{-\frac{t}{\rho_i}}, \qquad (1)$$

where E_{eqm} is the equilibrium modulus and E_i and ρ_i are the relaxation strengths and relaxation times respectively. A three-term Prony series ($n = 3$) was shown to adequately represent the relaxation modulus of trabecular bone with different volume fractions, BV/TV [3]. The values determined in this above cited study are used to consider the influence of time-dependent material properties on impact response.

6.1 Implementing viscoelasticity in LS-DYNA

In TI-E model, single value of elastic modulus, E is taken as $E(\infty)$ or E_{eqm} from equation (1); a long-term modulus representing the material resistance value when the load stabilises hence treated as in statically-applied manner in equilibrium. On the other hand, the viscoelastic material model in LS-DYNA, e.g. Mat076: General Viscoelastic requires elastic bulk modulus, K. Similar to the equilibrium modulus in TI-E, K in VE is defined as a long-term modulus. The Poisson's ratio was defined as 0.33. In order to exhibit relaxation phase in the VE model, shear relaxation moduli, $G_i$and decay constants, β_i for each term in Prony series are required. Hence, the general expression of the Prony series in LS-DYNA is

$$g(t) = \textstyle\sum_{i=1}^{n} G_i e^{-\beta_i t}. \qquad (2)$$

Comparing these two relaxation expressions in equations (1-2), the relaxation strengths and times parameters in equation (1) provided by Manda et al. [3] need to be converted onto shear moduli and decay constants in equation (2) respectively. Shear moduli are obtained from E and ν assuming isotropic behaviour while the decay constants (β_i) are simply inverse of the relaxation times (ρ_i) in which $i = 1, 2, 3$ for three terms of Prony series. In the case of three-term Prony series, the General Viscoelastic material model requires four input terms, in which the extra term is G_{eqm} with zero decay constant, fully expressed as

$$g(t) = G_{\text{eqm}} + G_1 e^{-\beta_1 t} + G_2 e^{-\beta_2 t} + G_3 e^{-\beta_3 t}. \quad (3)$$

Six (6) bone volume fraction models, i.e. BV/TV=19, 26, 33, 39, 42 and 43% are selectively chosen from Manda et al. [3] on the basis of signifying direct correlation between material's stiffness and the reported BV/TV. The required input parameters are given in table 2 by using moduli conversions into equation (3). All these conversions are applicable in this study assuming the material as having isotropic behaviour.

Table 2. Parameters of time-independent elastic (TI-E) and viscoelastic (VE) material models; Units: modulus [MPa], decay constant [ksec^{-1}]

BV/TV [%]	TI-E	VE adopting Prony series				
	E_{eqm}	K_{eqm}	G_{eqm}	i	G_i	β_i
				1	5.34	684.93
0.19	250.43	417.38	89.44	2	3.27	42.39
				3	5.90	3.75
				1	7.80	1111.11
0.26	317.18	528.63	113.28	2	5.55	119.76
				3	6.61	8.22
				1	12.18	1086.96
0.33	498.53	830.88	178.05	2	6.51	125.47
				3	6.83	10.24
				1	15.17	1351.35
0.39	545.65	909.42	194.88	2	10.55	129.03
				3	11.99	6.76
				1	16.32	3030.30
0.42	572.05	953.42	204.30	2	10.74	184.84
				3	10.66	9.10
				1	24.34	1010.10
0.43	904.68	1507.80	323.10	2	12.40	109.53
				3	19.95	7.15

6.2 Force response

The bar is discretised into 4-noded axisymmetric 10×40 solid elements impacted by a 2.5-kg rigid impactor. The length, l of the bar is 21 mm and its diameter, $2r$ is 10.6 mm. The impactor was assigned with axial velocity of 0.99 m/s, which is equivalent when it is dropped from 50 mm height. In order to terminate the simulation, the pulse duration is estimated [1, 5] as

$$t_{\text{pulse}} \approx \frac{1}{r}\sqrt{\frac{m_{\text{impactor}}\pi l}{E}}, \quad (4)$$

then the termination time was set to be slightly longer than the estimated pulse width in equation (4).

Two main outputs are investigated from the resulting force response at the end struck of axisymmetric bar, namely the peak load as the highest magnitude of compressive load experienced by the end struck and the pulse width, which is the contact duration of the impact incidence. Fig. 6 shows graphical elastic and viscoelastic responses of three different BV/TVs, clearly depicts that viscoelastic models have stiffer response than their TI-E counterparts.

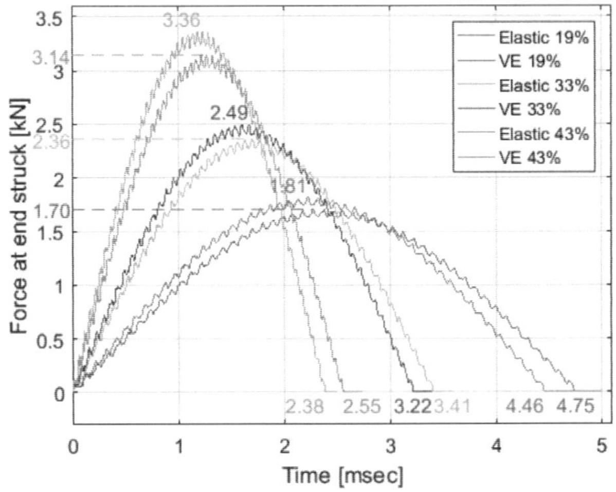

Figure 6. Effect of varying time-independent and viscoelastic properties towards peak load and pulse width.

The key numerical results for all models are tabulated in table 3 with *E*(*t*) represents E_{eqm} and *E*(0) for TI-E and VE respectively. The modulus function *E*(*t*) represents well the contribution of BV/TV to bone sample's stiffness for both material models. In addition, the stiffness of VE models, which is *E*(0) is consistently higher than TI-E (E_{eqm}).

Table 3. Elastic modulus input of time-independent elastic (TI-E) and viscoelastic (VE) material properties with the corresponding peak load (F_{max}) and pulse width (t_{pulse})

BV/TV	Model	E(t) [MPa]	Peak load [kN]		Pulse width [msec]	
			Value [kN]	% diff.	Value [kN]	% diff.
19%	TI-E	250.4	1.7037		4.7549	
	VE	291.1	1.8150	6.53	4.4629	6.14
26%	TI-E	317.2	1.9036		4.2455	
	VE	373.1	2.0428	7.31	3.9591	6.75
33%	TI-E	498.5	2.3580		3.4111	
	VE	570.0	2.4889	5.55	3.2212	5.57
39%	TI-E	545.7	2.4641		3.2646	
	VE	651.2	2.6461	7.39	3.0310	7.16
42%	TI-E	572.1	2.5223		3.1919	
	VE	677.7	2.7074	7.34	2.9777	6.71
43%	TI-E	904.7	3.1434		2.5544	
	VE	1063.4	3.3615	6.94	2.3839	6.67

Based on peak load values in table 3, the corresponding maximum compressive stresses at end struck fall in the range of $19.3 \leq \sigma_{max} \leq 38.1$ MPa as simulated by FE method. As comparison, trabecular bone was reported to have yield strength of 17.45 ± 6.15 MPa and $5 - 20$ MPa by Morgan and Keaveny [15] and Shim et al. [2], who used extensometer and split Hopkinson pressure bar respectively.

Maximum compressive stresses from table 3 recorded higher values than these reported strengths, nonetheless they were proven experimentally to be realistic in drop tests [5].

6.3 Modulus-to-output variable relations

The parametric study in [2] points to the requirement of using linear direct best fit regression for relating $\sqrt{E}$ to the peak load i.e. $F_{\max} \propto \sqrt{E}$. On the other hand, its effect on the pulse width is inversely proportional or negative power law i.e. $t_{\text{pulse}} \propto 1/\sqrt{E}$. The mathematical relations obtained from this numerical analysis is shown in figure 7.

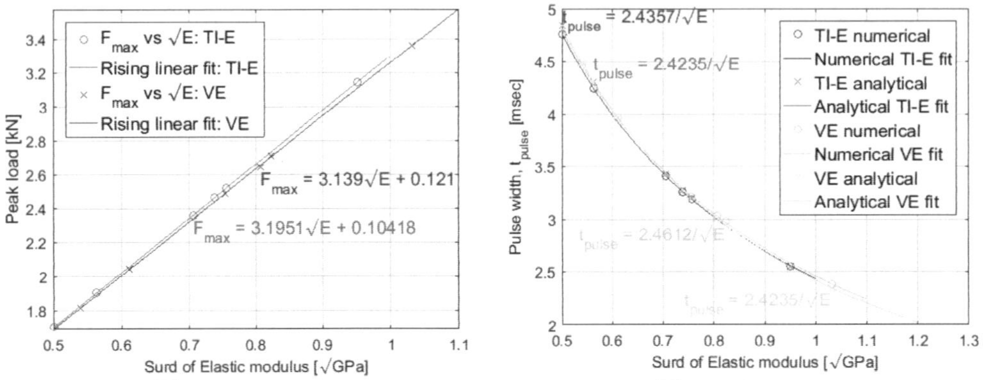

(a) $F_{\max} - \sqrt{E}$ direct proportionality (b) $t_{\text{pulse}} - \sqrt{E}$ inverse proportionality

Figure 7. Output variables relations with elastic modulus for varying material properties.

The output variable-elastic modulus relations in figure 7 show very close resemblance to the mathematical relations derived from the closed-form solution provided by Rosli [5] i.e. similar regression models return almost identical constants. Hence, it can be concluded that the current FE models well replicate the theoretical solutions, hence they are benchmarked and could be used for more complex axial bar impact problems.

7. SUMMARY

This numerical work replicates an axial bar subjected to rigid mass impactor problem as described in previous studies. Sufficiently-refined volume-weighted Galerkin axisymmetric elements model is used which is available in LS-DYNA as a cost-effective geometrical model, while linear elastic and time-dependent viscoelastic material models were evaluated. The study is successfully validated by theoretical solutions in which comparable results were obtained when analysing: 1) the effect of Poisson's ratio (difference with one-dimensional model is less than 3%); 2) parametric study on impactor's mass, drop height and structural stiffness; and 3) output variable relations (peak load and pulse width) with square-root of elastic modulus. The output variable-input parameter relations are useful to explain empirical mathematical relationships. Hence, the FE models developed in this study are successfully benchmarked and can be further developed for incorporating more complex inputs from axial bar impact problems.

ACKNOWLEDGEMENT

The first author is indebted to the Ministry of Education, Malaysia for the award of PhD sponsorship.

REFERENCES

[1] Timoshenko S, Goodier JN. (1951) Theory of Elasticity, 2nd Ed. McGraw-Hill Book Company, Inc., New York.

[2] Shim V, Yang L, Liu J, Lee V. (2005) Characterisation of the Dynamic Compressive Mechanical Properties of Cancellous Bone from the Human Cervical Spine. International Journal of Impact Engineering, 32(1-4):525-540.

[3] Manda K, Xie S, Wallace RJ, Levrero-Florencio F, Pankaj P. (2016) Linear Viscoelasticity – Bone Volume Fraction Relationships of Bovine Trabecular Bone. Biomechanics and Modeling in Mechanobiology, 15(6):1631-1640.

[4] Park SW, Schapery RA. (1999) Methods of Interconversion between Linear Viscoelastic Material Functions. Part I-a Numerical Method Based on Prony Series. International Journal of Solids and Structures, 36(11):1653-1675.

[5] Rosli AHB. (2019) Characterisation of Trabecular Bone Behaviour under Impact, PhD Thesis, The University of Edinburgh.

[6] LSTC. (2017) LS-DYNA® Theory Manual, Technical Report, Livermore Software Technology Corporation (LSTC), Livermore, California.

[7] Stronge WJ. (2000) Impact Mechanics, Cambridge University Press.

[8] Xie S, Wallace RJ, Callanan A, Pankaj P. (2018) From Tension to Compression: Asymmetric Mechanical Behaviour of Trabecular Bone's Organic Phase. Annals of Biomedical Engineering, 46(6):801-809.

[9] Lotz JC, Gerhart TN, Hayes WC. (1990) Mechanical Properties of Trabecular Bone from the Proximal Femur: A Quantitative CT Study. Journal of Computer Assisted Tomography, 14(1):107-114.

[10] Keller TS. (1994) Predicting the Compressive Mechanical Behaviour of Bone. Journal of Biomechanics, 27(9):1159-1168.

[11] Goulet RW, Goldstein SA, Ciarelli MJ, Kuhn JL, Brown MB, Feldkamp LA. (1994) The Relationship between the Structural and Orthogonal Compressive Properties of Trabecular Bone. Journal of Biomechanics, 27(4):375-389.

[12] Li B, Aspden RM. (1997) Composition and Mechanical Properties of Cancellous Bone from the Femoral Head of Patients with Osteoporosis or Osteoarthritis. Journal of Bone Mineral Research, 12(4):450-456.

[13] Ciarelli TE, Fyhrie DP, Schaffler MB, Goldstein SA. (2000) Variations in Three-dimensional Cancellous Bone Architecture of the Proximal Femur in Female Hip Fractures and in Controls. Journal of Bone and Mineral Research, 15(1):32-40.

[14] Kaneko TS, Bell JS, Pejcic MR, Tehranzadeh J, Keyak JH. (2004) Mechanical Properties, Density and Quantitative CT Scan Data of Trabecular Bone with and without Metastases. Journal of Biomechanics, 37(4):523-530.

[15] Morgan EF, Keaveny TM. (2001) Dependence of Yield Strain of Human Trabecular Bone on Anatomic Site. Journal of Biomechanics, 34:569-577.

Adaptive solar panel - overview and conceptual design

Wan Luqman Hakim Wan A. Hamid [1], Nurul 'Ilya Basharuddin[1,*], Nor Aiman Sukindar[2]

[1] Department of Mechanical and Aerospace Engineering, Kulliyyah of Engineering, International Islamic University Malaysia, Kuala Lumpur, Malaysia.
[2] Department of Manufacturing and Materials Engineering, Kulliyyah of Engineering, International Islamic University Malaysia, Kuala Lumpur, Malaysia.

[*] Corresponding author e-mail address: nurulilya98@gmail.com

Abstract. Renewable energy should reach its maximum usage capacity, around 20%, by 2025, as laid out by the Malaysian government to achieve the seventh Sustainable Development Goals (SDG): Affordable and clean energy. Many actions have been implemented to support the idea, including the development of solar energy panels in large photovoltaic installation. However, the installations require a wide area of land, and the efficiency offered is limited due to their fixed position. The efficiency of solar panels in absorbing the light from the sun is important to increase the energy supply. Several designs of the adaptive solar panel were developed by researchers, but the installation costs are high due to the components used. Commonly used actuators to move the solar panels are motor (DC motor, stepper motor, servo motor) that actuate according to the signals from photosensor, programming logic control (PLC), or microcontroller. Herein, the development of an Adaptive Solar Panel (ASP) is studied. SMA with high recoverability when it is strained is introduced as an actuator that capable of generating large actuation when heated to move the solar panel. In discovering the potential of SMA wire, a modern environmental-friendly design with six layers of solar panels with a rotational support system was designed in this study. In future work, the SMA wire actuation will be evaluated by computational simulation method using LS-Dyna. **Keywords:** Actuator, shape memory alloy, solar panel.

1. Introduction

The current energy sources for Malaysia's power generation are natural gas and coal. Only 2% of the total power generation is produced from renewable energy sources [1]. Thus, an ambitious target has been set by the Malaysian Government to utilize the renewable energy up to 20% by 2025. Due to our major dependence on fossil fuels, an alternative renewable energy source that is locally available is required to replace the depleting fossil fuels. Hence, utilizing solar energy as one of renewable energy sources is good for the environment as it is non-polluting. Besides, the Malaysian weather is suitable for solar panels made of photovoltaic (PV) cells as the average annual solar irradiance level is mostly above 1500 kWh/m^2 for every state [2]. However, the current solar panels offered limited efficiency in

[*] *Corresponding author.*
E-mail address: nurulilya98@gmail.com (Nurul 'Ilya Basharuddin)

sun-light absorption because of its static installation. Even with the existing solar tracking devices designed by some researchers, it still requires external power to sense the sun's movement and to actuate the solar panel. So, the complexity of the technology somehow increased the cost. Thus, simple solar tracking using shape memory alloy (SMA) for affordable and clean energy is designed. In this study, two thin-film solar modules manufactured by First Solar Company with a dimension of 1200 mm x 60 mm x 6.4 mm, weight of 30 kg, and two shape memory alloy (SMA) wires manufactured by SAES Getters Group will be used in the modelling.

2. Adaptive Solar Panel - Benefits and Potential

Based on previous studies in several countries having different climates, the performance of the PV arrays with and without solar tracking systems is investigated. Vieira (2016) compared the PV performance between the static and mobile systems and he found that the performance in power generation was increased by 11%. Nagy et al. (2016) found that the total energy could be saved up to 25% when they used a single axis of rotation solar tracking system with a soft pneumatic actuator in their adaptive solar facade. Compared to the fixed solar panel, Akbar et al. (2017) showed that their sun tracker design increased the efficiency of the output power generation by 25-30%. Seme et al. (2017) proposed a design of a bi-axis solar tracking system, and it generated 27% higher electrical energy compared to the fixed system. Alinejad et al. (2020) assessed the thermo-environomic of an integrated greenhouse with an adjustable solar photovoltaic blind system. They found that the SBG24 performed well in all studied parameter as it controlled the shading, it reduced the annual natural gas consumption by 3.57%, it supplied 45.5% electricity demand to the grid, and it generated 42.69 kWh/m^2 of energy supply compared to the existing greenhouse (GR). In a nut shell, the use of an Adaptive Solar Panel does not only increase the harvested energy but can also meet certain demands in form of shading and electricity supply. However, the performance may vary due to the types of tracking systems, the country's geographical locations, and the weather conditions.

3. Adaptive Solar Panel - Overview of the Existing Design

Solar tracking systems can be categorized based on the drivers, control system, tracking strategy, or degree of freedom of the movement (Awasthi et al., 2020). Overview of solar tracking systems with various mechanisms is summarized in Table 1.

In brief, based on the explored tracking systems, a certain amount of electrical energy is required as input for the controlling units (i.e. FLC, PLC and microcontroller), for the actuators (i.e. electrical motor) and the sensors (photosensor and LDRs). Therefore, electrical energy is required as an external energy source to activate them. By replacing those with smart materials (i.e. SMA wires) that can be activated by solar heating, the use of electrical energy as input is no longer required.

4. Design

4.1. Mechanical

The design of the ASP has three elements as shown in Figure 1; two thin-film PV modules, an actuator, a supporting frame with a stand. The two solar panel modules have a weight of 15 kg. In the following sections, the design considerations for the thin-film PV modules, the actuator, and the supporting frame and stand are described.

4.2. Solar Panel – Thin Film Solar Panel Module

Solar technology is rapidly expanding. In Malaysia, PV research and development explore cutting-edge technologies, cell design, and novel approaches in the development of solar materials and products [19]. Conventional PV cells consist of crystalline silicon which atoms are arranged in a 3D array [20]. Commercially available solar panels have their benefits and drawbacks [20-21]. Generally, PV modules can be categorized into three groups, which are amorphous, monocrystalline, and polycrystalline silicon.

Table 1: Solar tracking system

No.	Author's Name	Design Implemented
1.	Ganesh N. J. et al. (2011)	Sun tracking system with Ni-Ti shape memory alloy (SMA) spring having an average diameter of 5 mm and 52 coils that acts as sensor and actuator.
2.	Degeratu et al. (2014)	Sun tracking system controlled by SIEMENS program logic control (PLC) that requires external power sources actuated by Ni-Ti SMA.
3.	Azwaan Zakariah et al. (2015)	Sun tracking system using a combination of Arduino UNO ATmega328 microcontroller, motor driver (MD30C), two power window motors, maximum power point tracking (MPPT), four light dependent resistors (LDRs), and fuzzy logic controller (FLC). The microcontroller was programmed for control of the overall performance; the MPPT was used to adjust the current and voltage while the FLC provides inference decision to the microcontroller.
4.	Rahimi et al. (2015)	Sun tracking system using two photosensors, and two DC motors with a supply voltage of 6V DC. The sun's movement was tracked by the photosensor and the solar panel was actuated by the two DC motors.
5.	Chakrabory S. et al (2015)	Sun tracking system consists of stepper motor (4SHG-050A 51S 5V, 5Ω), AT89S52 microcontroller and five LDRs. The microcontroller with the help of LDRs was programmed to sense the sunlight for actuation of the servo motor.
6.	Das K. et al. (2016)	Sun tracking system consists of ATmega328 microcontroller, two LDRs, and servo motor. The microcontroller was programmed to sense the sunlight with the help of LDRs, for actuation of the servo motor.
7.	Canbay C. A. et al. (2016)	Sun tracking system consists of three Cu-Al-Mn-Ni SMA to coat as a Schottky contact on a p-Si substrate to produce four quadrants sensor using vacuum coating.
8.	Hussain S. Akbar et al. (2017)	Dual-axis sun tracking system using a combination of ATmega328 microcontroller, two direct currents (DC) motors, two LDRs, and two relays. The microcontroller was programmed to sense the sunlight with assistance from LDRs and relays, for decision of DC motor rotation.
9.	Riad et al. (2017)	Sun tracking system that works like a sunflower that was actuated by SMA springs with an average diameter of 5 mm and 52 coils.
10.	Riad et al. (2020)	Sun tracking system that used six SMAs in spring form with 10 mm diameter to actuate and move the solar panel.

With the rising development in solar energy production, thin-film technologies are currently improved to achieve higher efficiency by using materials such as cadmium telluride (CdTe), copper indium diselenide (CIS), and gallium arsenide (GaAs) [21]. This is proven by Nowshad Amin et al. (2009) and Sreenath et al. (2020) who conducted an experiment to analyse the performance of several types of solar panels under Malaysia's weather. The study found that crystalline solar modules gave better performance under the hot sun, while the thin-film and amorphous silicon solar panels gave higher performance under cloudy and diffused sunshine weather conditions. Moreover, the efficiency of crystalline silicon solar panels dropped when the temperature increased, unlike thin-film and amorphous silicon solar panels. Debbarma et al. (2017) and Manoj et al. (2019) found that Cadmium

Telluride (CdTe) was suitably and widely used for large-scale PV plant installation. Therefore, the Adaptive Solar Panel was modelled with the CdTe thin-film from First Solar Sdn. Bhd, with a dimension of 1200 mm × 60 mm × 6.4 mm.

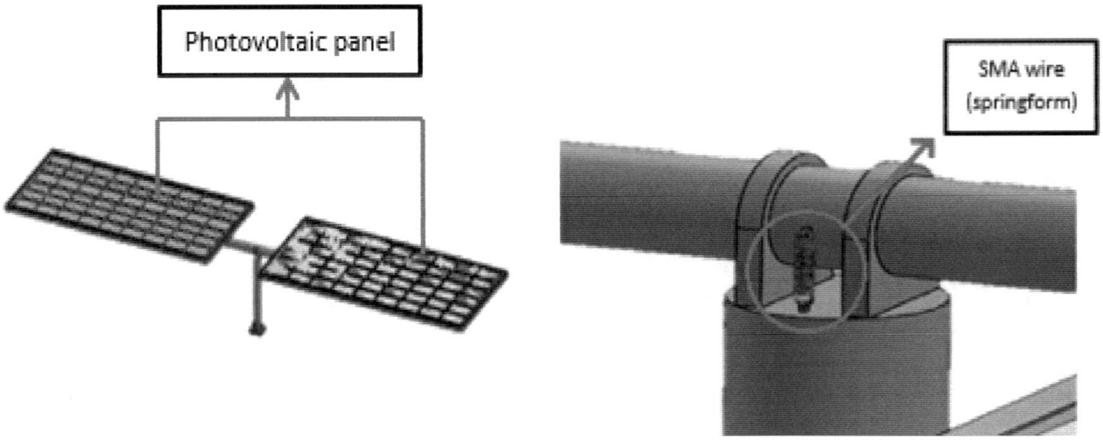

Figure 1. Current adaptive solar panel

4.3. *Actuator - Mechanical and Smart Actuation*

Solar trackers moved the solar panels so that the surfaces of the panels are oriented towards the sun. By utilizing actuators, the solar panels are able to face the sun throughout the day to capture the optimum energy and to maximize its output. According to section 3, there are two types of actuators: motors (i.e. DC motors, servo motors, and stepper motors) and SMAs.

A motor, which is a device that converts electrical energy into mechanical energy, operates via the interaction between the motor's magnetic field and electric current in wire winding to generate force or torsional moment. Once the motor is powered by external power sources, it generates torque, which actuates the solar panels to the aimed angle, capturing photons from the sun.

Shape memory alloy (SMA) is a unique material that can return back to its original shape after being deformed, when it is heated above the Austenite finishing temperature [17,26]. The SMA possesses excellent thermo-mechanical properties, such as shape memory effect (SME), super elasticity (SE), good damping, and good fatigue and corrosion resistance [27]. The SMAs have two main phases, which are a Martensite phase at low temperatures, and an Austenite phase at high temperatures [17,26]. The shape memory effect occurs when the SMA in the Martensite phase is deformed, and the residual deformation can be recovered by heating the SMA above the Austenite finishing temperature. The SMA also exhibits super elasticity properties when it is in the Austenite phase and is deformed due to the external load, and the deformation can be fully and spontaneously recovered when the load is removed [27].

SMA actuators are promising due to their large recoverable strain and specific energy output [28]. Therefore, the actuator chosen for the design is SMA wire manufactured by SAES Getters Group, with a 0.5 mm diameter and a maximum strain recovery of 8%, because of the simplicity of the smart actuation that requires no external power. The SMA wire was designed in spring form to achieve higher stroke [30]. However, details on the SMA design will be further evaluated based on the actuation, whether the designed SMA can actuate the 30 kg solar panels.

4.4. *Numerical Analysis for Support Stand and Beam*

According to First Solar Company, the mass of the solar panel is approximately 10-15 kg, depending on the glass thickness. Hence, for safety analysis, the maximum mass considered in the analytical calculation was 15 kg. Based on the maximum distortion energy theory, the structure is safe. The solar panels were assumed in equilibrium condition where the net forces were equal to zero. The formulas used in the analysis are listed below.

i - Equilibrium equations:

$$\Sigma \mathrm{F} = 0, \Sigma \mathrm{M} = 0 \tag{1}$$

where *F* and *M* are the forces and moments in the *x*, *y*, and *z* directions.

ii – Normal/axial stress:

$$\sigma = \frac{\text{Force}}{Area} \tag{2}$$

iii - Shear stress:

$$\tau = \frac{VQ}{It} \tag{3}$$

where *V* is the internal shear force, *Q* is the statical moment of area, *I* is the moment of inertia and *t* is the thickness (width) perpendicular to the shear.

iv - Bending stress:

$$\sigma = \frac{M \cdot y}{I} \tag{4}$$

where *M* is the internal bending moment, *y* is the distance from neutral axis, and *I* is the moment of inertia.

5. Results

The solar panel is rotated by 30° angle in 0.2 seconds. The result of the angle of rotation is shown in Figure 2. The adaptive solar panel rotates by 30 degrees as shown in the figure, when angular displacement was applied to the panel.

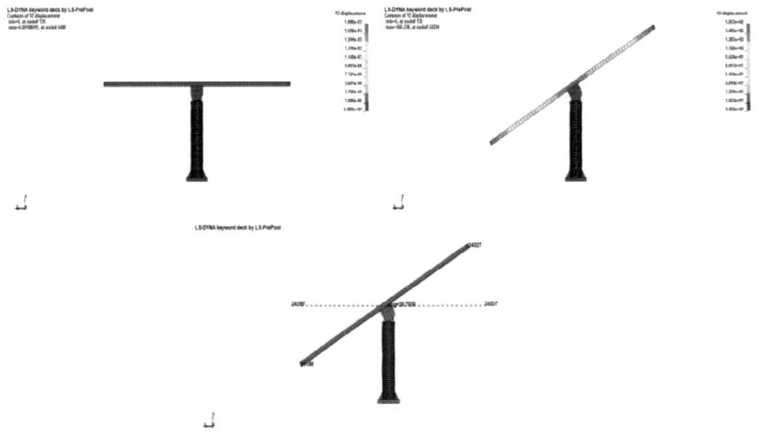

Figure 2. The adaptive solar panel before and after 30° rotation.

6. Conclusion

In conclusion, the design of an adaptive solar panel was made and finalized. The process of the design is mainly divided into three parts which are a selection of the solar panel, selection of actuator and analysis of support stand and beam. In future work, the actuation of the SMA wire will be evaluated by computational simulation method using LS-Dyna finite element software.

References

[1] Abdullah, Wan Syakirah Wan, Miszaina Osman, Mohd Zainal Abidin Ab Kadir, and Renuga

Verayiah. "The potential and status of renewable energy development in Malaysia." *Energies* 12, no. 12 (2019): 2437.

[2] Vaka, Mahesh, Rashmi Walvekar, Abdul Khaliq Rasheed, and Mohammad Khalid. "A review on Malaysia's solar energy pathway towards carbon-neutral Malaysia beyond Covid'19 pandemic." *Journal of cleaner production* (2020): 122834.

[3] Vieira, R. G., F. K. O. M. V. Guerra, M. R. B. G. Vale, and M. M. Araújo. "Comparative performance analysis between static solar panels and single-axis tracking system on a hot climate region near to the equator." *Renewable and Sustainable Energy Reviews* 64 (2016): 672-681.

[4] Nagy, Zoltan, Bratislav Svetozarevic, Prageeth Jayathissa, Moritz Begle, Johannes Hofer, Gearoid Lydon, Anja Willmann, and Arno Schlueter. "The adaptive solar facade: from concept to prototypes." *Frontiers of Architectural Research* 5, no. 2 (2016): 143-156.

[5] Akbar, Hussain S., Abulrahman I. Siddiq, and Marwa W. Aziz. "Microcontroller based dual axis sun tracking system for maximum solar energy generation." *American Journal of Energy Research* 5, no. 1 (2017): 23-27.

[6] Seme, Sebastijan, Gregor Srpčič, Domen Kavšek, Stane Božičnik, Tomislav Letnik, Zdravko Praunseis, Bojan Štumberger, and Miralem Hadžiselimović. "Dual-axis photovoltaic tracking system–Design and experimental investigation." *Energy* 139 (2017): 1267-1274.

[7] Alinejad, T., M. Yaghoubi, and Amir Vadiee. "Thermo-environomic assessment of an integrated greenhouse with an adjustable solar photovoltaic blind system." *Renewable Energy* 156 (2020): 1-13.

[8] Awasthi, Anshul, Akash Kumar Shukla, Murali Manohar SR, Chandrakant Dondariya, K. N. Shukla, Deepak Porwal, and Geetam Richhariya. "Review on sun tracking technology in solar PV system." *Energy Reports* 6 (2020): 392-405.

[9] Ganesh, N. Jeya, Subramanian Maniprakash, L. Chandrasekaran, S. M. Srinivasan, and A. R. Srinivasa. "Design and development of a sun tracking mechanism using the direct SMA actuation." Journal of Mechanical Design 133, no. 7 (2011).

[10] Degeratu, Sonia, Laurentiu Alboteanu, Sabin Rizescu, Daniela Coman, Nicu G. Bizdoaca, and Costel Caramida. "Active solar panel tracking system actuated by shape memory alloy springs." In *2014 International Conference on Applied and Theoretical Electricity (ICATE)*, pp. 1-5. IEEE, 2014.

[11] Zakariah, Azwaan, Jasrul Jamani Jamian, and Mohd Amri Md Yunus. "Dual-axis solar tracking system based on fuzzy logic control and Light Dependent Resistors as feedback path elements." In *2015 IEEE Student Conference on Research and Development (SCOReD)*, pp. 139-144. IEEE, 2015.

[12] Rahimi, Masoud, Meisam Banybayat, Yaghoub Tagheie, and Peyvand Valeh-e-Sheyda. "An insight on advantage of hybrid sun–wind-tracking over sun-tracking PV system." *Energy Conversion and Management* 105 (2015): 294-302.

[13] Chakrabory, Shreyasi, Nilanjana Mukherjee, Rashmi Biswas, Tanushree Saha, Astika Mohinta, Neha Kumari Modi, and Dip Prakash Samajdar. "Microcontroller based Solar Tracker system using LDRs and Stepper Motor." *International Journal of Computer Applications* 975 (2015): 8887.

[14] Das, Krishanu, Hridi Ghosh, and Maitrayee Sengupta. "Single axis solar tracking system using microcontroller (atmega328) and servo motor." *International Journal of Scientific and Research Publications* 6, no. 6 (2016): 486-489.

[15] Canbay, C. Aksu, A. Tataroglu, A. Dere, Ahmed Al-Ghamdi, and F. Yakuphanoglu. "A new shape memory alloy film/p-Si solar light four quadrant detector for solar tracking applications." *Journal of Alloys and Compounds* 688 (2016): 762-768.

[16] Akbar, Hussain S., Abulrahman I. Siddiq, and Marwa W. Aziz. "Microcontroller based dual axis sun tracking system for maximum solar energy generation." *American Journal of Energy Research* 5, no. 1 (2017): 23-27.

[17] Riad, Amine, Mouna Benzohra, Mohamed Mansouri, and Abdelelah Alhamany. "Design of Sunflower System Based on Shape Memory Alloy Actuator." In *First International Conference on Real Time Intelligent Systems*, pp. 45-54. Springer, Cham, 2017.
[18] Riad, Amine, Mouna Ben Zohra, Abdelilah Alhamany, and Mohamed Mansouri. "Bio-sun tracker engineering self-driven by thermo-mechanical actuator for photovoltaic solar systems." *Case Studies in Thermal Engineering* 21 (2020): 100709.
[19] Islam, M. R., Rahman Saidur, N. A. Rahim, and K. H. Solangi. "Usage of solar energy and its status in Malaysia." *Engineering e-Transaction* 5, no. 1 (2010): 6-10.
[20] Ab Kadir, Mohd Zainal Abidin, Yaaseen Rafeeu, and Nor Mariah Adam. "Prospective scenarios for the full solar energy development in Malaysia." *Renewable and Sustainable Energy Reviews* 14, no. 9 (2010): 3023-3031.
[21] Boxwell, Micheal. *Solar Electricity Handbook.* 6th. Warwickshire: Greensteam Publishing 2012 (2012): 19-21, 130-132.
[22] Amin, Nowshad, Chin Wen Lung, and Kamaruzzaman Sopian. "A practical field study of various solar cells on their performance in Malaysia." *Renewable Energy* 34, no. 8 (2009): 1939-1946.
[23] Sreenath, S., K. Sudhakar, A. F. Yusop, E. Solomin, and I. M. Kirpichnikova. "Solar PV energy system in Malaysian airport: Glare analysis, general design and performance assessment." *Energy Reports* 6 (2020): 698-712.
[24] Debbarma, Mary, K. Sudhakar, and Prashant Baredar. "Thermal modeling, exergy analysis, performance of BIPV and BIPVT: a review." *Renewable and Sustainable Energy Reviews* 73 (2017): 1276-1288.
[25] Manoj Kumar, Nallapaneni, K. Sudhakar, and Mahendran Samykano. "Techno-economic analysis of 1 MWp grid connected solar PV plant in Malaysia." *International Journal of Ambient Energy* 40, no. 4 (2019): 434-443.
[26] Pretorius, Van Reenen. U.S. Patent and Trademark Office. United States of America Patent 9,816,492. November 14 (2017).
[27] Kang, Liping, Hui Qian, Yuancheng Guo, Chenyang Ye, and Zongao Li. "Investigation of Mechanical Properties of Large Shape Memory Alloy Bars under Different Heat Treatments." *Materials* 13, no. 17 (2020): 3729.
[28] Vollach, Shahaf, and D. Shilo. "The mechanical response of shape memory alloys under a rapid heating pulse." *Experimental Mechanics* 50, no. 6 (2010): 803-811.
[29] SAES Getters Group. "SAES® Smart Materials (SSM) Brochure." SAES Getters Group (2008).
[30] W.L.H. Wan A. Hamid, L. Iannucci, and P. Robinson. "Finite-element modelling of NiTi shape-memory wires for morphing aerofoils." *The Aeronautical Journal* 124, no. 1281 (2020): 1740-1760.

IOP Conf. Series: Materials Science and Engineering 1244 (2022) 012017 doi:10.1088/1757-899X/1244/1/012017

An investigation on structural dynamic modification of a machine casing

Zahir Hanouf[1,4], Ahmed Abdelrahman[2] and Kartini Ahmad[3]

[1] *Department of Mechanical, Faculty of Engineering University of Bahrain*
[2] *Polytechnic Bahrain*
[3] *Department of science in Engineering Faculty of Engineering IIUM.*
E-mail: zhanouf@uob.edu.bh

Abstract. Investigation on structural dynamic characteristic is a crucial step in the noise, vibration and harshness (NVH) analysis in order to enhance the structural vibration, which subsequently contribute to improving the acoustical environment. The excitation of the structure can produce high vibration which could be transferred to the other parts of the body, then lead to high noise radiating components. Structural dynamic modification is one of the important techniques in the design of mechanical structures. It is an application of experimental modal analysis technique to investigate the effect of the physical and geometrical parameters on the dynamic properties of the structures such as frequencies and mode shapes.This paper investigates the dynamic properties of a machine casing structure which is extensively used in industrial applications . This investigation had successfully produced natural frequencies and their associated mode shapes of such open box for the original structure and modified structure. The study establishes the step-by-step procedures to carry out experimental modal analysis on the structures, using single input excitation and single output responses (SISO) technique. To ensure the validity of the results obtained an inverse method was done by fixing the response and moving the excitation. The results found were absolutely the same. Finally, comparison between results obtained from analysis showed significant differences on both frequencies and mode shapes.

1. Introduction

Vibration is a key factor to consider when constructing mechanical structures. As a result, it is critical to change the structure's responsiveness to load in order to avoid failure. The term structural dynamic modification is frequently used to regulate the vibration level. Changing the dynamic properties of a structure in order to address a vibration issue is difficult. This is because the majority of systems were not created with this necessity in mind. Furthermore, vibration was not the primary emphasis of the design effort, even though it became the focal point of the problem when the system was constructed. Solving the vibration problem towards the end of the project is usually a complexity and an intricate difficulty, since adjustments and modifications to the main portion are not always permitted. Additionally, while attempting to resolve a vibration problem in an operating system such as an engine or a power plant, the knowledge about the principal dynamic parameter is limited, and the system must be treated as a box, the dynamics of which may be determined experimentally by modal testing. Additionally, experimental modal analysis is utilised to determine the structure's mode shape and natural frequency. These dynamic qualities are critical for the analysis and design of any structure, since they aid engineers in comprehending a structure's dynamic behaviour.

Understanding a structure's dynamic properties, such as natural frequencies, mode shapes, and damping ratios, is critical for assessing structural safety, extending fatigue life, and monitoring structural health. By understanding a structure's dynamic behaviour at its natural frequencies, we may identify the location of damaged components before they cause fatalities, as damaged components display excessive deformation when subjected to the same level of excitation. Additionally, by understanding a structure's dynamic properties, engineers may design more effective structures. Weight, stiffness, and damping are critical characteristics that have a significant influence on a structure's natural frequencies and mode shapes, as well as on the locations of these natural frequencies and mode shapes. As a result, we must investigate these frequencies and determine how they impact the structure's vibrational behaviour when it is excited by a force. This can assist engineers in designing more vibration-resistant structures. Hence, most of the improvements in structural vibration analysis tools throughout the new era of engineering occurred as a result of the necessity to solve vibration problems. Modal analysis is an important technique for analysing structural dynamic features, operating conditions, and performance requirements, which enables the design of optimal dynamic behaviour or the resolution of structural dynamics issues in the current designs.

The main purpose of structural dynamic modification is to determine the influence of mass, stiffness, or/and damping modifications on the dynamic response of a structure using information from both the original system and the modifying component. There are two types of structural modification: direct structural structural modification and inverse structural modification techniques [1]. The direct structural modification approach is used to determine the modified structure's dynamic responses. Inverse structural modification, on the other hand, is used to identify the changes required to obtain desired dynamic properties, usually expressed in terms of desirable natural frequencies and mode shapes. Several researchers, including Park and Park [2,3], Wang et al. [4], Mottershead et al. [5], and Kyprianou et al. [1,6], have concentrated on the inverse structural modification approach. There are very few studies devoted to the topic of direct structural modification. By ignoring damping, Weissenburger [7] investigate a local eigenvalue modification approach for lumped mass and spring adjustments. Pomazal and Snyder [8], as well as Hallquist and Snyder [9], enhanced this concept by including damping. Sestieri and D'Ambrogio [10], as well as D'Ambrogio [11], utilized raw data of measured frequency response functions (FRFs) and lumped modifications to obtain the optimal structural modifications necessary to modify the structural dynamic features. Abrams, Ryan [12] considered the engine as a structural element of a Formula SAE race car chassis. Modal analysis was utilised in conjunction with a chassis beam mesh to offer a comparative study with and without engine structural support. Modal analysis is also utilised to justify a chassis stiffening increase. Moreover, a static analysis was conducted to determine chassis deflection, which revealed a reduction in deflection when structural support was added to the chassis' engine mounting locations. Another technique, based on the direct determination of frequency response functions using experimental or theoretical methods, has gained popularity in determining the impacts of structural modification [12–13,14–15]. Modal system synthesis techniques were utilised to analyse the effect of structural modifications such as mass, spring-damper, and tuned absorber on the dynamic behaviour of a mechanical structure. Calculations of the modified structure may be made using the modal model (scaled mode shapes discovered by experimental or finite element modal analysis) [16]. Other studies compared the FEM and experimental methods and discovered that the discrepancy between the FEM and experimental methods grows larger at higher frequencies, while the experimental method's modal frequencies are greater. [17,18]

2. Basic theory of SDM

The resolving of eigenvalue issue is the fundamental process of structural alteration. If the mass, damping, and stiffness properties of the structure are known, the modal properties of the modified structure can be determined by first changing the mass, damping, and stiffness coefficient matrices of the equations of motion and then solving for the modified equations' eigenvalues and eigenvectors. This would include resolving an eigenvalue issue in terms of physical coordinates. The procedure of modifying the local eigenvalues begins with a dynamic model of the unmodified structure. This model can be expressed in terms of the structure's mass, damping, and stiffness characteristics, or in terms of the structure's modal properties, including frequencies, damping, and mode shapes. Modal data can be derived from either a modal test of the structure or an Eigen solution of the differential equations of motion, which are commonly developed using finite element modelling techniques. Modifications to the structure's mass, stiffness, and damping coefficient matrices are made by adding to or subtracting from the mass, stiffness, or damping coefficient matrices of the differential equations of motion. The unmodified structure's equations of motion are represented as a set of linear second-order differential equations. They are written as follows for a structural model with n degrees of freedom:

$$\mathrm{M}(\ddot{x}(\mathrm{t})) + [\mathrm{c}]\{\dot{x}\,(\mathrm{t})\} + [\mathrm{K}]\{\mathrm{x(t)}\} = \{\mathrm{f(t)}\} \tag{1}$$

Similarly, the altered structure's equations of motion are stated as:

$$[\mathrm{M} + \Delta M\,]\{\,\ddot{x}(\mathrm{t})\} \; + \; [\mathrm{C} + \Delta C]\{\,\dot{x}\,(\mathrm{t})\} + [\mathrm{k} + \Delta k[\{x(t)] = \{\mathrm{f(t)} \tag{2}$$

Since the equations of motion are linear, they may be converted to the frequency domain without losing any information using the Laplace transform. Thereafter, the equations become:

$$S^2[\mathrm{M}]\{\mathrm{X(s)}\} + \; \mathrm{s}[\mathrm{C}]\{\mathrm{X(s)}\} + [\mathrm{K}]\{\mathrm{X(s)}\} = \{\mathrm{F(s)}\} + \{\mathrm{ICs}\} \tag{3}$$

Physical properties of the structure are on the left side of the equations, while the forces that were applied, and the initial conditions are on the right side. The initial conditions can be regarded as a specific case of the applied forces and so omitted from consideration in the subsequent evolution without compromising generality. To highlight the three fundamental components of every linear dynamic system, namely external disturbances (inputs), responses (outputs), and the physical system (linear filter), the modified equations of motion can be restated as:

$$[\mathrm{B(s)}]\ \{\mathrm{X(s)}\} = \{\mathrm{F(s)}\} \tag{4}$$

Where:

$$[\mathrm{B(s)}]=\mathrm{s}\ [\mathrm{M}] + \mathrm{s}[\mathrm{C}]+[\mathrm{K}] \tag{5}$$

[B(s)] is defined as the system matrix.

Since the early 1990s, the structural vibration approach has been used to modify and improve a structure's dynamic characteristics. Fabrizo Pimazio and Lucion Ricciardiello investigated the influence of structural variation on a system's dynamic behaviour in 1985 [19]. Two structures have been examined. The initial structure was straightforward, since it was used to connect the energy of a sector with the efficiency of modifying that sector. The second structure was more intricate in order to facilitate the study of perturbation formulas. The investigation found that even a little change in the size of a structure with a high energy content resulted in a significant change in the frequency, and that if the energy of the changed area is greater, the forecast and reference frequencies are less inaccurate. This error rose in proportion to the degree of modification.

A. Sesteiri examined the uses of SDM [20]. The approach is illustrated using two structures. First, the kinematic vibration transmission path is reduced in the case of engines, and secondly, the peaks of the spindle drive point on a flexible manufacturing cell are lowered. The engine part is chosen for modification since it is developed for static requirements. Thus, adding small masses to a few points on the engine demonstrated great sensitivity, i.e. it decreased vibration transmission. It was discovered that the high frequency range is mass-controlled, and that stiffness adjustment had little effects. The latter scenario involved the use of a structural modification approach on the Flexible Manufacturing Cell. The purpose of this example was to modify the machine's dynamic behaviour without affecting normal operation. The sensitivity analysis was done by looking at the effects of adding a single mass of increasing magnitude.

Initially, masses were placed to the machine's backside. There was just a very minor influence on the spindle-related FRFs. Following that, masses on the spindle housing were inspected. The result was

IOP Conf. Series: Materials Science and Engineering 1244 (2022) 012017 doi:10.1088/1757-899X/1244/1/012017

undesirable. Finally, the impact of a dynamic absorber on the spindle housing was investigated, and it was shown that the spindle FRFs improved significantly. However, the cost of making changes to the back of the machine was thought to be ten times less than the cost of making changes to the spindle housing. This shows the efficacy of the FR-based optimization approach on sophisticated structures such as machine tools.

A structural system's dynamic response is essentially determined by its natural frequencies and mode shapes. Thus, formal modification techniques may be utilised to create the required dynamic behaviour by modifying the natural frequencies and mode shapes through design variables. The design factors vary according to the sort of change required. When designing structural components like stiffened panels and cylinders, the design parameters include the stiffener spacing, the stiffener size and form, and the skin thickness. If the skin or/and stiffeners are constructed from layered composites, the orientation and fraction of the fibres become additional factors. The dimensions of the components are design variables in a fixed configuration structural system (frames, trusses, wings, fuselages, etc.). The thickness of plates, the cross-sectional area of bars, the area, moment of inertia, and torsional constants of beams all indicate the element's size. A structural dynamic modification has been applied to an open box structure in this paper. The structure has three sides and a front side, that are open. The box is made of mild steel, which was chosen for its low young modulus value, which makes it easier to bend and mould.

3.Methodology

To provide a detailed dynamic description of an open box, experimental modal analysis was undertaken. The structure was excited at a fixed position using an impact hammer, and an accelerometer was moved over each point. FRFs were generated by inputting the force and accelerometer signals into a multichannel analyzer. Data collection was performed using B&K Pulse Labshop Experimental Modal analysis software and then uploaded to ME'scope VES software for further analysis. It is essential to examine the measurement quality of the frequency response functions, as this directly correlates to modal parameters' quality. Thus, it is necessary to conduct certain tests to guarantee that the quality of the obtained frequency response functions is enough for evaluating the structure's dynamic properties. The first check is the coherence function; this is a frequency-dependent that indicates how much of the output signal is due to the true input signal and how much is due to the added measurement noise. If the coherence function value is close to one, there is a strong linear relationship between the input and output, and noise has a small effect. If the coherence function value is large compared to one unit, then there is no linearity between the input and output, as seen in figure 1.

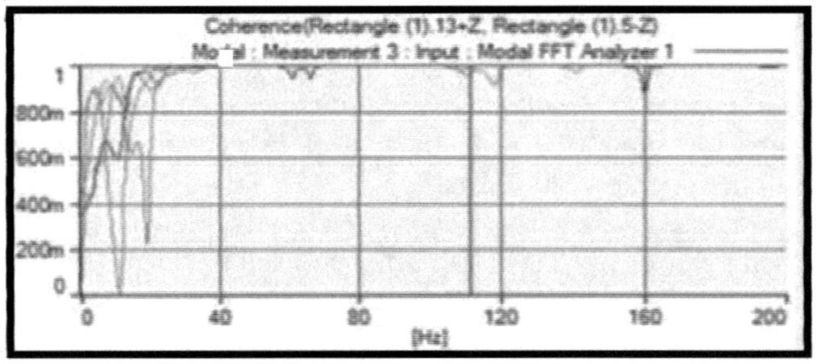

Figure 1. Coherence function

In the second check, a number of frequency response functions were tested for their reciprocity and reproducibility. The reciprocity and reproducibility of all measured FRFs should be same in principle. The approach involves determining the frequency response functions of a few points on the structure. First, we begin by testing the first couple by exciting the first point and take the response at the second point, and vice versa. Then we test the second couple by exciting the first point and taking the response at the second point, and vice versa. The results indicated that the frequency response functions derived from the first couple of points were identical to those showed in Figure 2. Furthermore, the second pair of points illustrate the same frequency response functions as in Figure 3. We conclude that the reciprocity and reproducibility of the frequency response functions assessed are acceptable.

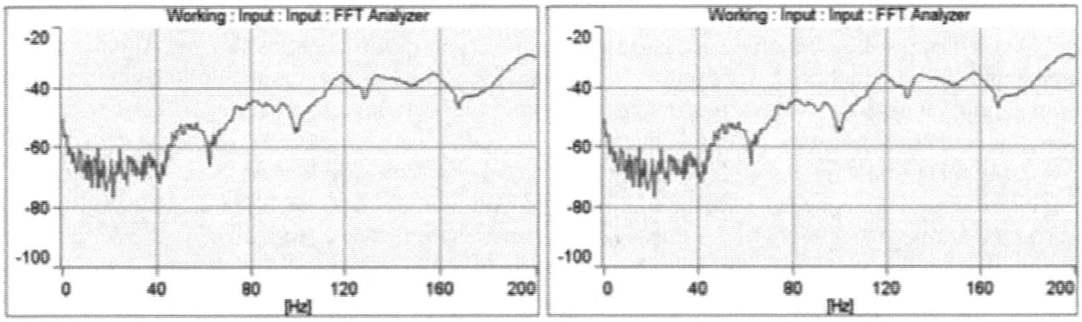

Figure 2. Frequency response of the first couple

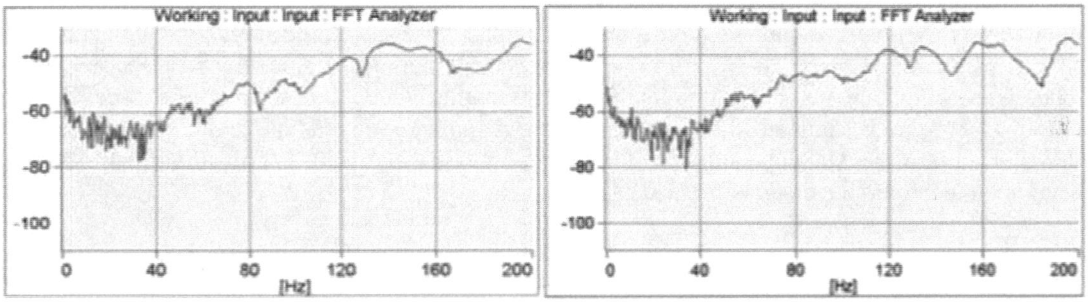

Figure 3. Frequency response of the second couple

The study is to determine the modal parameters which are natural frequency and mode shape. Experimental data was obtained on a steel open box placed on a soft support to ensure free boundary conditions and subjected to dynamic structural modifications. The equipment needed for this method are the impact hammer, accelerometer, vibration analyser, computer and also the specimen itself as shown in figure 4. The results obtained from modified structure will be compared with the original structure.

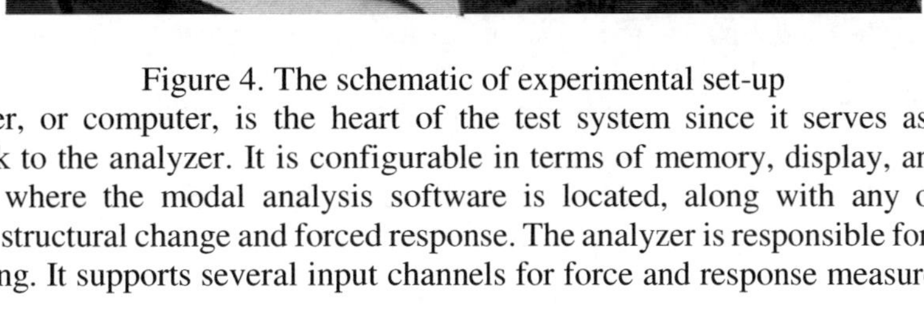

Figure 4. The schematic of experimental set-up

The controller, or computer, is the heart of the test system since it serves as the operator's communication link to the analyzer. It is configurable in terms of memory, display, and data storage. Typically, here is where the modal analysis software is located, along with any other analytical capabilities such as structural change and forced response. The analyzer is responsible for data collecting and signal processing. It supports several input channels for force and response measurements, as well as one or more excitation sources. Typically, measurement functions such as windowing, averaging, and FFT computation are performed within the analyzer.

This process needs minimal hardware. As indicated previously, the signal source might be a component of the analyzer. The desired force and response are measured using transducers in conjunction with a power supply for signal conditioning. The piezoelectric varieties are the most often utilized for modal testing since they measure force and acceleration. The signal conditioning power supply may be voltage or charge mode and is occasionally included as a component of the analyser, thus attention should be given while configuring and matching this component of the test system.

To achieve the desired impact force, several sized hammers are necessary. Due to the fact that the force is an impulse, the amplitude of the energy applied to the structure is a function of the hammer's mass and velocity. This is because linear momentum is defined as mass multiplied by the velocity of mass. The linear impulse is equal to the linear momentum's incremental change. Due to the difficulty of controlling the hammer's velocity, the force level is often regulated by modifying the mass. Weights for impact hammers range from a few ounces to several pounds. Additionally, most hammers may have mass added or subtracted, making them suitable for testing objects of varied sizes and weights. Impact testing must meet three requirements: pre-trigger delay, force and exponential windowing, and accept or reject capability.

The MeScope software was used to create the structure, which consist of twenty points. As seen in Figure 4, the dBFA suite software was configured for impact testing, and the analyzer was linked to the impact hammer, accelerometer, and laptop. The accelerator's other end was positioned at the steel plate. The plate is measured and marked with a total of twenty points. Three knocks were made at each point with an impact hammer equipped with a plastic tip to get an average reading of the impact. The data was transferred into the MeScope software for further analysis. By using curve fitting method in the software, the modal parameters are obtained.

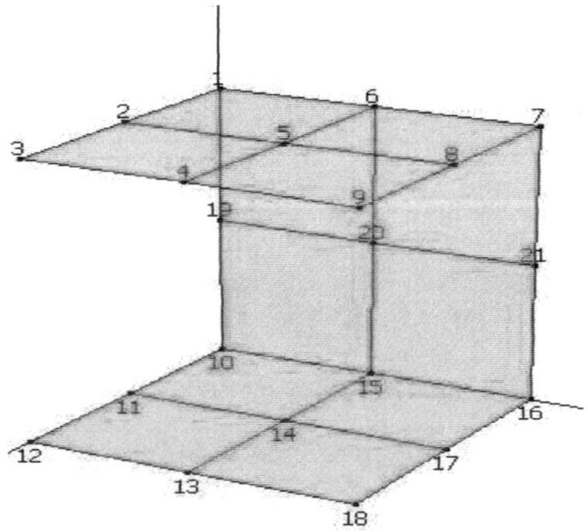

Figure 5. Structure with 20 points

4.Results and discussion

The first experimental Modal Analysis was conducted on the unmodified structure. Data was taken from the impact hammer testing by using dBFA software and then exported to Mescope for further processing to get the modal parameters from the experimental frequency Response Function as shown in figure 6. The results will be used for further comparison with both mass modification and stiffness modifications. Table 1 shows the results of the frequency values of unmodified structure and Table 2 shows the corresponding mode shapes.

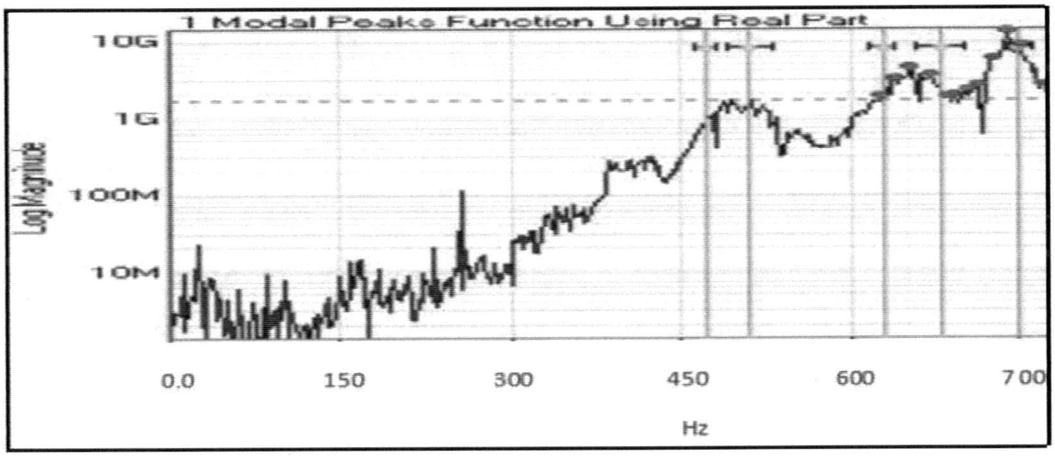

Figure 6. Frequency Response Function of unmodified structure

The main aim of this investigation is to improve the dynamic characterisation by using the structural dynamic modifications by adding mass or stiffness to the structure. At first mass was added to the box structure in the form of a cable around the box, and the structure was examined which showed a significant changes in the frequency response function as shown in figure 7.The results of the values of natural frequencies are shown in Table 1 and the corresponding mode shapes are shown in Table 2.

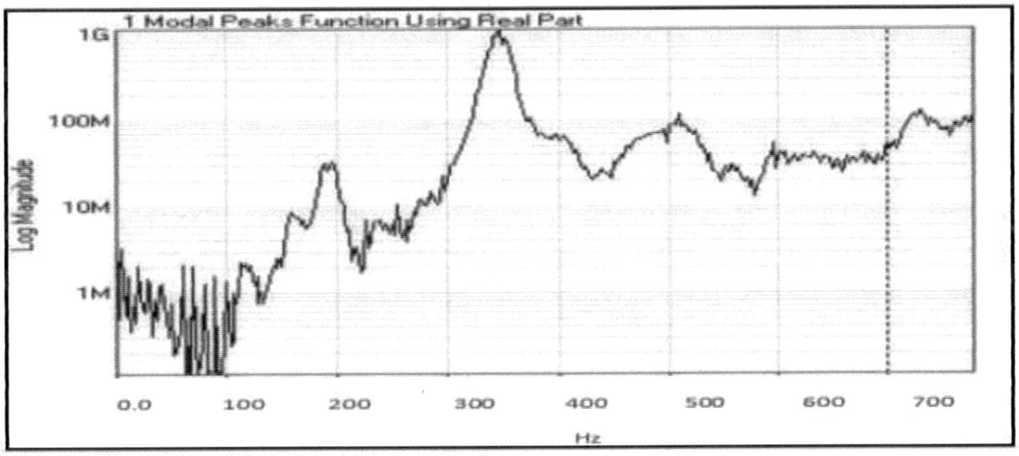

Figure 7. Frequency Response Function of Mass modification

The third modification applied to the original open box is to add a stiffness by adding spring to one corner of the open box as shown in the figure 4.The stiffness of the spring is 2N/m with 0.13 meter long and the material is stainless steel. The results are shown in Table 1 and Table 2.

4.1 Comparison between unmodified and modified structure

Comparison of natural frequencies and the corresponding mode shapes before and after modification are shown in the following tables 1 and 2. The aim of this study was to improve the dynamic behaviour of the machine by some minor modifications to it without effecting its normal operation. Hence adding a small amount of mass showed significant shifting of the natural frequencies and different mode shapes.

Table 1. Comparison of Natural Frequencies of the original, Mass and Spring modification structure.

Mode Numbers	Unmodified (Hz)	Mass modification (Hz)	Percentage decrease %	Spring modification (Hz)	Percentage increase %
1	180	24	**87**	250	**39**
2	240	64	**73**	380	**58**
3	424	156	**63**	500	**18**
4	564	216	**62**	548	**-3**
5	632	348	**45**	1100	**74**
6	732	632	**14**	1320	**80**

Mass modification reduces the modal frequencies by significant shifting down at all lower order modes as observed in the results. However, at lower order modes were less reduced. In other word the percentage decrease is gradually decreased from 87% to 14%. This is as expected, as the natural frequency is inversely proportional with the mass. The second comparison is performed between unmodified and stiffness modification structure by adding a spring with stiffness to the structure. The results show shifting up of natural frequencies for all modes. In addition, there is a significant percentage increase at higher frequencies compared to low frequencies. The percentage increase at the first and second modes are 39% and 58% respectively, however, the percentage increase at fifth and sixth modes are 74% and 80% respectively. In conclusion, the natural frequency of the structure is more sensitive to the mass at low frequencies, on the contrary, it is more sensitive to the stiffness at high frequencies

Table 2. Mode shapes of original, Mass modification and Spring modification structure.

Mode	Unmodified	Mass modification	Spring modification
1	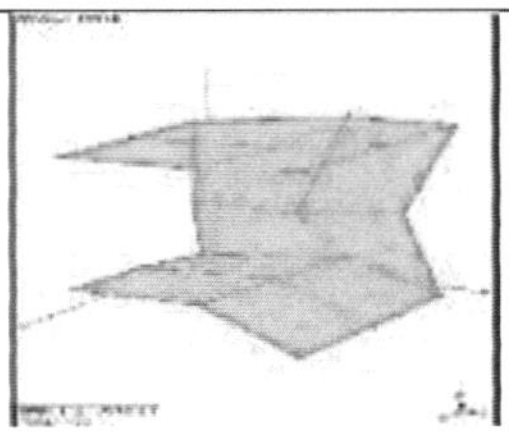		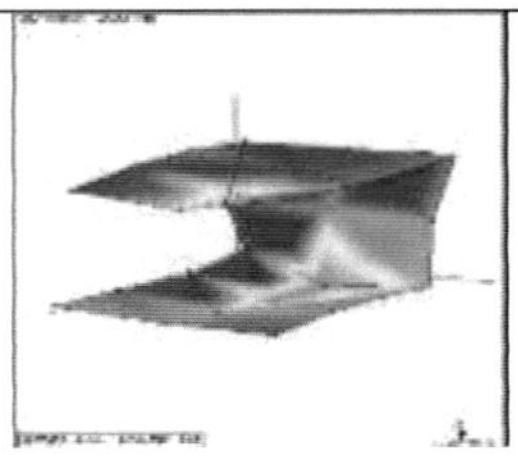
2	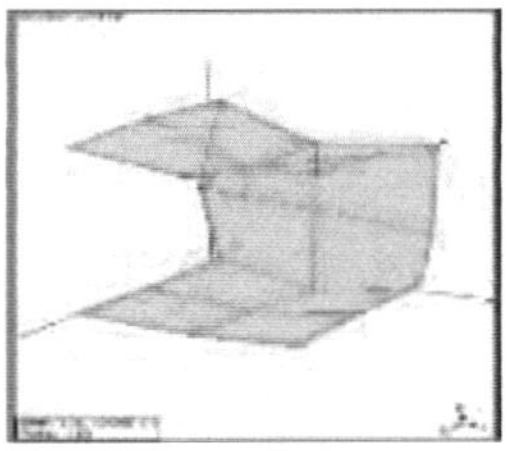	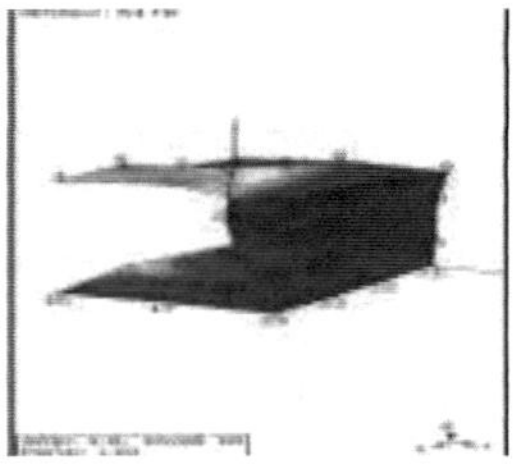	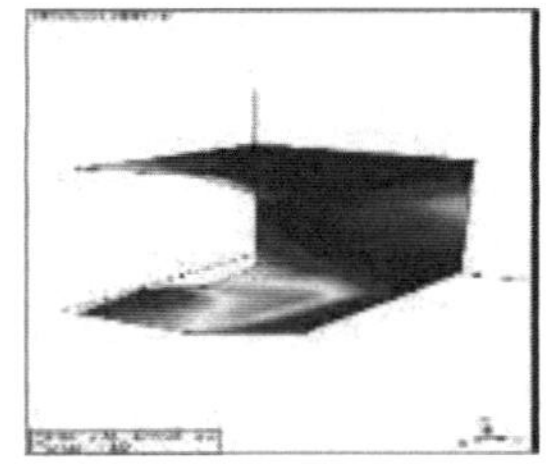

3	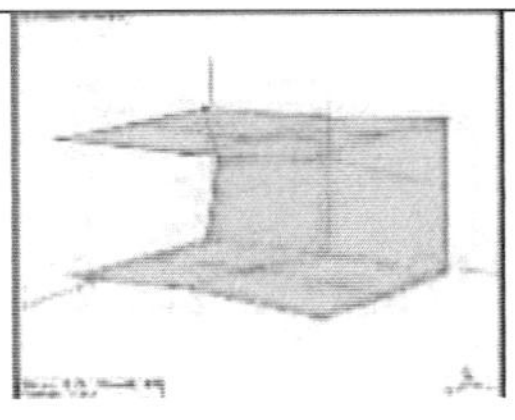	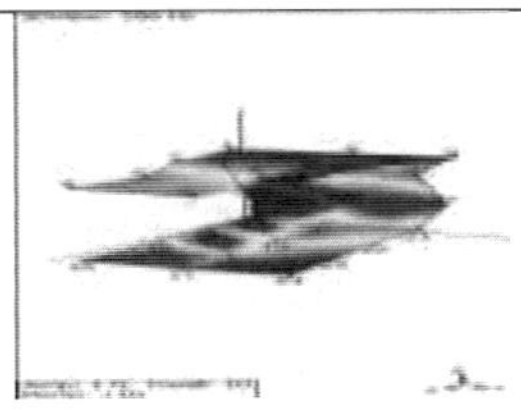	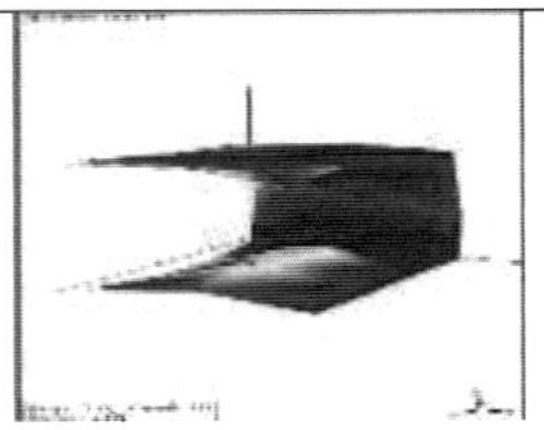
4	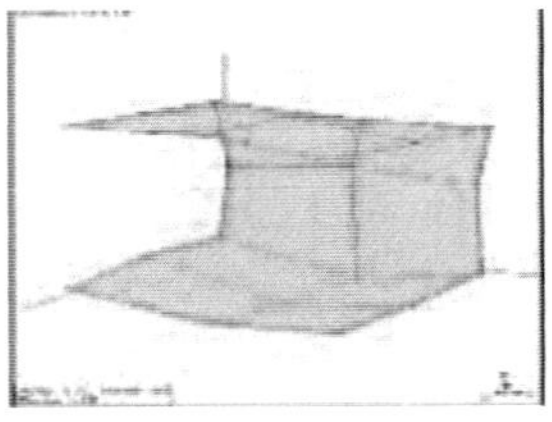	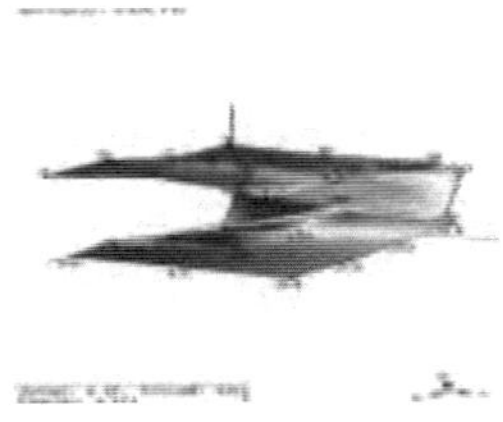	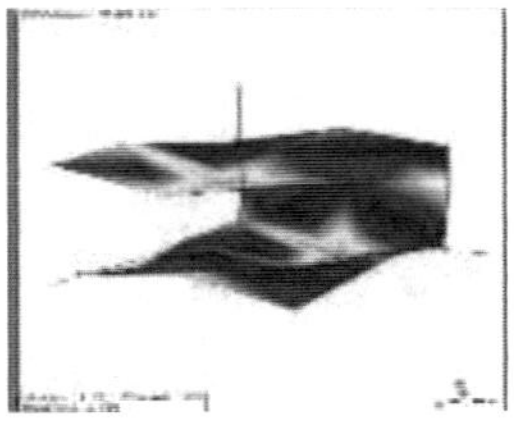
5	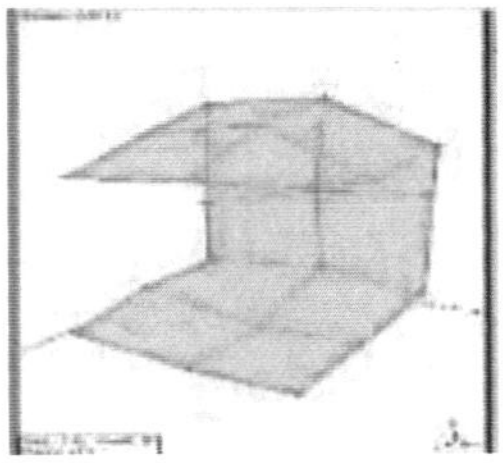	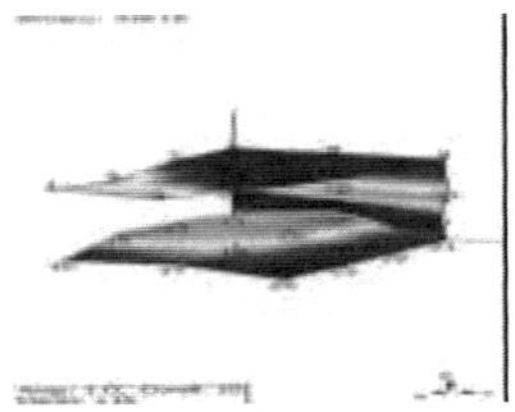	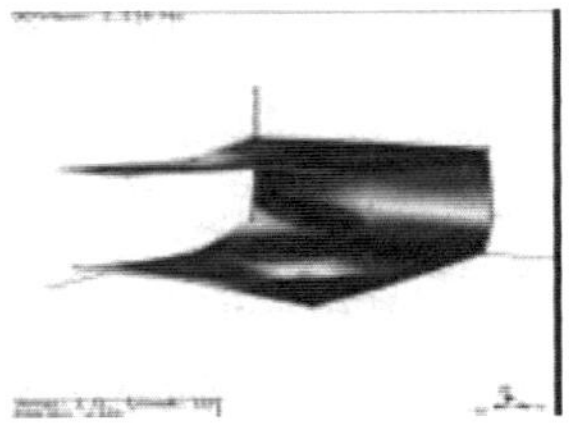
6	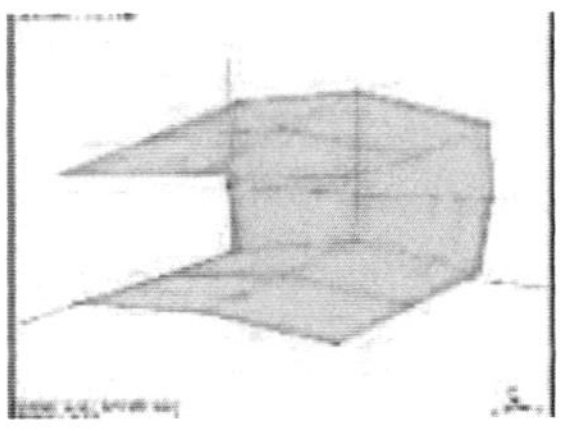	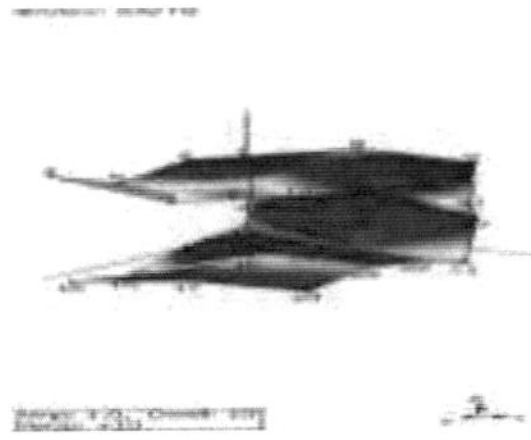	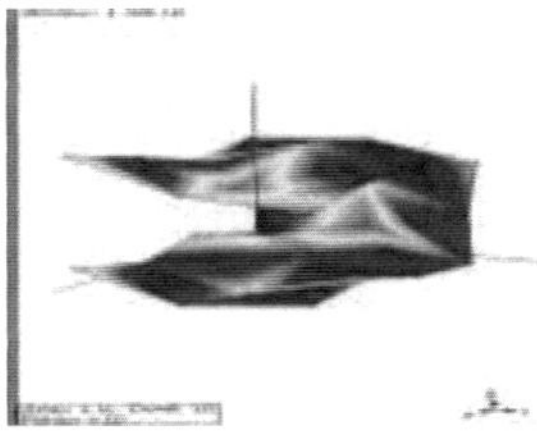

To have a good understanding of the structural dynamic modification method we need to compare the mode shapes as well. Table 2 shows the comparison between the experimental mode shapes of unmodified and modified structure. At the first mode both mass and stiffness modification modes show quite similar bending mode behaviour like the original bending mode. At the second mode both modification modes show tortional vibration, whereas the original mode shows bending mode. The third mode shows quit similarity between the original and stiffness modification modes as both showing first bending mode, however, mass modification mode shows tortional mode shape. For the fourth mode, both modification modes showing similar behaviour as both having second bending mode whereas the original mode shows first bending mode. Mode five shows similarity between original mode and mass modification mode. The last mode does not show significant similarity between the three mode shapes. The MSD can be considered as the best method by which the dynamic behaviour of a structure is improved by predicting both the natural frequencies and the mode shapes of the modified structure which lead to reduce the vibration by shifting up or shifting down the natural frequencies in order to prevent the machine from the phenomena of resonance.

5. Conclusion

Structural dynamic modification has been done efficiently and successfully on the open box to improve the dynamic characteristics of the structure. Experimental Modal Analysis step by step has been made and the results of natural frequencies and their corresponding mode shapes are compared. It was found that the measured modal frequencies are sensitive to mass modification at low frequencies and less sensitive at higher frequencies, on the contrary, it is more sensitive to the stiffness at high frequencies and less sensitive at lower frequencies. Therefore, the study of dynamic characteristics of any machine casing is a crucial step in machine design to enhance the structural vibration, which subsequently contribute to improving the acoustical environment.

6. Recommendation

In the future work, it is highly recommended to use the Finite Element Analysis (FEA) so that the result can be compared between experimental and Finite element method for more validation of results. Also recommend further work on damping modification.

7. References

[1] A Kyprianou, J E Mottershead and H Ouyang Structural modification. Part 2: assignment of natural frequencies and antiresonances by an added beam,

[2] Y H Park and Y S Park 2000 Structural modification based on measured frequency response functions: an exact eigen properties reallocation, *Journal of Sound and Vibration* 237 (3) 411–426.

[3] Y H Park and Y S Park 2000 Structure optimization to enhance its natural frequencies based on measured frequency response functions, *Journal of Sound and Vibration* 229 (5) 1235–1255.

[4] B P Wang et al 1982 Structural modification to achieve antiresonance in helicopters, *Journal of Aircraft* 19 (6) 499–504.

[5] J E Mottershead, C Mares and M I Friswell 2001 An inverse method for the assignment of vibration nodes, *Mechanical Systems and Signal Processing* 15 (1) 87–100.

[6] A Kyprianou, J E Mottershead and H Ouyang 2004 Assignment of natural frequencies by an added mass and one or more springs, *Mechanical Systems and Signal Processing* 18 (2) 263–289.

[7] J T Weissenburger 1968 Effect of local modifications on the vibration characteristics of linear systems, *Journal of Applied Mechanics, Transactions of the ASME* 35 327–332.

[8] R J. Pomazal and V W Snyder 1971 Local modifications of damped linear systems, *AIAA Journal* 9 (11) 2216–2221.

[9] J Hallquist and V Snyder 1973 Synthesis of two discrete vibratory systems using eigenvalue modification, *AIAA Journal* 11 (2) 247–249.

[10] A Sestieri and W D'Ambrogio 1989 A modification method for vibration control of structures, *Mechanical Systems and Signal Processing* 3 (3)

[11] W D'Ambrogio 1990 Some remarks about structural modifications involving additional degrees of freedom, *Mechanical Systems and Signal Processing 4* (1) 95–99.29–253.

[12] Abrams and Ryan 2008 Formula SEA Race car analysis: Simulation & testing of the engine as a structural member. F2008-SC-005.

[13] M Tahtali 1992 Vibration analysis of damped structures and structural reanalysis using a new structural modification method, in: *Mechanical Engineering, Middle East Technical University, Ankara, Turkey*, p. 78.

[14] H Hang, K Shankar and J Lai 2008 Prediction of the effects on dynamic response due to distributed structural modification with additional degrees of freedom, *Mechanical Systems and Signal Processing* 22 (8)1809–1825.

[15] G W Skingle 1989 Structural dynamic modification using experimental data, in: Imperial College of Science, Technology and Medicine, *University of London, London*, UK, p. 259.

[16] D Li, Q Zhou, G Chen, and Y Li 2017 Structural dynamic reanalysis method for transonic aeroelastic analysis with global structural modifications. *Journal of Fluids and Structures*, vol. 74, pp. 306–320.

[17] F J Herrada, J García-Martínez, A Fraile, L K H Hermanns, and F J Montáns 2017 A method for performing efficient parametric dynamic analyses in large finite element models undergoing structural modifications *Engineering Structures*, vol. 131, pp. 625–638.

[18] P Calka and K Lis 2017 Eksperymentalna analiza modalna profilu stalowego wypelnionego polimerobetonem," *STAL Metale & Nowe Technologie*.

[19] Fabrizo Pinazzi And Luciana Ricciardiello1985 The Effect Of Structural Variations On The Dynamic Behaviour Of A System, Computers & Structure Vol. 20. No. 4, pp. 659-667.

[20] A. Sestieri 2007 Structural dynamic modification of vibrating systems *Applied and Computational Mechanics* Vol1 203 – 214.

The Effect of Cutting Speeds on Tool Wear and Surface Roughness when Milling Carbon Fiber Reinforced Polymer

Aisyah Madihah Mustafa[1], Najlah Sakinah Shahruddin[1], Nor Farah Huda Abdul Halim[1*], Alya Naili Rozhan[1] and Maziati Akmal Hattiar[2]
[1]Department of Manufacturing and Materials Engineering,Kulliyyah of Engineering, International Islamic University Malaysia,53100 Gombak
[2]Department of Science in Engineering,Kulliyyah of Engineering, International Islamic University Malaysia,53100 Gombak

Email: farahudahalim@iium.edu.my

Abstract. Carbon fibre reinforced polymeric material (CFRP) is increasingly being used to replace metallic materials, particularly in the automobile and aerospace industries. The mechanical properties of the CFRP such as excellent weight to strength ratio give beneficial outcome in improving the performance of the part made from CFRP. However, machining of CFRP such as milling is difficult and challenging due to the abrasiveness of the CFRP. Therefore, the influence of cutting speeds on tool wear and surface roughness when milling CFRP in a dry condition is the aim of this paper. The milling test was carried out using 6 mm WC-Co end mill tool having 30° helix angle. Milling of CFRP was performed with three different cutting speeds, V_c (94, 113 and 132 m/min) while feed rate, f (1800 mm/min) and radial depth of cut, a_e (2 mm) were remained constant. Milling CFRP with V_c=132 m/min resulted in increasing of tool wear by 15 % and increasing of surface roughness by 10% when compared with V_c= 94 m/min. CFRP surface damages such as matrix smearing, delamination and fibre pull out was observed when milling with higher cutting speed.

1. Introduction

Attractive properties of Carbon Fiber Reinforced Polymer or CFRP such as excellent damage tolerance, fatigue resistance and high wear resistance made them extensively employed in high end industry especially in automotive and aerospace industry. More recently, CFRP has been used as a main material for aircraft main components such as door, wings and fuselage. This is attributed to its superior strength-to-weight ratio when compared to other materials such as metal, which improves fuel efficiency and load bearing reinforcement, both of which are determined by the aircraft's weight[1]. Also, CFRP can be manufactured nearest to its final product shape where post machining processes are still necessary to meet final dimension of the product. However, machining CFRP is challenging due to the anisotropy and non-homogenous characteristics of the CFRP. Edge trimming and drilling are still necessary for final part accuracy and hole making purposes. Selections of cutting parameters during machining of CFRP are crucial. Application of high cutting speed for example may lead to increasing of cutting temperature as well as increasing the tool wear rate. This can be explained by the abrasive nature of the carbon that may lead to the aggressive rubbing of the tool cutting edge on the machined surface[2-4]. Unlike machining metallic material where plastic deformation occurs during the removal process, CFRP undergoes brittle fracture where dust-like residual is formed, hence, minimizing the amount of heat removed [5,6]. Besides, machining beyond the glass transition temperature (Tg) will resulted in softening of the matrix resin, thus induced internal damage and degraded the quality of the CFRP [6,7]. Also, selection of cutting parameters during milling of CFRP such as cutting speed, feed rate, and depth of cut significantly affect the tool wear as well as surface damage. Rashid et al. [8] found that milling CFRP can be machined up to 13 m with low cutting speed (670 RPM) whereas at high cutting speed (6300 RPM) can only machined up to 8.4 m before the tool failed. This indicates that the increasing of contact time between the cutting tool and CFRP as the cutting speed increased. Moreover, studies found that increasing of cutting speed significantly increased the tool wear and shorten the tool life [7]. Sundi et al. [9] found that at lower cutting speed (2506 RPM), fibre pull out was not observed as compared with machining at high cutting speed (5012 RPM). They concluded that the fiber pull out occurred was due to the thermo-mechanical loads resulted from the increasing of cutting speed and cutting temperature along the milling process. Nguyen-Dinh et al. [10] observed that the effect of cutting parameters on surface roughness is more prominent when compared with cutting speed and machining distances during milling CFRP. They found that the measured surface roughness (Ra) increased up to 5.5 μm when V_c is increasing to 250 m/min, compared to machining at lower V_c= 150 m/min (Ra=1.7 μm). Therefore, in this research, an experimental investigation was performed to identify the effect of the cutting speed on the tool wear and surface roughness of the CFRP.

2. Experimental set-up and procedures

Mazak Nexus 410A-II Vertical Machining Centre was employed to conduct an experimental investigation during milling of CFRP. Three flutes of uncoated tungsten carbide (WC-Co) solid milling tool having 6 mm diameter was used as a main cutting tool. Two dimensions of CFRP which are 200x200x5 mm and 200x50x5 mm were applied to observe the tool wear and surface roughness development, respectively. Special fixture was used to clamp the CFRP panel and strip on the machine table as shown in Figure 1. Cutting parameters employed in this experimental investigation was tabulated in Table 1. All machining tests were performed dry.

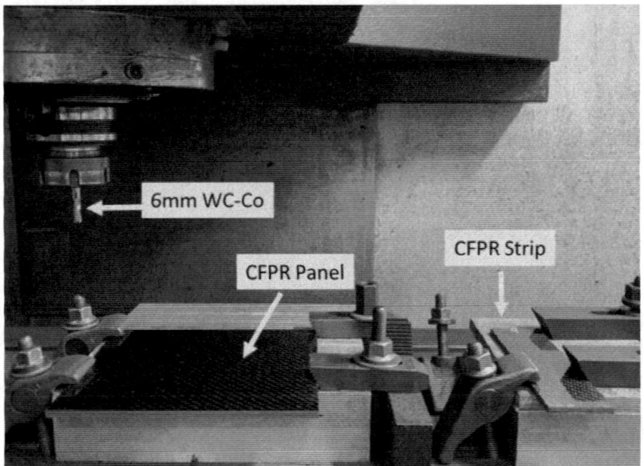

Figure 1. The experimental set up.

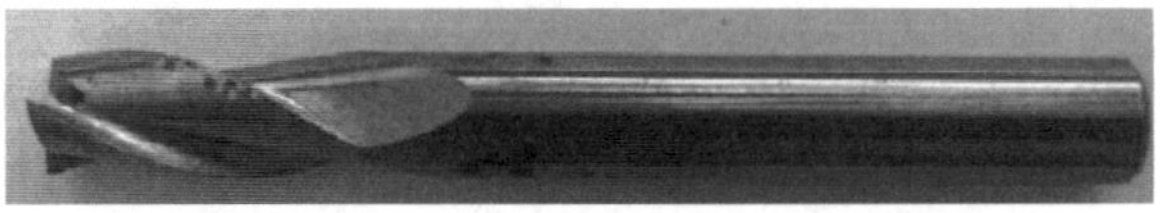

Figure 2. 6 mm diameter of K3EPEN 060C end mill

Table 1. Cutting parameters during milling CFRP

Cutting Parameters	Value
V_c (m/min)	94, 113, 132
f (mm/min)	1800
a_e (mm)	2

The tool wear and surface roughness was measured for every 400 mm distance travelled. The image of tool wear at each flutes was captured using Dino-Lite Premier and processed by using the digital imaging and measurement software. The average surface roughness (Ra) of the CFRP was measured using stylus type surface profiler, Mahr Federal Model 6910431 MarSurf M300 C set with 4 mm traverse length and 0.8 mm cut-off length. To confirm the carbon chip does not adhere on the CFRP surface and influence the measurement value, they were cleaned using an alcohol before Ra measurement. The CFRP machined surface was also observed and analysed using Alicona SL 3D profiler to further inspect the surface topography and condition after machining. The machining was conducted until 3200 mm machining length before the test was stop.

3. Results and discussion

3.1. Tool wear

The effect of different cutting speed during milling of CFRP over 3200mm machining length is shown in Figure 3. The tool wear value for all cutting speed increased as the machining length increased. Milling of CFRP with V_c=132 m/min resulted in highest tool wear with maximum 0.072µm of flank wear, followed by cutting tool with V_c= 113 m/min (0.064 µm) and V_c= 94 m/min (0.061µm) at the 3200 mm machining length. This indicates that the increasing of tool wear was caused by the increasing of contact length between the cutting tool and the workpieces during machining. Figure 4(i) shows the fresh tool whereas Figure 4(ii) shows the worn out tool after milling of CFRP.

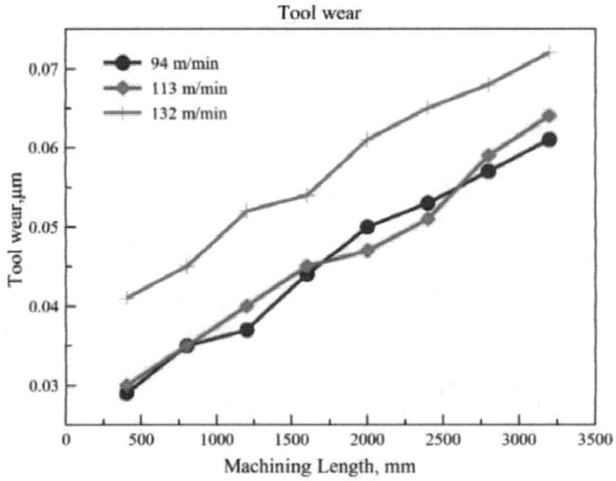

Figure 3. Tool wear against machining length.

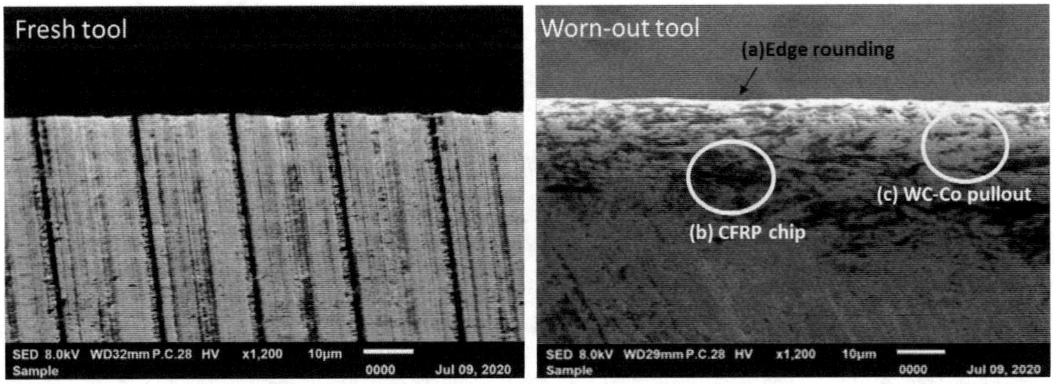

Figure 4. SEM images of (i) new cutting tool and (ii) worn out cutting tool after milling CFRP with 132 m/min after 3200 mm machining length.

Figure 4 (ii) shows that the worn-out cutting tool experiences edge rounding. It was also observed that adhered CFRP chip and matrix resin was observed at the cutting edge although the it has been cleaned with alcohol before the SEM examination was conducted suggesting that the matric resin have been thermally degraded during the milling operation. Furthermore, void was observed on the worn-out cutting tool due to the fractured chips of CFRP that abraded the tool surface causing dislodged of WC particles. This evidence proves that abrasive wear taking place as main wear mechanism during machining.

3.2. Surface roughness

The comparison of surface roughness progression when milling with V_c= 94, 113 and 132 m/min is shown in Figure 5. With regard with the effect of cutting speed, it was observed that high cutting speed of 132 m/min resulted 15 to 20% higher value of Ra when compared with lower cutting speed of 113 m/min and 94 m/min. At 3200 mm length, Ra value for V_c= 132 m/min was 1.45μm whereas, V_c= 94 m/min produced 1.25 μm Ra. Based on the surface roughness data presented in Figure 5, it was confirmed that the increasing of tool wear value as in Figure 3 affect the CFRP surface condition. This is an effect from the increasing of wear on the cutting tool radius that decreases the smoothness of the machining[7]. Therefore, the reduction in cutting tool sharpness resulted in degradation of the CFRP surface condition.

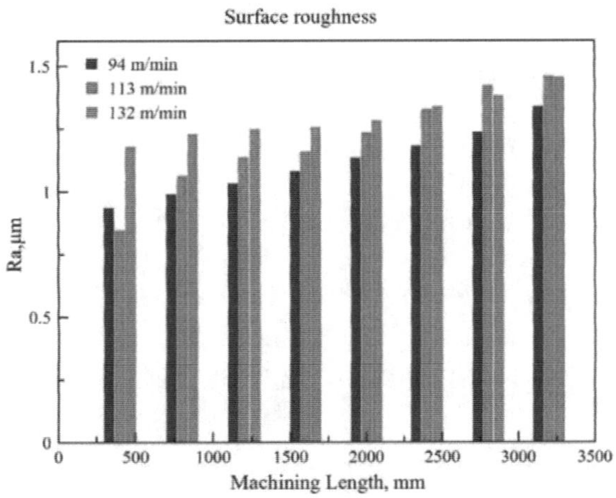

Figure 5. Ra against machining length.

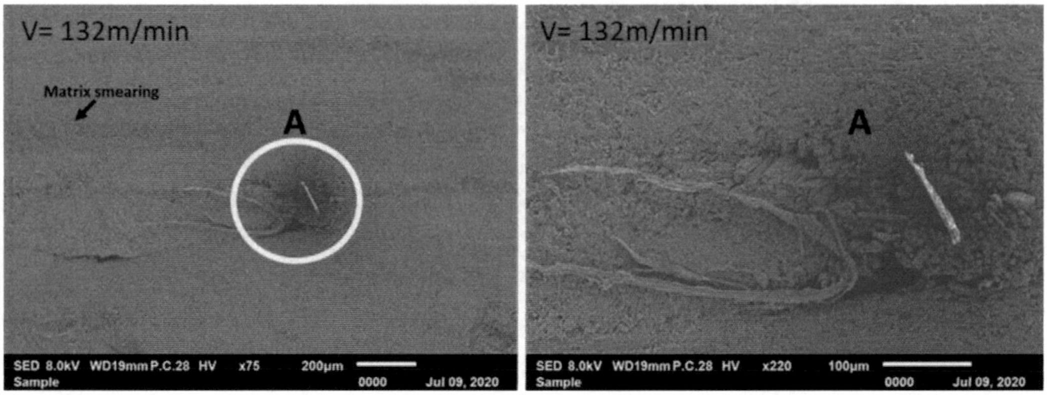

Figure 6. Surface damage that was observed after milling.

Matrix smearing and fibre pull-out was severely found on the CFRP machined surface when milling with higher V_c= 132 m/min as compared with other cutting speeds. This indicates that due to the worn out tool it is expected that the cutting force during milling was increased, thus resulted in deterioration of the surface condition as shown in Figure 6. Consequently, it is predicted that the heat generated caused by the worn out tool was then thermally degrade the matrix resin thus affecting the surface condition. It was also observed that fibre pull-out shown in Figure 6 occurred due to the thermo-mechanical failures between tool and workpiece. It is expected that high cutting temperature will be generated during milling operation at the cutting edge of WC-Co tool caused matrix interfacial shear strength decreased, that can lead in accelerate fibre pull-out. However, it was observed that the surface damage was less occurrence when milling with lower cutting speed.

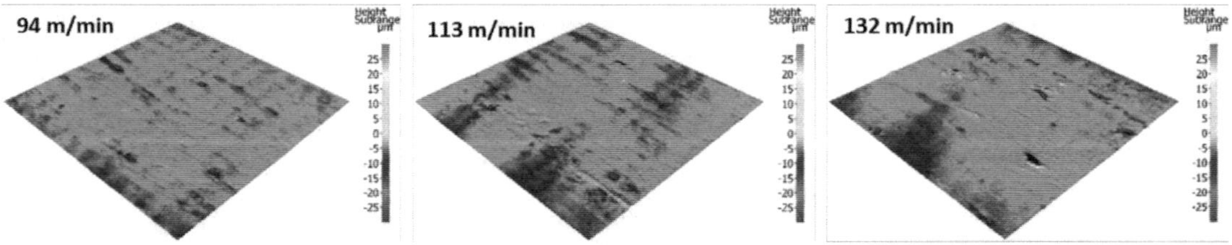

Figure 7. Image of surface profile observed on Alicona SL 3D profiler for 94, 113 and 132 m/min after 3200mm machining length

Figure 7 shows the surface profile observed by Alicona after milling with different cutting speed at 3200 mm machining length. The voids were severe on the machined surface with high cutting speed (132m/min) as compared with lower cutting speed. It shows that the machined surface for V_c= 132 m/min experienced the low surface profile quality after 3200 mm machined length therefore supported that the high cutting speed does resulted to fiber pull out that caused by the degrades of the matrix resin due to the heat generated.

4. Conclusions

The result of the effect of the cutting speed during milling of carbon fibre reinforced polymer using 6 mm end mill have been discussed. Some major observations are:

- Tool wear was measured up to 15% higher during milling CFRP with V_c= 132 m/min as compared with lower V_c = 113 and 94 m/min.

- The occurrences of delamination and fibre pull-out was less in lower cutting speed as compared with high cutting speed indicates that lower cutting speed is recommended to produce less damage.
- It was observed that milling of CFRP with lower V_c= 94 m/min resulted better surface roughness as compared with higher cutting speed where Ra value for V_c= 132 m/min was higher 10% than low cutting speed.
- The surface damage such as matrix smearing and fibre pull out on the CFRP machined surface was significantly observed when the value of tool wear increased.

Acknowledgement
This project is funded by International Islamic University Malaysia(IIUM) under RMCG20-027-0027 grant. A special gratitude to the supervisors and staffs of production lab, metrology lab and workshop of IIUM.

5. References

[1] "Composites: Airbus continues to shape the future - Commercial Aircraft - Airbus." https://www.airbus.com/newsroom/news/en/2017/08/composites--airbus-continues-to-shape-the-future.html (accessed Jun. 07, 2021).

[2] S. Ashworth *et al.*, "Varying CFRP workpiece temperature during slotting: Effects on surface metrics, cutting forces and chip geometry," in *Procedia CIRP*, 2020, vol. 85.

[3] A. N. Dahnel, S. Barnes, C. Kibbler, and H. Ascroft, "Study of cutting forces and surface roughness in milling of Carbon Fibre Composite (CFC) with conventional and pressurized CO2 cutting fluids," in *ASME International Mechanical Engineering Congress and Exposition, Proceedings (IMECE)*, 2015, vol. 2B-2015.

[4] S. Ghafarizadeh, G. Lebrun, and J. F. Chatelain, "Experimental investigation of the cutting temperature and surface quality during milling of unidirectional carbon fiber reinforced plastic," *Journal of Composite Materials*, vol. 50, no. 8, 2016.

[5] J. Y. Sheikh-Ahmad, F. Almaskari, and F. Hafeez, "Thermal aspects in machining CFRPs: effect of cutter type and cutting parameters," *International Journal of Advanced Manufacturing Technology*, vol. 100, no. 9–12, 2019.

[6] J. Sheikh-Ahmad and J. Mohammed, "Optimization of process parameters in diamond abrasive machining of carbon fiber-reinforced epoxy," *Materials and Manufacturing Processes*, vol. 29, no. 11–12, 2014.

[7] N. F. H. A. Halim, H. Ascroft, and S. Barnes, "Analysis of Tool Wear, Cutting Force, Surface Roughness and Machining Temperature during Finishing Operation of Ultrasonic Assisted Milling (UAM) of Carbon Fibre Reinforced Plastic (CFRP)," in *Procedia Engineering*, 2017, vol. 184.

[8] M. Akmam Noor Rashid, Z. Mohd Zain, M. K. Nor Khairusshima, W. Ibne Noor, M. Mullah, and S. Ahmed Khan, "Analysis and modelling of surface roughness in milling of JFRP composite using central composite design," *IOP Conference Series: Materials Science and Engineering*, vol. 1092, no. 1, 2021,

[9] S. A. Sundi, R. Izamshah, and M. S. Kasim, "Effect of machining parameters on surface roughness in edge trimming of carbon fiber reinforced plastics (CFRP)," *Tribology Online*, vol. 15, no. 2, 2020,

[10] N. Nguyen-Dinh, R. Zitoune, C. Bouvet, and S. Leroux, "Surface integrity while trimming of composite structures: X-ray tomography analysis," *Composite Structures*, vol. 210, 2019,

3D finite element analysis of laminated composites under three-point bending

M Bellahkim[1,*] A Ouezgan[1], Y Benbouras[1], A Maziri[1], E Mallil[1], J Echaabi[1]

Department of Mechanical Engineering, ENSEM, Hassan II University of Casablanca, Casablanca, Morocco

* E-mail: mouad.bellahkim@ensem.ac.ma

Abstract. Delamination is considered as one of the most critical failure modes which occurs near the free edges in laminated composites due to high interlaminar stress. When the classical plate theory is used to determine the in plane stress, the interlaminar stresses cannot be assessed by this theory, as it adopts the plane stress assumption. For that, a three-dimensional finite element analysis is adopted to calculate in-plane stresses as well as out-of-plane stresses of laminated composites under three-point bending. 8-node linear brick elements C3D8R with reduced integration available in ABAQUS element library is used for meshing the laminated composite models. The flexural numerical response of two main lay-up arrangements is modelled: Unidirectional laminates ($[0]_{12}$,$[90]_{12}$, $[45]_{12}$) and multidirectional laminates (cross-ply $[0/90]_{3S}$, angle-ply $[\pm 45]_{3S}$ and quasi-isotropic laminates $[0/90/\pm 45_2]_S$). Comparing the current results to experimental methodologies accessible in the literature demonstrates the predictive capability of the present model. Therefore, this study shows that the present formulations can predict with a good accuracy the deflection of the test specimens. Whilst, a significant dispersion is only observed for the unidirectional laminates$[0]_{12}$.

1. Introduction

Composite materials are now widely used in a variety of industries, and the design and dimensioning of composite structures are generally mastered and integrated into many computational codes thanks to the classical laminate theory. However, delamination failure problems were poorly predicted at near free edges because of interlaminar shear stress due to differences in the elastic properties of adjacent plies [1]. The free edges region is subjected to a three-dimensional stress field which cannot be predicted by the classical laminate theory which assumes a state of plane stresses.

The stress distribution in laminated composites has been treated using a variety of three-dimensional numerical analyses. A finite-difference solution for the free edge stresses has been obtained by Pipes and Pagano [2]. They demonstrated the singularity of interlaminar shear stress for an angle-ply laminate at the edge region under uniform axial extension. Afterwards, an approximate theory was developed by Pagano [3] [4] based on assumed in-plane stresses and the use of Reissner's variational principle. Similar work examined the interlaminar shear stress at free edges using finite element analysis [5] [6] [7], the boundary layer theory by Tang [8] and Tang and Levy [9] and the Layer-wise theory (LWT) [10]. Kant and Swaminathan [11] established a good review up to year 2000 on the analytical and numerical methods to calculate the interlaminar shear stress for the free-edge problem.

The stacking sequence is among the main parameters that influence the appearance of delamination near free edges. CZARNEK [12] and Herakovich [13] used the moiré technique at the free edge side of a symmetrical laminate designated where n = 2 and θ = 10°,25°, 30°, 45°. They highlight important displacement gradients at the +θ/-θ interface. Herakovich [14] and [15] compared the tensile behavior of the clustered and the alternating laminates with θ= 10°, 30°, 45°. They showed that the stress required to initiate the delamination of the alternating laminates is greater than that in which the clustered laminates. This is commonly referred to as the thickness effect, where two layers stacked successively with the same orientation behaving as a single layer of twice the thickness [16]. In the $(10° \leq \theta \leq 30)$ laminates; shear stresses appear at the interlaminar interfaces on the free edges and lead to quasi-pure mode III delamination [17]. The delamination is the main mode of observed damage for this type of stacking sequence with, in most cases, an absence of transverse cracks [16][18]. For where θ = 10°, 20° and n = 1, 2, 3, 4, Diaz and Caron [19] observed that depending on the fiber density at the interface, microcracks and matrix plasticity due to shearing may appear. This sliding is observed before delamination onset.

The objective of this work is to study the effect of stacking sequence of CHRP laminated composite on its flexural response by using 3D finite element method. This analysis is based on the experimental data of the literature [20]; where Caminero et al [20] characterized and described the damage in CFRP composite laminates with different stacking sequences subjected to flexural loading. The 3D FEA and 3D version of a Classical Laminate Theory (CLT) were adapted to examine the stress distribution under bending loads. Benbouras et al. [21, 22] used a recently established approximation formalism for isotropic materials to model the nonlinear behavior of laminated composite beams [21]. The global average stresses approach and the apparent flexural modulus method are used to obtain the laminate's equivalent longitudinal Young-modulus. A software program is developed to anticipate damage development using the most general failure criteria [22]. The authors [23] studied the successive failures and the effect of the thickness ratio on nonlinear behavior of a composites laminates reinforced with carbon/epoxy woven fabrics, It was discovered that successive failures are dependent on the stacking sequence, with orthogonal sequences having a considerable impact on plies delamination.

This work is focused on the bending behavior of Carbon/Epoxy laminated composites. Indeed, 3D linear finite element analysis is developed to study the effect of the stacking sequence on the flexural response of laminated composites. Finally, the 3D finite model obtained was compared with the experimental approach available in the literature [20], where some specimens contain delamination failure due to high interlaminar stress. 3D failure criteria, such as Hashin, Maximum stress, Tsai-Hill, will be applied in further work to predict this main type of failure.

2. Finite element analysis

The stacking sequences have a significant effect on flexural behavior of composite laminates [20, 23]. In fact, this effect was modeled by a numerical model using the ABAQUS software. Therefore, The predictive capability of the model obtained is demonstrated by comparing with experimental work of Caminero et al [20]. Two principal stacking sequence have been treated: Unidirectional laminates ($[0]_{12}$, $[90]_{12}$, $[45]_{12}$) and multidirectional laminates (quasi-isotropic laminates $[0/90/\pm45_2]_S$, angle-ply $[\pm45]_{3S}$ and cross-ply $[0/90]_{3S}$).

2.1. *Geometry and mesh generation*

The geometrical parameters used in this study are based on the experimental data [20]. The specimen's width and length were 10mm and 65mm respectively, with 9.25 of span-to-thickness ratio (L/t) see Figure 1.

8-node linear brick elements with reduced integration (C3D8R) were used to mesh the structure and hourglass control was employed for the specimen. Reduced integration declines the number of constraints introduced by an element when there are internal constrains such as Kirchoff transverse shear constraints in bending problems. There were 16800 elements and 19383 nodes in total (Figure 2). One element was modeled through the thickness of each of the 12 plies. Each Ply was given the material attributes stated in TABLE 1 and an orientation utilizing the ABAQUS GUI.

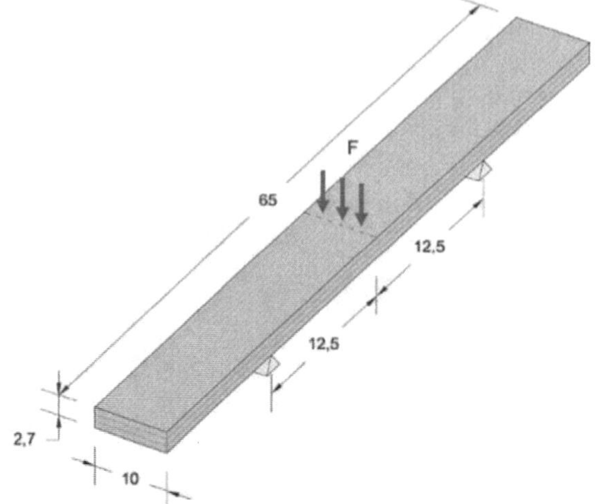

Figure 1. Geometrical parameters and applied load of the three-point bending test specimen

2.2. Boundary condition and loads

The loads applied in the FEA simulation were taken as the maximum loads given in the experimental load-displacement curves [20]. All of the specimens were simply supported. The Figure 2 shows the boundary conditions applied to the FE model, where , and are the displacement components in the x, y, z directions respectively, and , and are the rotations around the x, y, and z axis respectively. In the loading line, all the degrees of freedom of the nodes are fixed except the displacement along the z-direction, and for the support line all the degrees of freedom are constraint except the displacement along the x-direction and rotation around the y-axis.

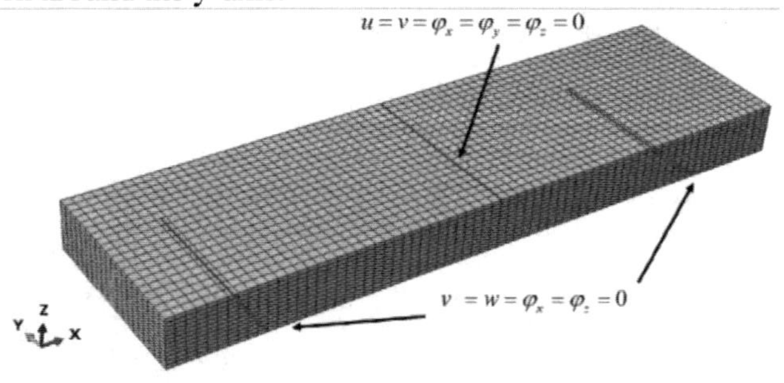

Figure 2. The boundary conditions and mesh

2.3. Material model and mechanical properties

Linear elasticity for an orthotropic material implemented in ABAQUS/Standard is used to simulate the elastic behavior of composites plates. This model defined by giving the "engineering constants": the three elastic moduli E_1, E_2, E_3; Poisson's ratios ν_{12}, ν_{13}, ν_{23}; and the shear moduli G_{12}, G_{13}, G_{23} associated with the material's principal directions. These moduli define the elastic compliance according to the following equation.

$$\begin{Bmatrix} \varepsilon_{11} \\ \varepsilon_{22} \\ \varepsilon_{33} \\ \gamma_{12} \\ \gamma_{13} \\ \gamma_{123} \end{Bmatrix} = \begin{bmatrix} 1/E_1 & -\nu_{21}/E_2 & -\nu_{31}/E_3 & 0 & 0 & 0 \\ -\nu_{12}/E_1 & 1/E_2 & -\nu_{32}/E_3 & 0 & 0 & 0 \\ -\nu_{13}/E_1 & -\nu_{23}/E_2 & 1/E_3 & 0 & 0 & 0 \\ 0 & 0 & 0 & 1/G_{12} & 0 & 0 \\ 0 & 0 & 0 & 0 & 1/G_{13} & 0 \\ 0 & 0 & 0 & 0 & 0 & 1/G_{23} \end{bmatrix} \begin{Bmatrix} \sigma_{11} \\ \sigma_{22} \\ \sigma_{33} \\ \sigma_{12} \\ \sigma_{13} \\ \sigma_{23} \end{Bmatrix} \quad (1)$$

Note that due to the symmetry of the compliance matrix, ν_{ij} is not equal to ν_{ji}, they are related by $\nu_{ij}/E_i = \nu_{ji}/E_j$ and only nine of the engineering constants are independent.

The same material was applied for all specimens only the orientation of the plies is changed according to the stacking sequences. The studied materials are provided by the materials manufacturer Hexcel Composite Ltd [24] and the given mechanical proprieties are listed in Table 1. The mechanical properties relating to tension are applied in the lower layers and those relating to compression are applied in the top layers.

Table 1 Mechanical properties (GPa) for the M21E/IMA carbon fibre/epoxy structure given by the manufacturer [24]

E_{11}^T	E_{11}^C	$E_{22}^T = E_{33}^T$	$E_{22}^C = E_{33}^C$	$G_{12} = G_{13}$	G_{23}	$\sigma_{12} = \sigma_{13} = \sigma_{23}$
178	146	118	7,4	5,2	1,2	0,39

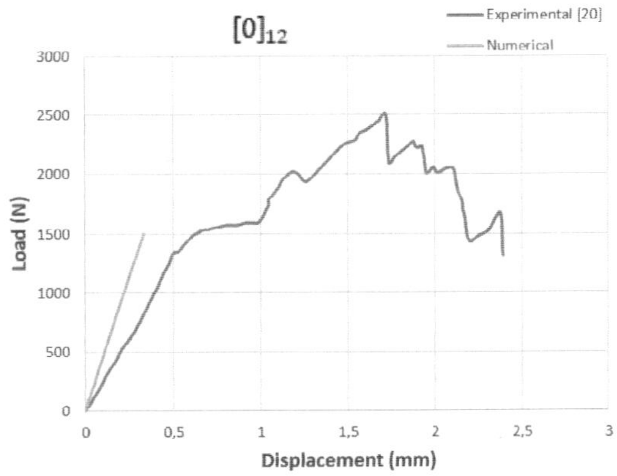

Figure 3. Load-displacement curve of $[0]_{12}$ laminate

3. Results and discussion

The numerical results found were compared with the experimental results of Caminero et al [20], where a series of three-point bending tests on unidirectional and multidirectional laminates are carried out to investigate the influence of stacking sequence on flexural parameters and failure modes. Generally, the comparison between the load-displacement curves is made before the first macroscopic failure. The typical load-displacement curves extracted from the central node of middle plane are shown in Figure 3 to Figure 8. All numerical curves of different stacking sequence present a good correlation about 3% with the experimental response except the $[0]_{12}$ laminate which show a 40% of disparity (Figure 3). An optical micrographs of $[0]_{12}$laminate indicates that premature interlaminar shear failure (delamination failure) is observed instead of flexural failure mode [20]. This flexural response of premature shear failure and local effects can explain the divergence between the calculated and measured displacement for the $[0]_{12}$laminate. In addition, a good similarity is observed for the multidirectional $[\pm 45]_{3S}$laminate in the linear part (Figure 8). However, the laminate presents a non-linear load-displacement experimental behavior leading to a pseudo-ductile performance.

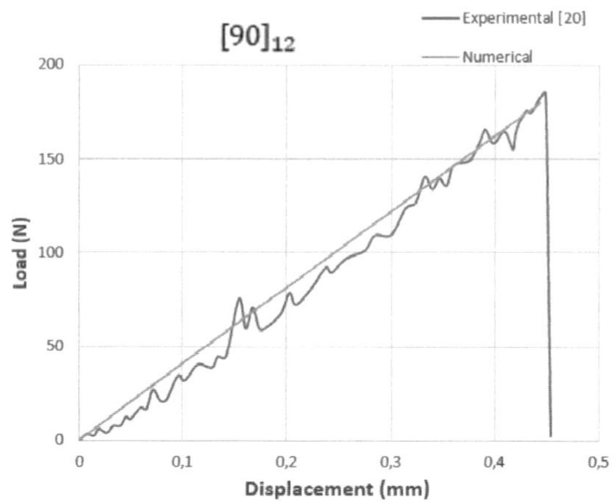

Figure 4. Load-displacement curve of $[90]_{12}$ laminate

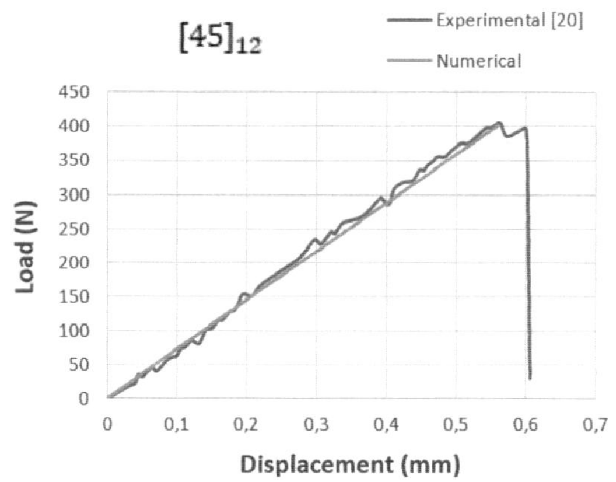

Figure 5. Load-displacement curve of $[45]_{12}$ laminate

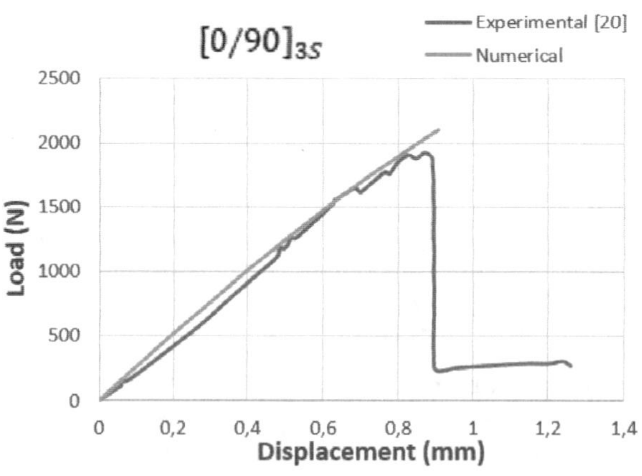

Figure 6. Load-displacement curve of $[0/90]_{3S}$ laminate

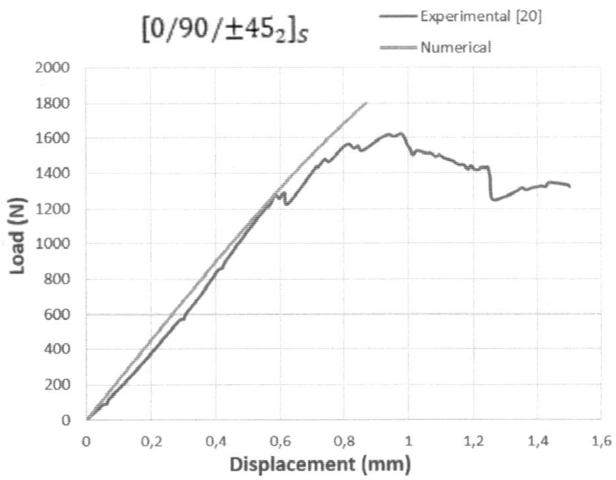

Figure 7. Load-displacement curve of $[0/90/\pm45_2]_S)$ laminate

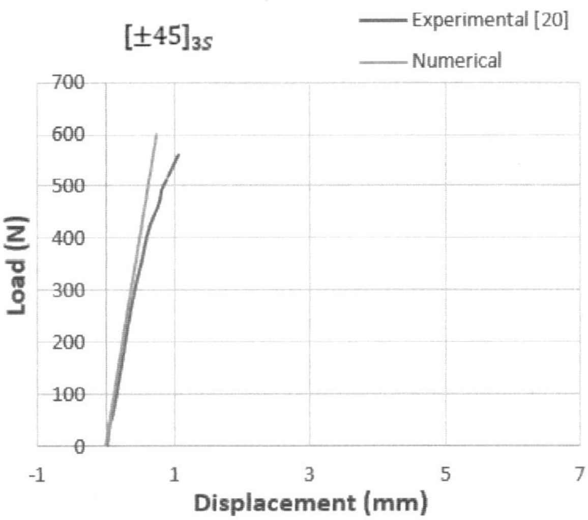

Figure 8. Load-displacement curve of $[\pm45]_{12}$ laminate

4. Conclusions

In the present paper, a numerical simulation of flexural response for the CFRP composite laminates with different stacking sequences has been proposed. Two main lay-up arrangements have been analyzed: Unidirectional laminates $[0]_{12}$, $[90]_{12}$, $[45]_{12}$ and multidirectional laminates (cross-ply$[0/90]_{3S}$, angle-ply $[\pm45]_{3S}$ and $[0/90/\pm45_2]_S$). This simulation is mainly based on experimental data. The comparison with the experimental results indicates that all stacking sequences presents an acceptable correlation (about 3%) except the $[0]_{12}$ unidirectional laminate (45%). In perspective, the interlaminar stresses will be calculated and some 3D failure criteria (Hashin, Maximum stress and Tsai-Hill) must be applied in critical areas of the laminated composites to identify the damage progression and failure modes for different orientations mainly delamination failure.

References

[1] Herakovich, Carl T. "On the relationship between engineering properties and delamination of composite materials." Journal of Composite Materials 15.4 (1981): 336-348.

[2] Pipes, R. B., and Pagano, N. J. (September 1, 1974). "Interlaminar Stresses in Composite Laminates—An Approximate Elasticity Solution." ASME. J. Appl. Mech. September 1974; 41(3): 668–672.
[3] Pagano, N. Jn. "Free edge stress fields in composite laminates." *International Journal of Solids and Structures* 14.5 (1978): 401-406.
[4] Pagano, N. J. "Stress fields in composite laminates." *International Journal of Solids and Structures* 14.5 (1978): 385-400.
[5] Wang, A. S. D., and Frank W. Crossman. "Some new results on edge effect in symmetric composite laminates." *Journal of Composite Materials* 11.1 (1977): 92-106.
[6] Kassapoglou, Christos. "Determination of interlaminar stresses in composite laminates under combined loads." Journal of reinforced plastics and composites 9.1 (1990): 33-58.
[7] Murthy, Pappu LN, and Christos C. Chamis. "Free-edge delamination: laminate width and loading conditions effects." *Journal of Composites, Technology and Research* 11.1 (1989): 15-22.
[8] Tang, S. "A boundary layer theory-Part I: Laminated composites in plane stress." *Journal of composite materials* 9.1 (1975): 33-41.
[9] Tang, S., and A. Levy. "A boundary layer theory-part II: extension of laminated finite strip." *Journal of Composite Materials* 9.1 (1975): 42-52.
[10] Lee, C. Y., and J. M. Chen. "Interlaminar shear stress analysis of composite laminate with layer reduction technique." *International journal for numerical methods in engineering* 39.5 (1996): 847-865.
[11] Kant, T. A. R. U. N., and K. Swaminathan. "Estimation of transverse/interlaminar stresses in laminated composites–a selective review and survey of current developments." *Composite structures* 49.1 (2000): 65-75.
[12] Czarnek, Robert, Daniel Post, and Carl Herakovich. "Edge effects in composites by moire interferometry." *Experimental Techniques* 7.1 (1983): 18-21.
[13] Herakovich, C. T., et al. "Influence of layer thickness on the strength of angle-ply laminates." Journal of Composite Materials, 16.3 (1982): 216-227.
[14] Herakovich, C. T. "Edge effects and delamination failures." *The Journal of Strain Analysis for Engineering Design* 24.4 (1989): 245-252.
[15] Herakovich, Carl T. "Failure modes and damage accumulation in laminated composites with free edges." *Composites science and technology* 36.2 (1989): 105-119.
[16] Brewer, John C., and Paul A. Lagace. "Quadratic stress criterion for initiation of delamination." Journal of composite materials 22.12 (1988): 1141-1155.
[17] T. Lorriot, G. Marion, R. Harry, and H. Wargnier. "Onset of free-edge delamination in composite laminates under tensile loading." *Composites Part B: Engineering* 34.5 (2003): 459-471.
[18] Y. Yuan, X. Yao, B. Liu, H. Yang, and H. Imtiaz. (2017). "Failure modes and strength prediction of thin ply CFRP angle-ply laminates." *Composite Structures* 176 (2017): 729-735.
[19] Diaz, Alberto Diaz, and Jean-François Caron. "Interface plasticity and delamination onset prediction." *Mechanics of Materials* 38.7 (2006): 648-663.
[20] Caminero, M. A., G. P. Rodríguez, and V. Muñoz. "Effect of stacking sequence on Charpy impact and flexural damage behavior of composite laminates." *Composite Structures* 136 (2016): 345-357.
[21] Benbouras, Y., Maziri, A., Mallil, E., & Echaabi, J. "A nonlinear analytical model for symmetric laminated beams in three-point bending." *Int J Technol* 8.3 (2017): 437-447.
[22] Benbouras, Y., Bellahkim, M., Maziri, A., Mallil, E., & Echaabi, J. "Nonlinear modeling of the failure of a graphite epoxy under a three-point bending test." *Polymers and Polymer Composites* 28.2 (2020): 119-139.
[23] Benbouras, Y., Bellahkim, M., Maziri, A., Mallil, E., & Echaabi, J "An analytical and experimental study of the nonlinear behaviour of a carbon/epoxy under a three-point bending test." Plastics, Rubber and Composites, . (2021) 1-9.
[24] H. Resources. Prepeg data sheets. hexply M21E epoxy matrix. Product data, www.hexcel.com

Effect of process parameters on the laser microdrilling performance of stainless steel, aluminium and copper

W. Noor[1], T. Saleh[1,3], M. A. N. Rashid[1] and A. M. Ibrahim[2]

[1]Autonomous Systems and Robotics Research Unit (ASSRU), Department of Mechatronics Engineering, Kuliyyah of Engineering, International Islamic University Malaysia, 53100, Gombak, Kuala Lumpur, Malaysia
[2] Department of Mechatronics Engineering, Kuliyyah of Engineering, International Islamic University Malaysia, 53100, Gombak, Kuala Lumpur, Malaysia
[3] Author to whom any correspondence should be addressed.

tanveers@iium.edu.my

Abstract. Micromachining techniques are being used regularly in various engineering and production sectors such as Micro Electromechanical Systems, Aerospace, Automotive, Electronics and Biomedical industries. For Laser Beam Micromachining, highly energized laser beam is focused on a small region of the workpiece surface. As a result, it is heated up rapidly to sufficiently high temperatures, then the material starts to melt and/or vaporize from the surface. This phenomenon of material removal is called laser ablation. The holes that are produced by one-dimensional laser beam drilling (LBD) can come with defects like taper, heat affected zone (HAZ) and Recast Layer. On the other hand, LBD process offers high control, high efficiency, precision, and production rate, particularly for drilling microscopic holes in a variety of materials. LBD performance is measured on different parameters such as taper and recast layer of the machined holes. Previous studies report many observations regarding the performance parameters and the process input parameters. But a thorough study of the performance parameters with respect to different material properties has not been reported yet. The research work for this paper focuses on the experimental investigation for different materials such as stainless steel (type SS304), Aluminum and Copper to observe the effects of Laser input parameters (namely laser power, scanning speed, and pulse repetition rate) on the performance of the LBD. The study also considers different thermophysical as well as optical properties such as thermal conductivity, specific heat, melting point, absorptivity and how it affects the outcome of LBD in a combined manner. Characterization of different output parameters such as Entry area, Exit Area and Recast Area is done by SEM (Scanning Electron Microscope) machine. For 75 Loop count, 90% Laser Power, 950 mm/s Scanning Speed and 10kHz pulse repetition rate, the highest amount of Recast Area is observed in SS, valued at 0.053 mm^2. The largest amount of taperness is found in Cu at 44.404°. Scanning Speed and Laser Power has been identified as the most significant process factors for LBD performance.

1. Introduction

Laser Micromachining has been at the core of various engineering applications in recent times. In cutting edge biomedical technology, microfluidic passageways, MEMS, myriad devices such as catheters, stents are manufactured with the aid of LBMM(Laser Beam Micro-Machining)[1]. The aerospace industry requires drilling of many microscopic cooling holes on the turbine blades and vanes using LBD, where the materials are mostly superalloys[2]. The automotive industry greatly depends on composite materials such as Carbon fibre-reinforced composites (CFC) due to its beneficial properties and laser machining is well suited for CFC due to difficulty associated with traditional machining processes[3].

LBMM offers control and ease in making recurrent microscopic patterns, shapes, electronic circuits, and electronic micro-appliances such as microreactors, microanalytic systems and optical devices[4]. LBMM is used in such cases to fashion out intricate details on the devices. These features are difficult or impossible to achieve using conventional machining processes. Tapered conical shapes can be replicated on material surfaces using LBMM, resulting in better wettability of the surface[5]. Fiber Lasers incorporate some key advantages such as high thermal efficiency and robustness of operation which negates any non-linear environmental effects[6].

In Laser Micro Drilling, highly energized concentrated ray of laser is impacted onto the surface of the material. The material is removed in layers. Most of the energy applied to the surface is absorbed by the material and it is heated up, while the rest is reflected. The state of the material changes to first molten and then vapor. This steady removal of material is called laser ablation.

Laser ablation phenomenon can be divided into three regimes: (1) melting (2) vaporization (3) deterioration of chemical bonds and material disintegration[7]. Every material has different threshold value for ablation. This value indicates to the amount of minimum energy required to set off the ablation[8].

The application of fiber lasers has been mostly restricted to experimental usage up until present times. The most notable component of the fiber laser setup is the doped fiber medium or 'gain medium'. The fiber is doped with elements such as Erbium or Ytterbium. The electrons within are pumped with a diode laser and excited to higher energy level. Once the electron collides with another photon and descends to lower energy level, it releases additional photons which is set on a collision course with other electrons. This bombardment of photons is confined within the fiber core by surrounding it with sheaths or claddings[9]. These claddings facilitate total internal reflection of the laser that is initially emitted from source. The resulting highly luminous output is then focused onto workpiece. Fiber lasers offer flexibility and produce high quality beams[10].

Even though fiber laser micromachining offers high speed operations and low operating costs, it has some downsides. The holes formed by this process consist of taper, spatter and recast layer. The overall performance of the machining process is gauged on some parameters such as the hole taperness, HAZ, material removal rate, surface roughness and so on. It has been observed that with the change in different process parameters, there are significant changes to the output parameters and machining performance. Harp et al.[11] illustrated that pulse width and pulse duration of the fiber laser directly impacts the MRR, geometry and shape of the holes. Ancona et al.[12] demonstrated the effect of pulse energy and repetition rate on drilling efficiency. Furthermore, it was discovered that the depth of ablation per pulse increases logarithmically with pulse energy[13]. In case of combined effects, the exact correlations between different input and output parameters are unknown. To derive these correlations many different methods of statistical analysis has been done. Okasha et al.[14] tinkered with laser power, assist gas type and gas pressure to understand the effect of these parameters on the recast layer thickness, HAZ, taper and diameter of the hole. All the input parameters had significant impact on the output parameters, be it individually or collectively. An Nd:Yag laser was used to microdrill the holes on Inconel-718. Goyal et al.[15] investigated the effect of pulse width, pulse repetition rate, assist gas pressure and laser trepanning speed on taper and circularity of holes.

Along with different process parameters, significant effects of material properties such as thermophysical and optical properties on the laser machining performance have been reported in previous studies. Ahmed et al.[16] illustrated the effect of input variables such as scanning speed and laser intensity for laser milling operation on alloys of titanium(Ti6Al4V), nickel(Inconel 718) and aluminum (AA 2024). They concluded that material properties such as thermal conductivity and emissivity are chief contributors to the difference in the value of MRR for the same machining parameters. For modelling such laser machining processes transient heat transfer equation is considered where physical properties such as density of the material is incorporated into the equation[17].

From the literature survey, it is evident that the combination of process parameters and material properties govern the ablation and characteristics of laser micro-drilling of holes. The quality and precision of the process are manipulated predominantly by tweaking the laser power, scanning speed, pulse repetition rate. It is observed that parametric study of fiber laser micromachining is scarce and

IOP Conf. Series: Materials Science and Engineering 1244 (2022) 012020 doi:10.1088/1757-899X/1244/1/012020

thorough exploration of the cause and combined effects from the holistic perspective of both property and parameters of laser machining is yet to be done. The study involves characterization of the holes and an attempt to explain the underlying causes of the outcome.

2. Materials and experimental method

This section details out the processing method of workpiece before and after laser machining operation, choice of the materials, schematic of the experimental setup and characterization of the different output parameters.

2.1. Experimental Setup

Aluminium, Stainless Steel and Copper Plates of identical shape and size were used to conduct the experiment. The thickness of the plates was 0.2mm each. The length and width were 26.7mm and 25.7mm, respectively. The plates were cleaned, preserved, and maintained in favorable temperature and humidity.

Each plate was machined one sample at a time. In total there were three samples for each material. First the plate was fastened to a fixture by screws and washers. This was done to eliminate the occurrence of errors due to accidental displacement of the workpiece. The screws were tightened but not too firmly as it might bend the material. The fixture was then carefully placed and aligned on the base of the laser setup. The base was made of Perspex and it could be moved horizontally and vertically in manual mode using screws. The laser head can also be adjusted accordingly. The focal point of the laser beam was focused on the workpiece by changing the focal length. The focal length is measured as the distance between the laser head and the workpiece. The spot diameter for the laser beam was set at 0.2 mm. The loop count was set at 75 for all passes and all samples.

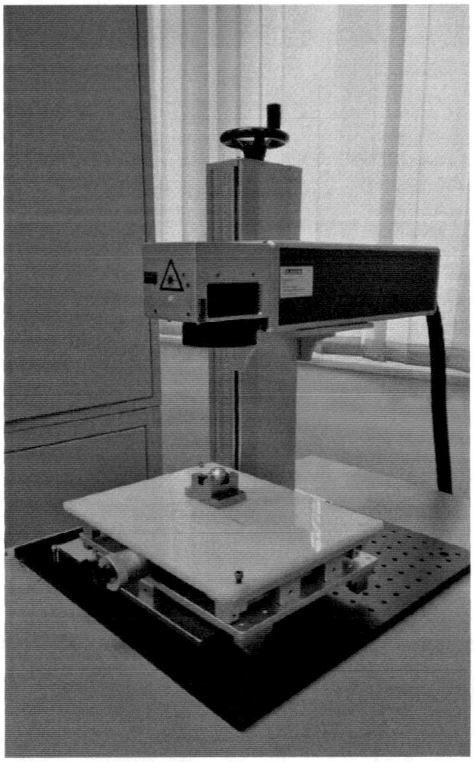

Figure 1. Experimental setup of the fiber laser

At first a pilot hole was drilled using the laser for marking the starting hole. Then a rectangular array of 25 holes were made, each row and column containing 5 holes for the first two samples. For the last sample 24 holes were machined. The holes were drilled by varying the laser parameters, such as, pulse repetition rate, scanning speed and percentage of power. All the parameters were varied in four stages. To observe the effect of a single parameter, other two were fixed and only the parameter of interest was varied. The parameters were varied in the following order: 1) pulse repetition rate, 2) scanning speed, 3) percentage of power. The holes were drilled 0.2 mm apart from each other in both vertical and horizontal directions. Time needed to drill each hole was measured using a stopwatch. After machining the samples were blow-dried for cleaning of debris. Then they were stored in a sample box wrapped in soft tissue papers.

Figure **1** illustrates the experimental setup.

The important components[18] of a Fiber Laser Machining device are:

i. An active medium, typically an optical fiber that has been doped in rare elements such as Erbium, Ytterbium, Neodymium, Dysprosium etc.

ii. A cladding that sits around the fiber core.

iii. Enhancement devices such as collimators, beam splitters and optical lenses.

A schematic diagram of a fiber laser is given in Figure 2:

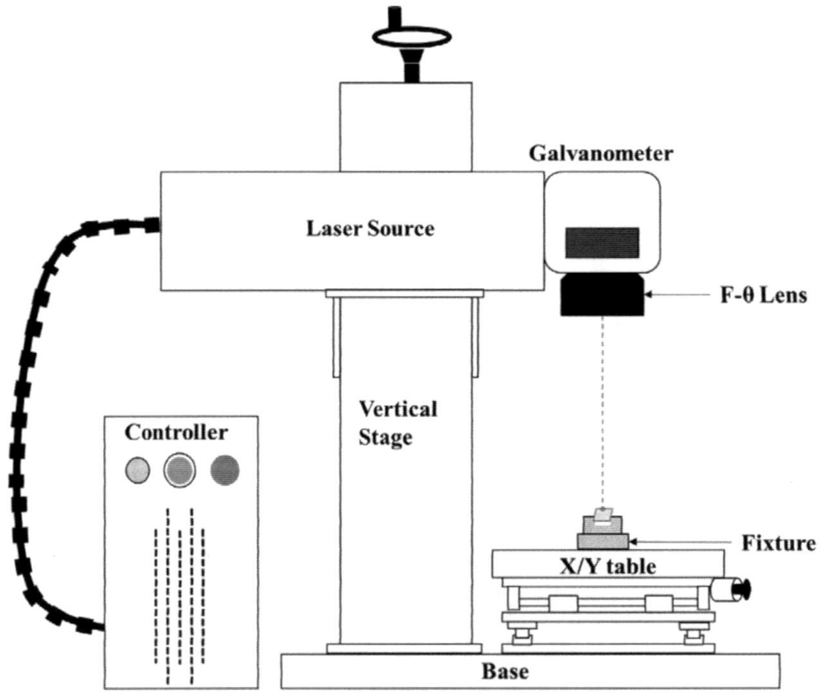

Figure 2. Schematic Diagram of a Fiber Laser Setup(Rashid et al.[19])

2.2. Characterization

After finishing the machining, the samples were ready for characterization. To aid in characterization, we use the SEM(Scanning Electron Microscope) machine. The SEM machine can magnify an image up to 30000 times and it is used to capture very detailed image of the holes. The images were taken in such a way that the recast layer would be easily discernible. For capturing the recast layer image, the workpiece was slanted at an angle of 40 degrees. SEM machine bombards the workpiece with electrons and forms an image using the scattered electrons. Before taking the pictures, the workpieces were further cleaned by submerging them in pure alcohol filled beaker and placing the beaker on an ultrasonic cleaner machine. All the pictures were taken with JSM-IT100 InTouchScope™ Scanning Electron Microscope. The images were magnified 150 times to get better visualization. After capturing the images, the samples were restored in the sample box for further study.

For acquiring the entry and exit diameter of the holes, the entry area or exit area, both denominated by A, were converted into diameter(D) in millimeters using the following Equation 1.

$$D = \sqrt{\frac{4\times A}{\pi}} \quad (1)$$

The recast layer is measured by measuring the area of the recast zone along the edge of the entry of each hole. Figure 3 depicts the recast layer of different materials at the same machining parameter.

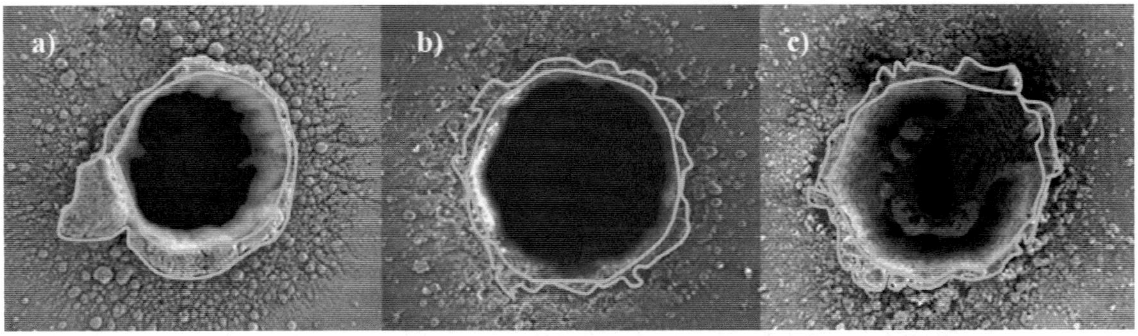

Figure 3. Characterization of Recast Layer for a) Stainless Steel, b) Aluminium and c) Copper sample. All holes were drilled at 5khz, 15.5W laser power, 500mm/s scanning speed at 75nos Loop Count.

The measurements were made using ImageJ[20] software. For measuring hole taper, the difference between the entry(D_1) and exit hole diameter(D_2) was divided by the thickness(t) of the sample. Figure 4 presents a two-dimensional front view of the hole taper. If θ is the taper angle then taperness in degrees is obtained by Equation 2.

$$\theta = \frac{D_1 - D_2}{2t} \times \frac{180}{\pi} \quad (2)$$

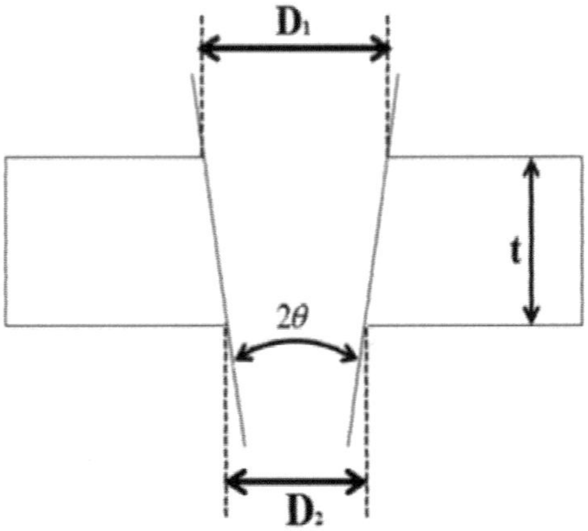

Figure 4. Characterization of hole taperness

3. Results and Discussion

In this section the parametric effect and the variation in output due to different material properties will be discussed. As indicated by previous studies, the performance parameters rarely vary in a linear manner with the change in input and material properties. This section will be divided and dedicated to the effect of each parameter as well as the effects of each material properties on the outcome of LBD. The parameters that are discussed here are scanning speed and laser power. The effect of pulse repetition rate is negligible so it has been averaged up for the study. The material properties in this study are thermal conductivity(k), specific heat(c_p), melting point(T_f) and absorptivity.

3.1. Study of Parametric effect

This section describes the effect of both laser power and scanning speed on the performance parameters of recast layer and taperness. It is obvious from **Figure 5**(a) that with the increase in power there is an increase in recast layer formation along the hole. In fact at 15.5W power and 950mm/s scanning speed the highest average value of recast layer is observed at 0.0455mm^2 for stainless steel. But at the same time for Copper and Aluminium the effect of laser power is almost negligible on recast layer formation. This can be attributed to the fact that stainless steel has much lower thermal conductivity and very high vaporization temperature compared to other materials so the heat distribution is poor. As the power is increased the amount of molten material that is ejected larger cause due to poor heat distribution the material is not properly vaporized, most of it is only melted away and resolidified again. For Taperness the effect is again different for different materials. We can see that taperness angle decreases significantly with the increase in power(**Figure 5**(b)) for stainless steel and Aluminium, while for Copper it shows very little change. High value of laser power ensures that energy density is high and the exit area increases with it. That is why the taperness goes down with increase in power. Copper has the lowest absorptivity of all the materials. Lee et al.[21] showed that optical absorbance of materials

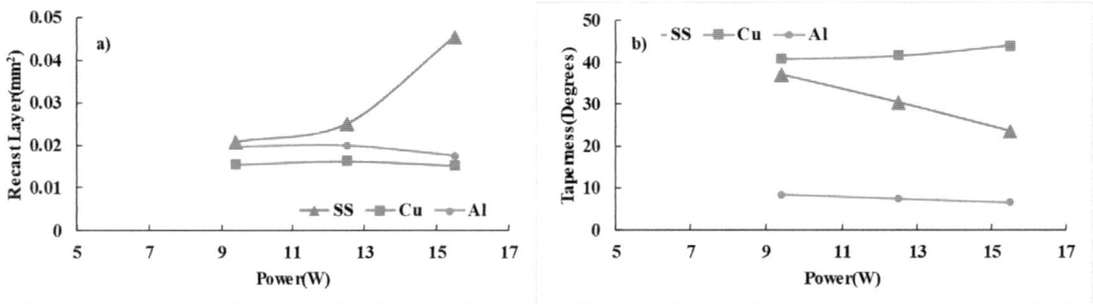

Figure 5. Effect of laser power on the a) recast layer and b) taperness of Stainless Steel, Aluminium and Copper. The pulse repetition rate has been averaged up while all the holes were machined at 75nos Loop Count and 950mm/s Scanning Speed.

has a positive correlation with the ablation depth of laser machining. Thus, owing to the lowest specific heat value, for Copper the value of taperness is higher at the same power compared to other materials. **Figure 5** omits the value of Taperness and Recast Layer at 6.4W power as there was no signs of material removal on either the entry or exit side of the Copper sample for this power setting.

Scanning Speed's effect is much more pronounced on the taperness compared to that on the recast layer. For stainless steel, recast layer decreases with the increase of scanning speed. Due to lower interaction time between the material and incident laser with higher scanning speed, the amount of ejected material is low, this may be the reason why recast layer decreases. However for the other materials the effect of scanning speed on recast layer is random. Taperness increases significantly with the increase in scanning speed as lower interaction time means not enough energy is propagated across the material. As a result the exit area becomes smaller. It is noteworthy that for Aluminium the value of taperness is the lowest among all the materials, for a fixed combination of input parameters. Benton et al.[22] illustrated that with the increase in specific heat c_p the depth of cut decreases for same amount of power. As the power increases the effect of ablation depth is even more profound. Aluminium has the highest specific heat value, so for same power and scanning speed the value of taperness is much lower than the rest. For lowest power(6.4W) the highest average value of taperness for Aluminium is found at 950mm/s scanning speed, which is 35.52° and for only this power setting we can see that taperness increases rapidly with scanning speed. For higher power settings the range of change in taperness is significantly decreased for Aluminium.

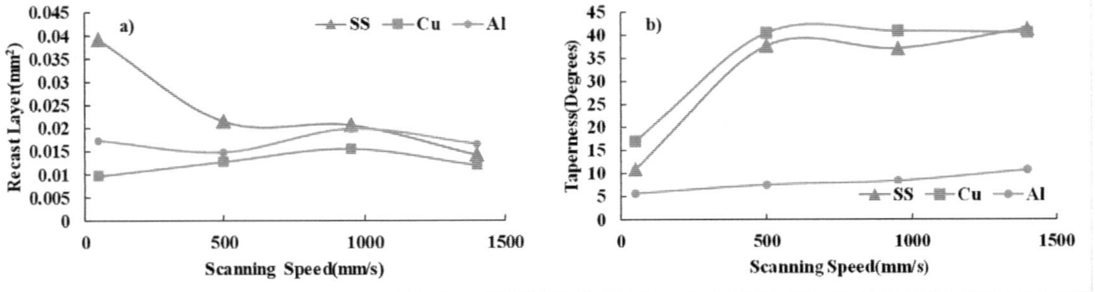

Figure 6. Effect of scanning speed on the a) recast layer and b) taperness of stainless steel, Aluminium and Copper. The pulse repetition rate has been averaged up while all the holes were machined at 75nos Loop Count and 9.4W laser power.

3.2. Study of the Material Index(I) Derived from Material Properties in terms of Performance parameters

3.2.1. Derivation of the Material Index(I). Apart from process parameters, material properties such as thermal conductivity(k), specific heat(c_p), melting point(T_s) and absorptivity(A) have significant effect on the output of the laser machining. In this section, a study on these effects have been presented. There are two types of material properties in the discussion. One being thermophysical properties(k, c_p, T_s) and the other one is optical property(A).

It is not possible to conduct an experiment where we can vary one material property at a time while keeping the rest of the material properties as fixed. Because all the material properties of one material are inherent and salient features. We have introduced a material index comprised of the material properties(k, c_p, T_s and A) considered for this study. The material index(I) is calculated according to Equation 3 which is given below:

$$I = \frac{k.A}{c_p.T_s} \tag{3}$$

This index(I) has been derived according to the assumption that the amount of material removed due to LBMM has a proportional relationship with thermal conductivity(k) and absorptivity(A) while having an inversely proportional relationship with specific heat(c_p) and melting point(T_s). Benton et al.[22] had conducted a simulation study that reported the relationship between thermal conductivity and depth of cut in material is proportional while for specific heat, the relationship with depth of cut is inversely proportional. Mishra et al.[23] denotes that for the same pulse energy, ablation volume would decrease for higher melting temperature. Wu et al.[24] discussed that the same amount of energy per pulse higher for higher absorptivity of material would result in higher levels of ablation. So the studies conducted in [22],[23],[24] confirm our assumptions. However, the dimensional analysis of the material index(I) is out of the scope of this study. Table 1 tabulates the value of various properties:

Table 1. Value of different material properties for Stainless Steel, Copper and Aluminium Alloys

Material Type	Thermophysical Properties			Optical property	Material Index, I(kg/(m.K.s))
	Thermal Conductivity, k(W/(m.K))	Specific Heat, c_p(J/(kg.K))	Melting Point,T_s(K)	Absorptivity, A	
Stainless Steel	14	490	1723.15	0.36	5.96913E-06
Copper Alloy	385	385	1356.15	0.05	3.68691E-05
Aluminium Alloy	138	900	878.15	0.15	2.61914E-05

All the values in Table 1 are thermal properties which vary with the ambient temperature and conditions. The values have been approximated for an ambient temperature of 20°C. As we are experimenting with metallic alloys, the melting temperature here is solidus temperature(T_s), the highest temperature at which an alloy remains solid. The wavelength of the incident laser is 1064nm. The value of absorptivity of Copper and Aluminium has been approximated at 1064nm from the study conducted by Deininger et al.[25]. Kuang et al.[26] denotes that for 1.06 μm wavelength the absorption coefficient is approximately 0.36 for SS304(stainless steel). Thermal conductivity and specific heat values for stainless steel has been procured from AZoM website[27].

3.2.2. Discussion on the Variation of Performance Parameters with respect to Material Index(I). The variation of performance parameters with respect to the material index(I) will be discussed with respect to both laser power and scanning speed. Figure 7 describes the variation of taperness and recast layer for varying power levels and scanning speeds. Graphs for Taperness and Recast Layer at 6.4W Power has again been omitted from Figure 7(a) and Figure 7(b) as it has no visible ablation effect for copper. As the material index(I) consists of all the material properties, the discussion will be done from the perspectives of thermal conductivity, specific heat, melting point, and absorptivity in a holistic manner. Generally, for higher thermal conductivity heat spreads radially as well as vertically from the spot where the laser beam impacts. Also the conduction of heat throughout the material is faster and that facilitates increased ablation of the material. Recast Layer is formed from the ejected molten material while the laser ablation takes place. The higher the amount of localized heat the higher the recast layer. That is why stainless steel has the highest recast layer and copper has the lowest(Figure 7(a) and (c)). Because of the lowest thermal conductivity and highest melting point of stainless steel, it results in accumulation of heat along the cross sectional area of the workpiece and not along the depth. Copper has the lowest average value of recast layer(0.007mm2) for all the penetrated holes at 12.5W. At higher power, lower scanning speed and higher values of material index(I) recast layer seems to be very low, which is desirable. Taperness is seen to first decrease slightly for different power levels and then increase rapidly with increase in material index. As thermal conductivity increases, generally the taperness shall decrease due to more heat being propagated radially as the heat is concentrated on the surface of the material. For high scanning speed the taperness also increases due to less interaction time between material and laser beam(Figure 7(d)).

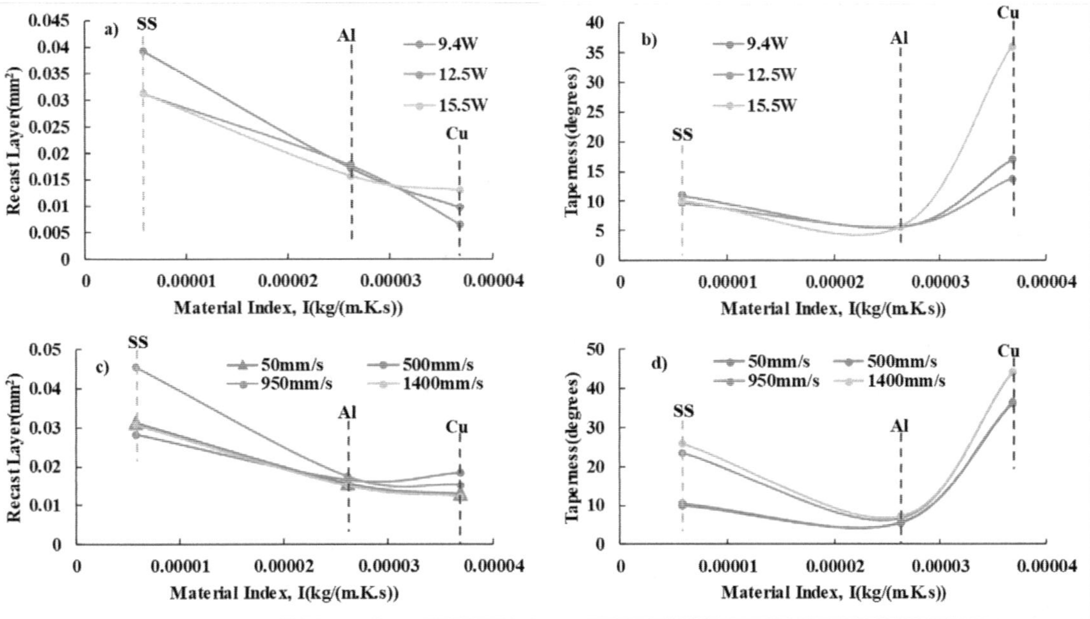

Figure 7. Effect of Thermal Conductivity on recast layer and taperness respectively, with respect to variable power((a) and (b)) and scanning speed((c) and (d)). The loop count has been kept constant at 75 nos and the pulse repetition rate has been averaged up for this graph. The machining parameters for(a) and (b) are 50mm/s, while for (c) and (d) it is 15.5W power.

Specific heat is the amount of heat a material absorbs to increase the temperature of 1kg of that material by 1K/1°C. Higher specific heat usually correlates to lower melting point of the material. Stainless steel has very low thermal conductivity so the recast layer increases even though the value of specific heat is higher for stainless steel. Stainless steel has the highest average value of recast layer(0.0455mm^2) at 15.5W power and 950mm/s scanning speed. The value of taperness also decreases with higher specific heat, as the material has lower melting point. Even though stainless steel has a higher melting point than copper, its absorptivity is far superior to both the materials, so the taperness value decreases for stainless steel. Higher melting point coupled with low thermal conductivity is the reason why stainless steel has very high values of recast layer. High values of melting point are associated with having higher vaporization point so for the same amount of energy there will be partial ablation, meaning the molten material will not reach vaporization temperature. This means the amount of molten material that is ejected will be more with higher melting point. For taperness the effect is reversed as it increases with increase in melting point. But for stainless steel it decreases due to high absorptivity of stainless steel. Copper has the lowest absorptivity so the heat is not enough to remove material along the cross section of the material. With the increase in power and scanning speed the range of change is also increased, as evident from Figure 7(b).

Absorptivity is the fractional value by which the measure of absorbed radiation is determined. Usually, higher values of absorptivity should indicate higher level of absorbed thermal energy. That being said, the combined effect of other thermal properties and process parameters have a hand in the final outcome of the machining. With the increase in absorptivity there is rapid increase in recast layer from Copper to stainless steel. That is because the effect of thermal conductivity here is dominant over absorptivity. The change in recast layer from Copper to Aluminium is negligible due to minuscule difference between their absorptivity. For higher scanning speed the amount of molten material decreases so the increasing trend of recast layer for stainless steel is only seen for highest power setting. The effect of material index on taperness is random as the value first decreases for Aluminium and then increases for stainless steel. Here the combined effect of low absorptivity, low melting point and high specific heat associated with Aluminium comes into play. For stainless steel the high melting and vaporization temperature coupled with low thermal conductivity results in poor penetration of workpiece.

4. Conclusion

The aim of this paper was to study the effect of different process parameters as well as thermophysical and optical properties on the output and performance of fiber laser micromachining on Aluminium, Copper and Stainless Steel. The selected process parameters were laser power, pulse repetition rate, and scanning speed. The variables were increased in four stages and for each pass only one variable was changed. Then the output parameters were measured using SEM. The discussed performance parameters are recast layer and hole taper. For most of the materials, there is no clear pattern as to how the process variables and material properties affect the output of the machining. It is evident that the governing factors of machining using fiber laser consist of the material properties such as thermal conductivity, specific heat, melting point, and absorptivity. For 75 Loop count, 90% Laser Power, 950 mm/s Scanning Speed and 10kHz pulse repetition rate, the highest amount of Recast Area is observed in SS, valued at 0.053 mm^2. The largest amount of taperness is found in Cu at 44.404°. Higher thermal conductivity led to lower recast layer but higher taperness. Materials with high specific heat value should exhibit low taperness for same combination of process parameters. The effect of melting point and absorptivity is too random and governed by process parameters as well as thermal conductivity and specific heat.

Acknowledgement

This project was funded by the Ministry of Higher Education Malaysia (MOHE Grant No: FRGS/1/2018/TK03/UIAM/02/2) and Asian Office of Aerospace Research and Development (AOARD Grant No: BAA-AFRL-AFOSR-2016-0007). Wazed Ibne Noor was paid as a graduate research assistant from the MOHE grant and Mir Akmam Noor Rashid was paid as a graduate research assistant (GRA) from the AOARD grant .

References

[1] Miller P R, Aggarwal R, Doraiswamy A, Lin Y J, Lee Y-S and Narayan R J 2009 Laser micromachining for biomedical applications *Jom* **61** 35–40

[2] Marimuthu S, Antar M and Dunleavey J 2019 Characteristics of micro-hole formation during fibre laser drilling of aerospace superalloy *Precis. Eng.* **55** 339–48

[3] Voisey K T, Fouquet S, Roy D and Clyne T W 2006 Fibre swelling during laser drilling of carbon fibre composites *Opt. Lasers Eng.* **44** 1185–97

[4] Mamilla V R and Chakradhar K S 2014 Micro machining for micro electro mechanical systems (MEMS) *Procedia Mater. Sci.* **6** 1170–7

[5] Worts N, Jones J and Squier J 2019 Surface structure modification of additively manufactured titanium components via femtosecond laser micromachining *Opt. Commun.* **430** 352–7

[6] Canning J 2006 Fibre lasers and related technologies *Opt. Lasers Eng.* **44** 647–76

[7] Parandoush P and Hossain A 2014 A review of modeling and simulation of laser beam machining *Int. J. Mach. Tools Manuf.* **85** 135–45

[8] Liu X, Du D and Mourou G 1997 Laser ablation and micromachining with ultrashort laser pulses *IEEE J. Quantum Electron.* **33** 1706–16

[9] Williams E 2014 Experimental and theoretical investigations of nanosecond fibre laser micromachining

[10] Kleine K F, Whitney B and Watkins K G 2002 Use of fiber lasers for micro cutting applications in the medical device industry *International Congress on Applications of Lasers & Electro-Optics* vol 2002 (Citeseer) p 70

[11] Harp W R, Dilwith J R and Tu J F 2008 Laser ablation using a long-pulsed, high-fluence, CW single-mode fiber laser *J. Mater. Process. Technol.* **198** 22–30

[12] Ancona A, Röser F, Rademaker K, Limpert J, Nolte S and Tünnermann A 2008 High speed laser drilling of metals using a high repetition rate, high average power ultrafast fiber CPA system *Opt. Express* **16** 8958–68

[13] Ancona A, Nodop D, Limpert J, Nolte S and Tünnermann A 2009 Microdrilling of metals with

an inexpensive and compact ultra-short-pulse fiber amplified microchip laser *Appl. Phys. A* **94** 19–24

[14] Okasha M M, Mativenga P T and Li L 2011 Sequential laser mechanical microdrilling of Inconel 718 alloy *J. Manuf. Sci. Eng.* **133** 11008

[15] Goyal R and Dubey A K 2014 Quality improvement by parameter optimization in laser trepan drilling of superalloy sheet *Mater. Manuf. Process.* **29** 1410–6

[16] Ahmed N, Rafaqat M, Ishfaq K, Ur Rehman A, Hassan A, Umer U, Ragab A E and Al-Zabidi A 2019 Comparison of laser milling performance against difficult-to-cut alloys: parametric significance, modeling and optimization for targeted material removal *Materials (Basel).* **12** 1674

[17] Elhadj S, Matthews M J and Yang S T 2014 Combined infrared thermal imaging and laser heating for the study of materials thermophysical and processing properties at high temperatures *Crit. Rev. Solid State Mater. Sci.* **39** 175–96

[18] Schaeffer R 2012 *Fundamentals of laser micromachining* (CRC press)

[19] Rashid M A N, Saleh T, Noor W I and Ali M S M 2021 Effect of laser parameters on sequential laser beam micromachining and micro electro-discharge machining *Int. J. Adv. Manuf. Technol.*

[20] Schneider C A, Rasband W S and Eliceiri K W 2012 NIH Image to ImageJ: 25 years of image analysis *Nat. Methods* **9** 671

[21] Lee J Y, Kang M C, Kim K H, Park W I and Cho S H 2016 Effects of optical absorbance with ablation characteristics in femtosecond laser irradiation of carbon reinforced Al2O3 composites *Adv. Appl. Ceram.* **115** 123–8

[22] Benton M, Hossan M R, Konari P R and Gamagedara S 2019 Effect of process parameters and material properties on laser micromachining of microchannels *Micromachines* **10**

[23] Mishra S and Yadava V 2015 Laser Beam MicroMachining (LBMM) – A review *Opt. Lasers Eng.* **73** 89–122

[24] Wu B, Liu P, Wang X, Zhang F, Deng L, Duan J and Zeng X 2018 Effect of laser absorption on picosecond laser ablation of Cr12MoV mold steel, 9Cr18 stainless steel and H13A cemented carbide *Opt. Laser Technol.* **101** 11–20

[25] Deininger C 2011 Copper machining with lasers *Laser Tech. J.* **8** 28–31

[26] Kuang J H, Hung T P, Lai K, Hsu C M and Lin A Der 2012 The Surface Absorption Coefficient of S304L Stainless Steel by Nd:YAG Micro-Pulse Laser *Adv. Mater. Res.* **472–475** 2531–4

[27] AZoM 2017 Stainless Steel - Grade 304 (UNS S30400)

Deep-learning-based detection of missing road lane markings using YOLOv5 algorithm

Nur Hanisah Azmi[1], Ali Sophian[1*], Ali Aryo Bawono[2]

[1] Mechatronics Department, Kulliyyah of Engineering, International Islamic University Malaysia, Selangor, Malaysia

[2] Rapid Road Transport Department, TUMCREATE Limited, Singapore, Singapore

*Corresponding author: ali_sophian@iium.edu.my

Abstract. Road lane markings are critical for ensuring road user safety. To improve their safety, there are even different types of road lane markings, such as single solid lines, double solid lines, dashed single line etc. Their colors generally are white and yellow. This road lane markings mainly used to provide guidance and information for road user to comply with the rule of the road. Unfortunately, these markings get worn out with time and may even disappear. In order to prevent this from happening, regular inspection and maintenance need to be conducted. Manual inspection is tedious, slow, and prone to human errors. With the recent technological advancement, especially in machine vision and artificial intelligence, automated or semi-automated missing road lane marking detection systems can potentially be developed. In this work, preliminary study of the implementation of one of the latest deep learning algorithms, i.e. YOLOv5, has been carried out in the detection and classification of missing road lane markings. This paper shows the preliminary results which look promising as the mean Average Precision (mAP@0.5) reaches 0.995.

1. Introduction

Traffic accidents have been identified as one of the most serious problems in Malaysia [1]. Road lane markings are one of the most important aspects of safe roadways as it is to delineate the roadway path and specific traffic lanes. They provide information including roadway alignment, vehicle positioning and other important driving-related tasks to road users. Uniformity of the lane marking is an important factor in minimizing confusion and uncertainty about their meaning among road users.

Unmarked or missing road lane markings are considered as areas of high potential hazards such as those new unmarked pavement roads in rural areas or those missing road lane markings that have been scraped away by either wind, water or other natural agents. These conditions can lead to fatal traffic accidents and missing road markings have been identified as one of causes of fatal road accidents in Malaysia [2].

Many road inspection activities, including survey on quality of road lane marking, are still done manually. The manual approaches are tedious and prone to human errors. Some automation that makes use advanced technology, such as computer vision and artificial intelligence (AI), can improve the performance and effectiveness of such inspection. Artificial intelligence is a replica of human intelligence programmed in a machine. While machine learning is a subset of artificial intelligence which is a technique used through trained data algorithms while deep learning is a subset of machine learning which is an approach of artificial neural networks. From the perspective of engineering, computer vision seeks to automate tasks that the human visual system can do [3].

There are several methods that have been reported in the research literature for the detection of the missing road lane markings, including those that use video cameras, mobile laser scanners and Light Detection and Ranging (LiDAR) [4]. Some development that uses the video cameras by using relatively low processing power computers, such raspberry pi, has also been discussed [5][6]. Many approaches typically use the same algorithm as Hough transform to generate lane marking by detecting straight edges according to the image features of the lane [7]. Another method suggested a road lane marking identification method based on the combination of fuzzy logic and random sample consensus (RANSAC) algorithm but with some drawback in the detection along a curved part of the road and the unstructured roads. The detection is more challenging due to factors such as bright light, shadows, water and so on. These factors degrade the detection performance [8][9].

A machine learning technique which is Support Vector Machine (SVM) was proposed for an accurate and efficient road marking detection system with the Principal Component Analysis on Histogram of Oriented Gradients (PCA-HOG) as its algorithm [10]. The paper comprises of three elements that are the vanishing point in the image is defined and is used to generate a bird's-eye view of the road, the local median binarization approach is used to segment the image and it can avoid the interference of light and shadow and the feature vector is generated by using a combination of PCA-HOG and SVM to evaluate the feature vector as a particular road marking. Further studies on the SVM algorithm for the Digital Highway Data Vehicle (DHDV) system was used to acquire a laser image which can then be analysed for lane marking identification and reconstruction. Four phases were implemented in this research method, the first phase is the binarization of the laser images with a new threshold method [4]. This SVM tool is normally used in the advanced driver-assistance system (ADAS) to monitor and receive road information in real time.

Deep Convolutional Neural Network (DCNN) is a category of deep neural networks which is commonly used for visual image analysis that outperforms the traditional methods. A study presented an approach for robust road lane marking identification quality using a deep convolutional neural network known as Lane Marking Detector (LMD) [11]. A dilated convolution is chosen for enhancing performance with lower complexity together with shallower and thinner architecture for low computational cost. In contrast with the DCNN, a discovery was made on a modified faster region based conventional neural network (Faster RCNN) and proposed a fast, deep convolutional neural network that surpasses the Faster RCNN but still restricted by the scale of its convolution-based map performance capability though doing good in both accuracy and efficiency in object detection [12]. A lot of network architecture was discovered later in recent years, but this deep convolutional neural network method was frequently used to outperform the traditional approaches on many applications.

A revolutionary approach to object detection is presented as YOLO which was first published in 2016 and turned out to be one of the major real-time object detection evolution known as YOLOv1 [13]. The next YOLO generation was introduced as YOLO 9000 or YOLOv2 that can detect over 9000 object classes [14]. The same developers again successfully launched YOLOv3 as the most popular and reliable version at that time [15]. YOLOv4 was then introduced with a high margin and surprising accuracy rate compared to the EfficientDet Family, which surpasses YOLOv3 [16]. The latest version in YOLO family is the PyTorch based YOLOv5 was launched by Glenn Jocher in June 2020 which outperforms previous versions in terms of COCO AP. High accuracy object detection is not the main aim of YOLOv5 as it was developed primarily for ease of use in real-time, high-quality, and persuasive object detection performance.

In the next sections, the methodology and preliminary results of our work in the development of automated detection of road lane markings of various types based on YOLOv5 will be presented. The conclusion and further works are to follow at the end of the paper.

2. Methodology

In this project, the focus is primarily on machine vision systems that can be used to identify a missing road lane marking for inspection purposes. Therefore, unlike advance driving assistance (ADAS), the deep learning model developed or used in this project does not have to work real-time. Although real-time capability is not required for this project, the detection speed is still crucial as the amount of data from tens of kms of roads will be huge. In addition to the detection of the missing line, the system should also detect and register road lane markings and classify them correctly.

The detection of the lane markings is developed by using YOLOv5 model which is one of the latest object detection deep learning models that have been developed. The use of this family is also deemed attractive as the community support is widely available and growing. YOLOv5's architecture consists of the backbone, the neck, and the head sections. The backbone is a convolutional neural network that receives the input image, extracts the image features, and generates a fixed-length representation despite the possible variation in the input image size. The neck consists of network layers that combine the image features to be fed to the next stage. The head makes the prediction of the categories of detected object.

Out of the YOLOv5 family, YOLOv5s is selected to be used in this project due to its superior speed while its detection performance is comparable to other YOLOv5 algorithms.

2.1. The Development of the Road Lane Marking Detection Algorithm

The stages of the development of the road lane marking detection algorithm based on YOLOv5s is shown in the flowchart in Figure 1. Started by building the dataset and the annotation of the images, the dataset was then used to train and test the deep learning model. Some performance metrics, which will be described later, are used to evaluate the performance of the model during training and testing. The training is done through many iterations (epochs) in order to improve the performance of the detection model. Improving metrics indicate that the YOLOv5 model would be potentially used in this application. On the other hand, if the metrics are stagnant, then it is likely that YOLOv5 is not suitable for this application.

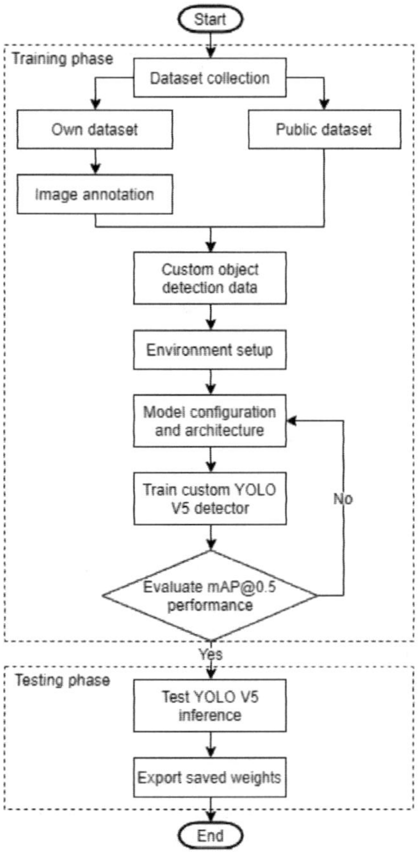

Figure 1. YOLOv5s-based model development for road lane marking detection.

2.2. Dataset and Object Classes

In this work, the dataset collection for this missing road lane marking detection project was built by using a recorded video of roads in Kuala Lumpur by using a dashboard camera. A total amount of 298 image frames were extracted from the video for building the dataset. Each image has 1280x1280 pixels. Figures 2 shows some typical samples of the images in the dataset.

Figure 2. Typical images grabbed from the video in the developed dataset collection.

2.3. Data Annotation

Annotation or labelling was performed to build the training image dataset by using a labelling tool called labelImg, which is recommended as it supports the annotation format that is compatible with the YOLOv5 model used in this missing road lane marking detection project. Figure 3 shows an annotated image to be used for training the deep learning model. In this particular image, the green box indicates a missing marking, while the yellow and purple are for single solid line and single broken line, respectively. The missing marking was purposely labelled inside the single broken white line to indicate the type of missing road lane marking.

Figure 3. An example of data annotation of different types of road lane marking. In this work, the tool called labelImg was used for annotating the images.

Table 1 shows the statistics of the annotated objects for each class. As can be seen, there are seven classes, which are single broken white line, single broken yellow line, single solid white line, single solid yellow line, double solid white line, double solid yellow line, and the missing line. The missing line is a part of the road where a road lane line marking is supposed to be present or line marking that is visibly worn out. The data will later be used for identification which parts of the road that need lane marking repainting.

Table 1. Number of Annotated Objects for All Classes

No.	Class	Number of Objects
1	SingleBrokenWhite	523
2	SingleSolidWhite	201
3	SingleSolidYellow	140
4	SingleBrokenYellow	40
5	DoubleSolidWhite	55
6	DoubleSolidYellow	58
7	MISSING	269

Custom object detection data were generated by uploading the 298-image dataset together with its annotations to an online tool called Roboflow. The dataset was split into 208 images for training, 60 (20%) for validation, and 30 (10%) for test. Then, a dataset health check was performed to make sure the class balance is relatively evenly distributed. Data augmentation is used in this study in order to increase the amount of training data Subsequently, the dataset were exported to the Google Colab in the YOLOv5 PyTorch format for training.

Figure 4 and Figure 5 show some of both the original ground truth and augmented ground truth data that are used in the training. The use of the augmented data improves the quality of the training of the model.

Figure 4. Samples of typical ground truth data used in the training.

Figure 5. Augmented training images

2.4. Model Training

The YOLOv5 repository was cloned and its dependencies were installed to set up the programming environment to be ready to run object detection training and inference commands. The minimum environment setup that is required to run the YOLOv5 lane detection training and inference command are PyTorch version 1.5, Python version 3.7 and CUDA version 10.2. Other libraries used are Numpy 1.17, Torch 1.5, Torchvision 0.6.0 and pyAML5.3.1.

The training was run with 1000 epochs. After each cycle or epoch of the training, generally metrics showed that the model became better and better. The mains metrics used are precision (P), recall ® and mAP@0.5, which is the mean Average Precision that takes the detection as positive if the intersection over union (IoU) is larger or equal to 0.5. Precision P and recall R are given by

$$P = TP/(TP+FP) \tag{1}$$

$$R = TP/(TP+FN) \tag{2}$$

, where TP = true positive, FP = false positive and FN = false negative. Recall tells us how many true positive predictions out of all actual positives, while precision indicates how many are true positives predictions out of all the positive predictions.

From the training the weights of the model that gave the best performance metrics, especially the mAP@0.5 were chosen and saved for the actual use of the model in detection of the lines and their classification.

3. Results

3.1. Training Output

The performance evaluation of the model during the training after each epoch can be seen in the plots shown in Figure 6. The plots of the recall, precision and mAP metrics show that the training has managed to develop the model to become an effective road lane marking detection model. Each of these metrics has been described briefly in the previous section. The mAP@0.5:0.95, shown in Figure 6(d), represents the average mAP at different IoU thresholds ranging from 0.5 to 0.95 in increments s of 0.05, i.e. the thresholds are 0.5, 0.55, 0.6, 0.65, 0.7, 0.75, 0.8, 0.85, 0.9, and 0.95. The growing mAP@0.5:0.95 plot also shows that the model is getting better in the detection and classification of the road lane markings or the missing ones as the training iteration is performed repeatedly.

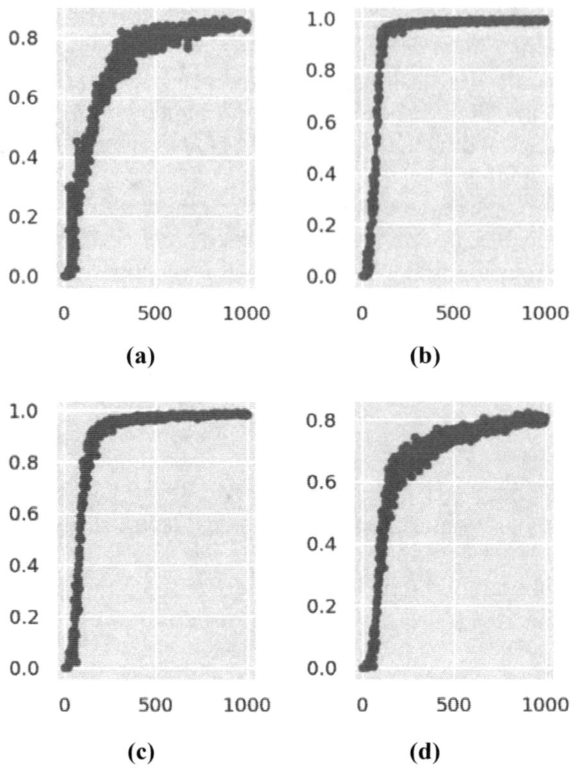

Figure 6. Performance metrics progress against the training epoch of the YOLOv5 road lane marking detection model. (a) Precision, (b) Recall, (c) mAP@0.5 and (d) mAP @0.5:0.95

The recall is specially very high at more than 0.95, while the precision is lower at just above 0.8. This means that the model can detect and recognize the types of most of the lines in the images, including if they are missing. However, the model also produces a rather high number of false positives. Nevertheless, the overall performance is still very good as shown by the high mAP, which is higher than 0.995.

Figure 7 shows the Precision-Recall Curve that shows the tradeoff between the precision and recall values when the threshold is varied. The resulting curve indicates that the model can reach a high performance, which is characterized by having both high precision and high recall values. The red box in the plot shows that it is possible to have both precision and recall that are higher than 0.9.

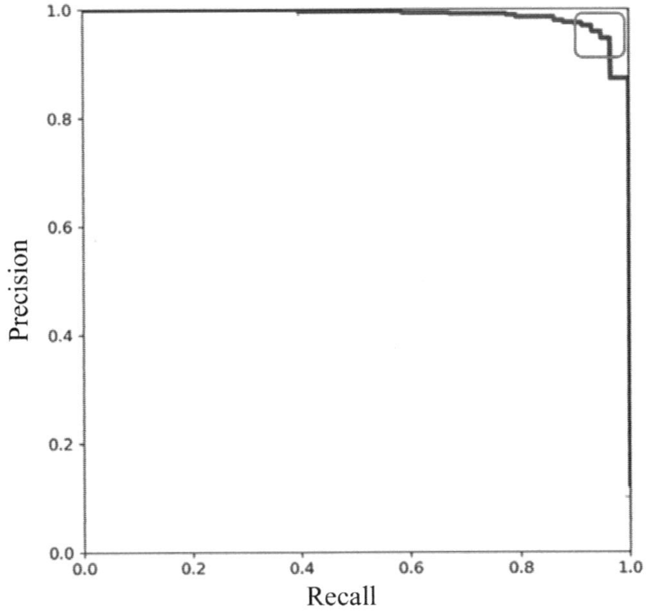

Figure 7. Precision-Recall Curve for all classes

3.2. Inference Testing on Image

The inference testing with the best weights was done by using images that had not been used in the training. Figure 8 shows some examples of the results. In these particular images, it can be clearly seen that the model has successfully detected the lines and the missing lines. It can also be seen that the missing marking is located inside of other lane marking types to indicate the type of missing marking.

This result is very promising in order for us to develop it further to design an automatic identification system of parts of the inspected road that need to have marking repainting. When this information is complemented with geolocation information, the system will be very useful and informative for the road inspection ana maintenance operators.

Figure 8. Inference testing results show that the model successfully classifies the missing line and the lines of different types.

4. Discussion

In this preliminary work, dataset was built by using a video of roads in Kuala Lumpur. A detection and classification model based on YOLOv5 was built and trained by using the dataset. From the results of the experiment that was done, the model evaluation performance of the custom YOLOv5 detector was very promising with mAP@0.5 reached higher than 0.99. This preliminary work will be continued by incorporating further algorithm in order to allow the quantification of missing lane marks along a give route and GPS data will be used for the positioning to the missing lines. It is expected that the final solution will help the implementation of automated and more informative road inspection system.

5. Conclusion

Road lane markings serve as an important navigation aid for road users during day and night which plays an important role in ensuring road in any country around the world. There is a demand on automatically detecting the missing or significantly faded road lane markings for improved road inspection and maintenance for road users' safety. The development of deep learning techniques has provided a great opportunity to make this happen. This paper has presented some preliminary results in the study of the use of YOLOv5, which is a very latest deep learning model, in the detection and classification of the missing road lane markings. The results are very promising, with a mean Average Precision (mAP@0.5) of 0.995, paving the way for the implementation of automated inspection of road lane markings by using an RGB camera. Further work will include optimization of the accuracy of detection and classification by improving the training dataset, among others.

Acknowledgement

The authors would like to thank the owner of KL Stuff channel on YouTube who has given the permission to use their video, titled "KL Drive | City Centre, 2020".

References

[1] Abu Mansor S N, Ahmad Saman M S, Tengku Razman, T M S, and Masnel H 2019 *IOP Conf. Ser. Mater. Sci. Eng.* **512** 12023
[2] Abdul Manan, M M, Várhelyi A, Çelik A K, and Hashim H H 2018 *IATSS Res.* **42**
[3] Aziz M V G, Hindersah H, and Prihatmanto A S 2017 *4th International Conference on Electric Vehicular Technology (ICEVT)* 149–153
[4] Li L, Luo W, & Wang K C P 2018 *Sensors (Switzerland)* **18**
[5] Mirunalini K, David VK 2019 *Int Journal of Research and Analytical Reviews* **6**
[6] Blaga B C Z, Deac M A, Al-Doori R W Y, Negru M, & Danescu R 2018 *Proc. - 2018 IEEE 14th Int. Conf. on Intelligent Computer Communication and Processing, ICCP 2018*, 229–236
[7] Hernández D C, Kurnianggoro L, Filonenko A, & Jo K H. 2016 *Sensors (Switzerland)*, **16**
[8] Xu Y, Shan X, Chen B Y, Chi C, Lu Z F, & Wang Y Q 2017 *7th Int. Conf. on Power Electronics Systems and Applications*
[9] Du X., & Tan K K 2016 *IEEE Transactions on Image Processing* **25** 2075–2088
[10] Ding L, Zhang H, Xiao J, Li B, Lu S, Klette R, Norouzifard M, & Xu F 2020 *Multimedia Tools and Applications* **79** 17193–17210
[11] Chen, P R, Lo S Y, Hang H M, Chan S W, & Lin J J 2019 *International Conference on Digital Signal Processing*
[12] Tian Y, Gelernter J, Wang X, Chen W, Gao J, Zhang Y, & Li X. 2018 *Neurocomputing* **280** 46–55
[13] Redmon J, Divvala S, Girshick R, & Farhadi A. 2016 *Proc. of the IEEE Computer Society Conference on Computer Vision and Pattern Recognition* 779–788
[14] Redmon J, & Farhadi A. 2017 *30th IEEE Conference on Computer Vision and Pattern Recognition, CVPR 2017* 6517–6525
[15] Redmon J, & Farhadi A 2018 *arXiv* 1804.02767
[16] Bochkovskiy A, Wang, C Y & Liao H Y M 2020 *arXiv* 2004.10934

IOP Conf. Series: Materials Science and Engineering 1244 (2022) 012022 doi:10.1088/1757-899X/1244/1/012022

Detection of unbalance and looseness faults in a ventilation turbine using vibration signature analysis

DJEMANA. Mohamed[1], BENYOUCEF. Rania[2], HRAIRI, Meftah[3]

[1]École Supérieure de Technologies Industrielles, Safsaf City, BP 218 23000 Annaba, Algeria

[2]Laboratoire D'automatique et de Productique, University of Batna, Algeria.

[3]Department of Mechanical Engineering, International Islamic University Malaysia, P.O. Box 10, 50728 Kuala Lumpur, Malaysia

meftah@iium.edu.my

Abstract. Heavy industry, which generally uses turbomachines, often uses both technologies to perform high-performance vibration monitoring of its production tool. Vibration analysis is one of the means used to monitor the health of rotating machinery in operation. This is part of a policy of forecast maintenance of the industrial production tool. This work is part of the monitoring and diagnosis of rotating machines by vibration analysis taking as an example the X205 circulation fan. Numerical simulation was done to test the capabilities and limitations of a dynamic simulation. The modelling of vibratory phenomena is developed using SolidWorks software to perform a dynamic simulation of the 3D model. The numerical simulations were performed to find the effect of different types of defects, such as defect of unbalance fin fan and bolt joint looseness, on the output of system. The numerical results were confirmed either by experimentally using employing results available in the open literature that existent at factory of cement plant (SCHS) and a good agreement was observed. The proposed and derived model has demonstrated the viability of dynamic simulation approach to rotating machines by vibration analysis and serves as a significant alternative approach to the direct experimentation on the same systems in terms of cost and time.

1. Introduction

Multiple fields of industry employ rotating machines. When these machines' working conditions are continuously monitored, the threat of catastrophic failures can be avoided. One method of condition monitoring is known as vibration analysis, and it is highly accurate and cost-effective [1]. The practice of vibration analysis is based on a solid theory foundation and has been in use long enough to have matured as a measurement tool [2]. It should be noted that rotating machinery typically generates a complex high frequency signal that is the result of many superimposed components, each of which can have different amplitude, non-linearity, and non-stationarity characteristics [3]. This complexity requires more robust and effective ways to separate individual vibration signals from the whole.

Fault diagnosis has recently become the focus of researchers. Gao et al. [4,5] performed a systematic review of four fault-diagnosis approaches namely, knowledge-based, signal-based, model-based and hybrid/active. The key to a method's success often relies on three steps: fault symptom determination, sensitive feature extraction, and condition pattern classification. Tayarani-Bathaie et al. [6] proposed a dynamic neural network that would perform the diagnosis of gas turbine faults. Their artificial neural network and empirical mode decomposition were used to identify automatic bearing faults apparent within vibration signals.

Qin et al. [7] proposed a concurrent fault diagnosis method based on Bayesian discriminating and time series analysis with dimensionless parameters. They applied their method to a centrifugal multi-level impeller blower with six states, including single and simultaneous defects in gear and bearing. Asr et al. [8] performed a combined bearing and gear fault diagnosis of the gearbox system using empirical mode decomposition (EMD), statistical features, and a non-Naive Bayesian classifier. Pan et al. [9] performed a compound fault detection of the bearing outer race and the cracked tooth of gear using the proposed symplectic geometry mode decomposition method.

The development of computer tools and numerical techniques has meant that numerical simulation now occupies a very important place and has established itself as a means that is both effective and economically inexpensive in comparison with experimental testing [1], numerical simulation is one of the tools for simulating real phenomena. It is a process whereby a computer program is run on a computer to simulate a complex physical phenomenon. It is to be noted that, most of the existing works have considered the rotating machine faults within the context of experimental set up but none have taken into consideration the effect of bolt joint looseness of fixed bearing and examine that numerically. Therefore, the results presented here are of unique significance and novelty.

In this work, this method is used in order to build a model that will be used to predict and detect efficiently the anomalies of a ventilation turbine and that will offer us the advantage of optimizing the potential of the latter. To this end, a dynamic simulation under the same conditions as the real model has been carried out, where the vibratory signal is the source of information retained to inform us about the state of this machine and the experimental validation of the numerical model. Firstly, we are interested in reproducing the unbalance fault which is the main cause of the faults affecting this equipment, and which is characterized by a mass imbalance around the axis of rotation which produces centrifugal forces generating vibrations at the level of the bearings likely to accelerate their degradation. Initially, any part has a greater or lesser degree of un-balance, the main causes of which may be machining or assembly faults or asymmetric heating of the rotor during operation [10]. Secondly, we are interested in reproducing the tightening defect, which can be the result of an initial unbalance [11]. In the present study, the vibratory information characterizing these faults is acquired for the un-balance by two methods, the experimental method and the numerical simulation, and for the loosening by a numerical simulation.

2. Vibration analysis

Vibrations are common phenomenon in rotating machinery, which could carry important information about condition of the rotating machinery. All specific failures in rotating machinery have their own characteristics of vibrations. Measurements and analysis of the vibratory behavior of the system and vibration changes leads to the detection of problems and faults [12,13]. Almost all machine vibrations are due to one or more reasons like unbalance, eccentricity, misalignment, looseness, belt-drive problems, gear defects, bearing defects, electrical faults, resonance.

Unbalance fault has been classified as one of the most common causes of vibration in rotating machines [14]. Machinery vibrations due to unbalance fault is usually characterized by a dominant peak at the fundamental rotational frequency (1x RPM), which usually changes in proportion to the square of the rotational speed and in the radial direction. The total elimination of unbalance fault in rotating machines is almost impracticable, due to the difficulties associated with achieving perfection in the manufacture of components as well as their installation. Based on this premise, a significant number of research have been centered on estimating unbalance and its correction.

Mechanical looseness can either occur in the form of internal looseness between the machine and its base mounting, internal looseness of the machine components or structural looseness. Internal looseness occurs due to lack of proper fit between the components of a machine (bearing-to-bearing housing, shaft-to-bearing, coupling-to-shaft, etc.), which consequently leads to the excitation of numerous harmonics of the machine running speed. These harmonics are usually generated due to combined effects of exciting forces from the rotor and the non-linear response from the loose components [15]. Looseness between machine and base mounting is commonly the resultant of a crack in the base structure or bearing pedestal, which may generate high 2x RPM component and some harmonics. Structural looseness on the other hand occurs when there is a weakness in the foundation or looseness in the supporting structure.

The presence of mechanical looseness in a machine train is often characterized by chaotic response, which have been observed to sometimes excite multiples of $\frac{1}{2}$x or $\frac{1}{3}$x RPM [16]. Other studies on mechanical looseness in rotating machines such as those related to bearing caps or supports have also been seen to have large numbers of harmonics and sub-harmonics, which were dependent on the analysis direction and point [17]. The vibration signatures produced by a rotating machine with mechanical looseness have also indicated that the vibration amplitude at the higher order frequency region will usually be greater than half of the vibration amplitude caused by the rotational speed, which will persist after machine balancing operations [18].

3. Methodology

In this work, the experimental and simulation analysis were carried out to predict and detect unbalance and looseness faults in an industrial fan. The modelling and simulation analysis were performed by SolidWorks software. On the experimental part, the vibration measurement on the rotating machinery were performed using accelerometer and tachometer. Validation of simulation and experimental results was carried out for the detection of unbalance fault.

3.1. Experimental procedure using hardware

For this study, we use the most common conditional maintenance approach for the diagnosis of rotating machines, vibration analysis. This technique provides a global view of the state of equipment and to detect possible malfunctions and monitor their evolution [19].

This application concerns the X205 industrial fan, which consists of an asynchronous motor (980 rpm), a coupling and an impeller with blades mounted on a shaft with two bearings, fitted with cylindrical roller bearings. The installation is shown in Figure 1.

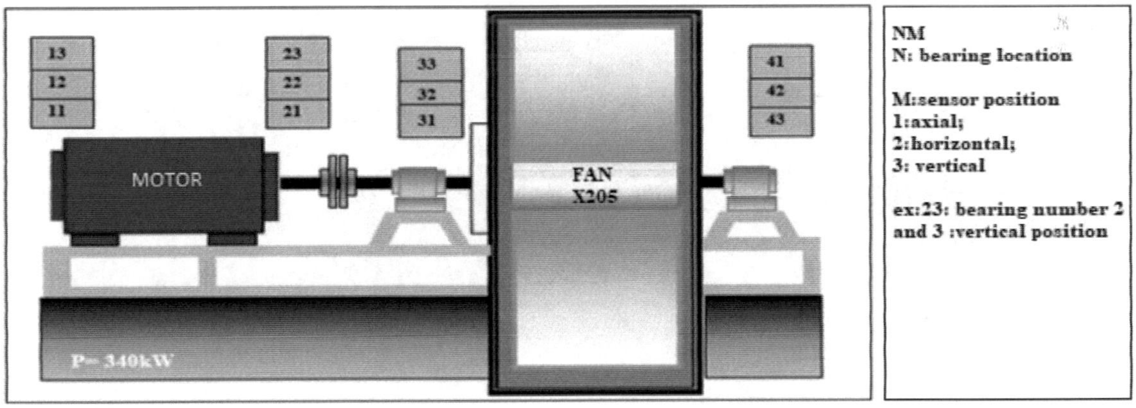

Fig. 1 Block diagram of the fan chain.

The acquisition of the vibration signals was carried out using a system consisting of a FALCON WLS wireless triaxial transducer associated with a FALCON analyzer collector. The software used for the vibration analysis was OneProd XPR300. Twelve measurement points were set up in all three directions in order to evaluate the behavior of the fan.

IOP Conf. Series: Materials Science and Engineering 1244 (2022) 012022 doi:10.1088/1757-899X/1244/1/012022

3.2. Numerical simulation

The numerical simulation was carried out using the SolidWorks Motion add-on, which is an integrated module in SolidWorks that allows the dynamic animation or simulation of motorized or externally forced assemblies by considering predefined contact conditions between the assembled components, which are assumed to be rigid.

The use of SolidWorks motion for the analysis of mechanisms consists of three steps: model generation, analysis (or simulation) and visualization of the results, as illustrated in Figure 2 [20].

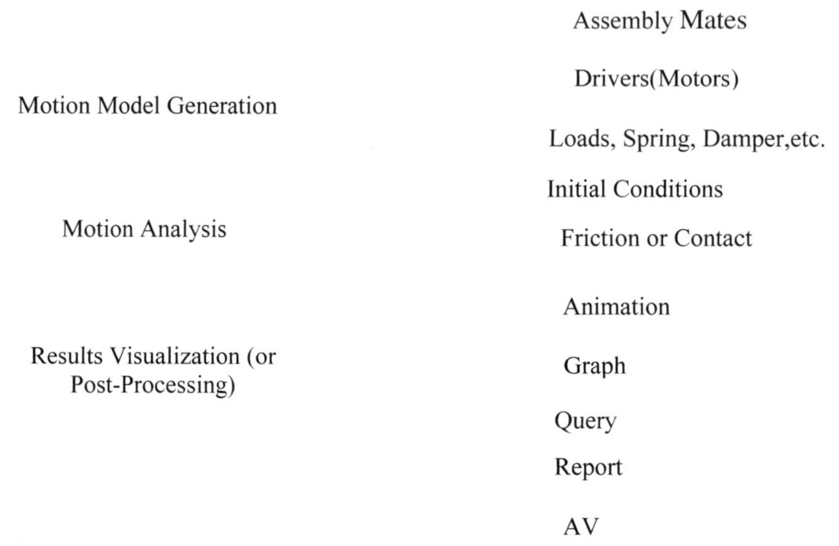

Fig. 2 General process of using SolidWorks Motion

The modelling concerns a part of the fan chain consisting of two roller bearings, a shaft, and an impeller. Figures 3 and 4 shows the modelling of the different parts of the fan chain.

Fig. 3 Exploded view of the fixed bearing

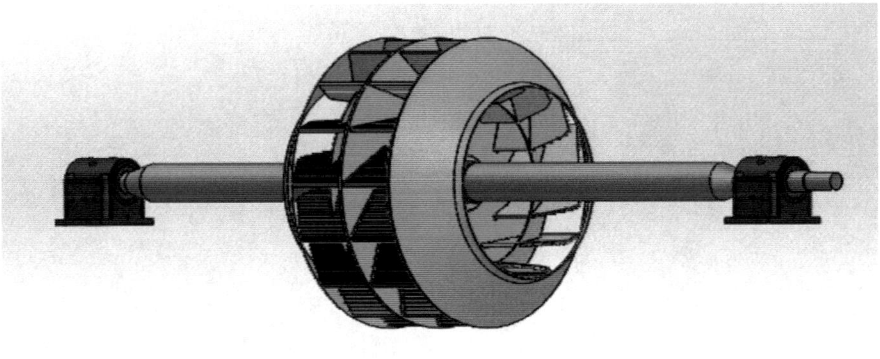

Fig. 4 Assembly of components

In the case of the fan, the unbalance phenomenon presented in the centrifugal impeller due to dust or wet flour and is superimposed non homogeneously on the impeller blades. The unbalance fault modelled by modelling dust bars on the impeller blades as shown in Figure 5.

Fig. 5 Dust distribution on the turbine

Time signals of linear acceleration were generated from a dynamic simulation of the motion performed with the assembled model, a circular motor was applied to affect the motion to the turbine with a progressively increasing rotational speed reaching a constant angular velocity of 102.63 rad/s, for an estimated total motion simulation time of 6s. Figure 6 shows the velocity curve applied to the motor.

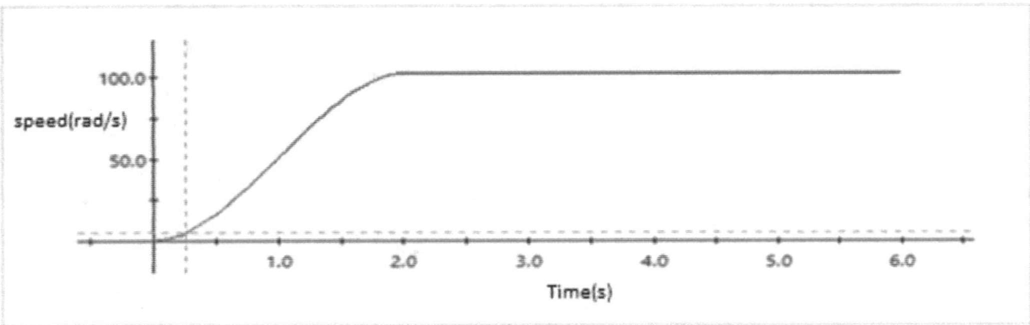

Fig. 6 Progressive speed curve

We have introduced the predefined contact conditions in SolidWorks motion for the lubricated contact of the bearing components Outer ring, Inner ring and rollers as shown in Figure 7.

IOP Conf. Series: Materials Science and Engineering 1244 (2022) 012022 doi:10.1088/1757-899X/1244/1/012022

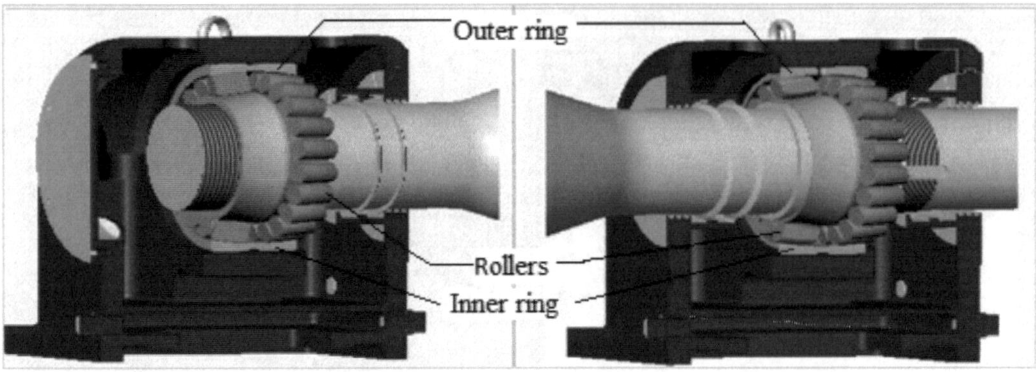

Fig. 7 Cross-section of contacting components

Table 1 Properties of contacts conditions

Elastic properties (Impact)	
Stiffness	100000.00 N/mm
Exponent	1.5
Max. damping	49.91566312 N/(mm/s)
Penetration	0.10 mm
friction	
v_k	10.16 mm/s
μ_k	0.05
Static friction	
v_s	0.10 mm/s
μ_s	0.08

Once the acquisition is complete, the signals are imported into MATLAB, so that their Fourier transform can be performed and compared with the experimental signals.

In order to model the clamping failure, we assumed that there is a clearance between the two bearing parts that occurs while the fan is in motion. We modelled this clearance by changing the stresses on the perfect bearing; to maintain a clearance between 0 mm and 0.5 mm when the fan is rotating, which in real life conditions embodies that the screws are loose (Figure 8).

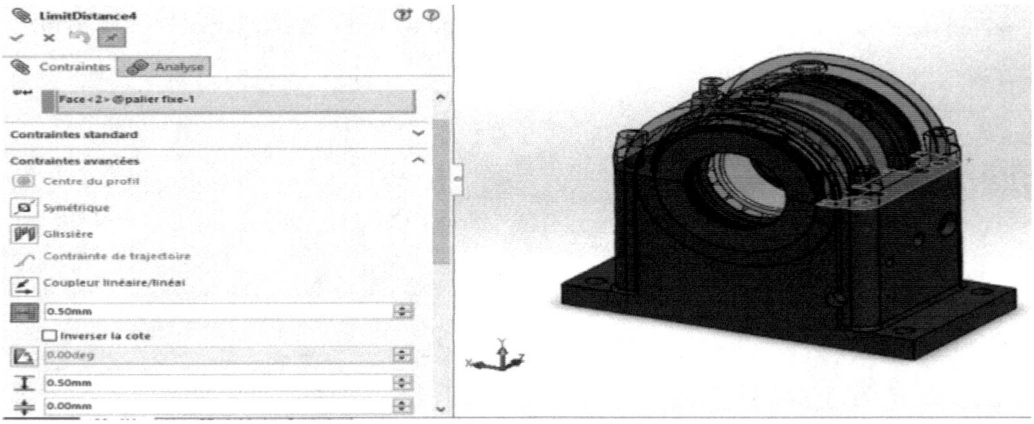

Fig. 8 The modelling of the clamping fault at the fixed bearing.

4. Results and discussion

4.1. Unbalance of fin fan

Figure 9 shows the spectrum recorded in point 32 (shown in fig.1) which represents the horizontal direction of bearing 3 showing a typical unbalance fault manifested by a high vibration amplitude at the rotation frequency (16.62 Hz).

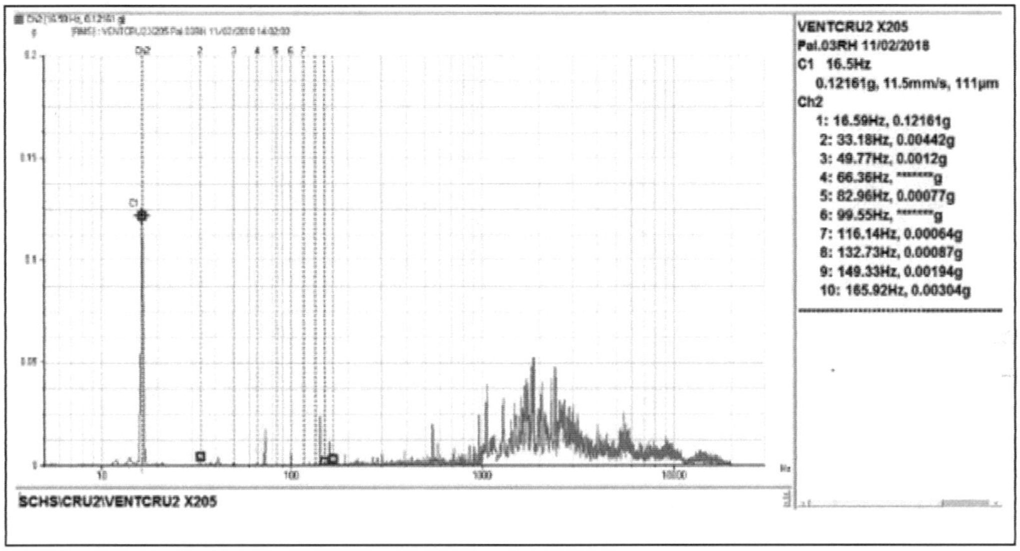

Fig. 9 The experimentally measured spectrum on bearing.

The results of the dynamic simulation are shown in figure 10 and 11 display the comparison between the experimental and simulated spectra.

From the results shown in Figure 11, it can be seen that the FFT of the experimental model and the simulated FFT, are correctly superimposed, the peaks at the rotational frequency (16.33 Hz) which characterize the unbalance defect coincide with a decrease in the vibratory amplitude visible in the simulation spectrum and the absence of some peaks. The differences observed can be explained by the external conditions that may affect the operation of the equipment (wind, temperature, etc.) and which are not considered in this study.

IOP Conf. Series: Materials Science and Engineering 1244 (2022) 012022 doi:10.1088/1757-899X/1244/1/012022

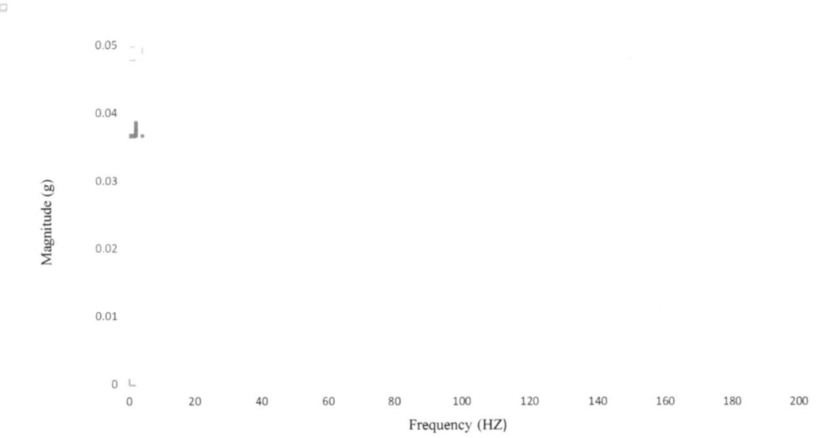

Fig. 10 Simulated spectrum of the unbalance fault.

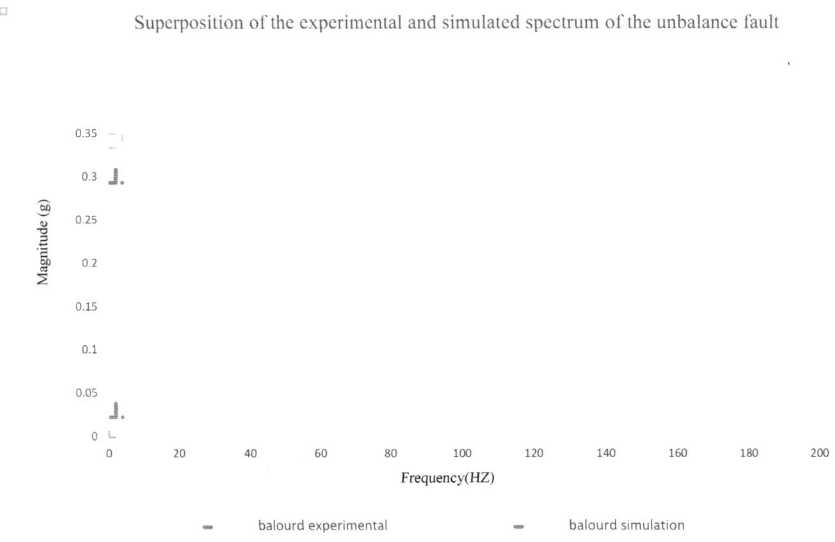

Fig. 11 Comparison of experimental and simulated FFTs.

one of the more commonly used indices include the correlation coefficient deviation metric (CCDM). When a fan chain is considered healthy, baseline, or reference, vibration signatures are taken. These are compared to later signatures to assess system health.

The correlation coefficient is the basis for the second index, CCDM, and is calculated by

$$CCDM = 1 - C_C$$

Where CC is the correlation coefficient calculated using vibration signatures for the fan chain under healthy and damaged conditions, as defined before, in the same frequency range. It's calculated using the following equation:

$$C_C = \frac{\sum_{k=\omega_i}^{\omega_F} \left[Z_{E,H}(k) - \bar{Z}_{E,H} \right] \left[Z_{E,D}(k) - \bar{Z}_{E,D} \right]}{\sqrt{\sum_{k=\omega_i}^{\omega_F} \left[Z_{E,H}(k) - \bar{Z}_{E,H} \right]^2} \sqrt{\sum_{k=\omega_i}^{\omega_F} \left[Z_{E,D}(k) - \bar{Z}_{E,D} \right]^2}}$$

Comparing the results by measuring the degree of intercorrelation between the two signals corresponding to the experimental and numerical models gave us a similarity rate of 85%. This level of correlation between the two models confirms the good modelling of the fan chain and the assumptions of the dynamic simulation model and is a good start for further test studies.

4.2. Bolt joint Looseness of fixed bearing

The typical spectrum of a mis-tightening measured on a machine in which there is a backlash contains a large number of peaks at frequencies multiple of the rotational frequency (up to 10th order). It is also sometimes possible to find peaks at ½ harmonic (1/2 x rotational frequency) and its multiples at a level that is much lower than the harmonic of the fundamental frequency. [21]

A clamping fault in fixed bearing represented in figures 12 and 13. An absence of some peaks which can be explained by external conditions such as environment (temperature, humidity, noise..etc.) that are not taken into account in this study.

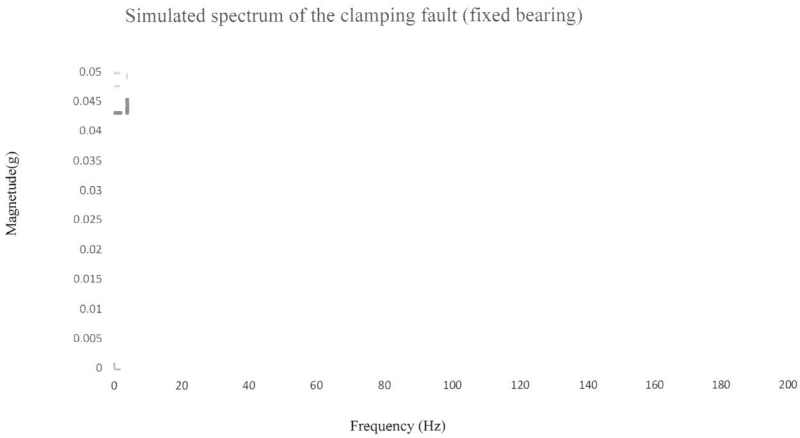

Fig. 12 Simulated loosening spectrum (Fixed bearing).

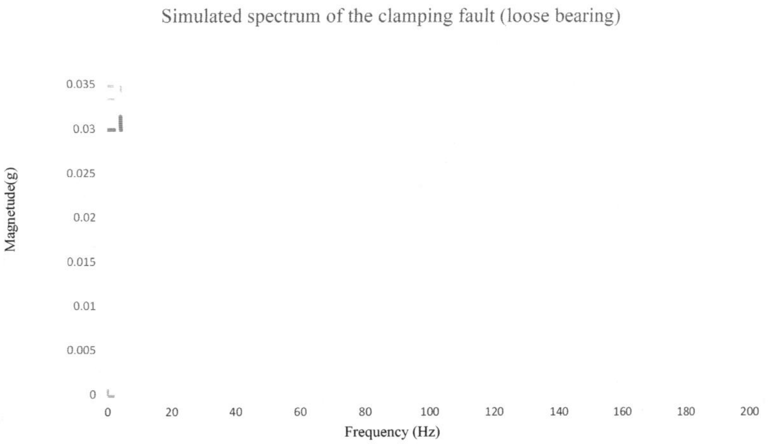

Fig. 13 Simulated loosening spectrum (Floating bearing).

IOP Conf. Series: Materials Science and Engineering 1244 (2022) 012022 doi:10.1088/1757-899X/1244/1/012022

5. Conclusion

The activity of validating models and simulations has become an essential activity, to ensure the credibility that can be given to the results provided by the simulation. We have created a dynamic simulation of the 3D model of the X205 circulation fan. So that it can be used as a decision support tool in the future, helping us to predict possible faults that may occur in the future.

The numerical simulation approach of the unbalance fault has given quite satisfactory results specific to the fan studied. Although this approach is far from covering all the physical phenomena that can exist in rotating machines, it remains quite promising and opens up an interesting perspective in terms of learning about the faults that can occur, saving time and money and making it easier to adapt to the various rotating machines that exist in industry. This approach, therefore, deserves to be explored for other faults and in other machines, and may one day replace learning by the classic methods (test bench and feedback).

References

[1] Claeys, M., Sinou, J-J., Lambelin, J-P., Todeschini, R. Modal interactions due to friction in the nonlinear vibration response of the “Harmony” test structure: Experiments and simulations, Journal of Sound and Vibration, vol., 376, August, 2016, 131-148.

[2] Ben Ali, J., Saidi, L., Harrath, S., Bechhoefer, E., & Benbouzid, M. (2018). Online automatic diagnosis of wind turbine bearings progressive degradations under real experimental conditions based on unsupervised machine learning. Applied Acoustics, 132, 167–181

[3] Zhao, M., Jiao, J., & Lin, J. (2019). A data-driven monitoring scheme for rotating machinery via self-comparison approach. IEEE Transactions on Industrial Informatics, 15(4), 2435–2445

[4] Gao, Z.; Cecati, C.; Ding, S.X. A survey of fault diagnosis and fault-tolerant techniques—Part I: Fault diagnosis with model-based and signal-based approaches. IEEE Trans. Ind. Electron. 2015, 62, 3757–3767.

[5] Gao, Z.; Cecati, C.; Ding, S.X. A survey of fault diagnosis and fault-tolerant techniques—Part II: Fault diagnosis with knowledge-based and hybrid/active-based approaches. IEEE Trans. Ind. Electron. 2015, 62, 3768–3774.

[6] Tayarani-Bathaie, S.S.; Vanini, Z.N.S.; Khorasani, K. Dynamic neural network-based fault diagnosis of gas turbine engines. Neurocomputing 2014, 125, 153–165.

[7] Qin, A., Hu, Q., Lv, Y., & Zhang, Q. (2019). Concurrent Fault Diagnosis Based on Bayesian Discriminating Analysis and Time Series Analysis with Dimensionless Parameters. IEEE Sensors Journal, 19(6), 2254–2265. https://doi.org/10.1109/ JSEN.2018.2885377.

[8] Asr, M. Y., Ettefagh, M. M., Hassannejad, R., & Razavi, S. N. (2017). Diagnosis of combined faults in Rotary Machinery by Non-Naive Bayesian approach. Mechanical Systems and Signal Processing, 85, 56–70. https://doi.org/10.1016/j. ymssp.2016.08.005.

[9] Pan, H., Yang, Y., Li, X., Zheng, J., & Cheng, J. (2019). Symplectic geometry mode decomposition and its application to rotating machinery compound fault diagnosis. Mechanical Systems and Signal Processing, 114, 189–211. https://doi.org/10.1016/j. ymssp.2018.05.019.

[10] Boulenger, A. Pachaud, C. Analyse vibratoire en maintenance : surveillance et diagnostic des machines, Dunod/L’Usine Nouvelle, Paris : 2007.

[11] Fouzi, B. (2019). Diagnostic et détection des défauts mécaniques affectant les systèmes électromécaniques (Doctoral dissertation, UNIVERSITÉ BADJI MOKHTAR-ANNABA).

[12] Piotrowski, J. Shaft Alignment Handbook, third ed., Taylor and Francis Group, Boca Raton, 2007.

[13] Friswell, M. Models for the condition monitoring of rotating machinery. Proceedings of the International conference "Condition Monitoring", Cambridge, UK, 2005, pp.385-392.

[14] Sinha, J.K., Lees, A.W., Friswell, M.I. Estimating unbalance and misalignment of a flexible rotating machine from a single run-down, J. Sound Vib., vol. 272, no. 3–5, pp. 967–989, 2004.

[15] Scheffer, C., Girdhar, P. Practical machinery vibration analysis & predictive maintenance, First Edition. Oxford: Elsevier, 2004.

[16] Lu, W.X., Chu, F.L. Experimental investigation of pedestal looseness in a rotor-bearing system, Key Eng. Mater., vol. 413–414, pp. 599–605, 2009.
[17] Betta, G., Liguori, C., Paolillo, A., Pietrosanto, A. A IMP-based FFTanalyzer for the fault diagnosis of rotating machine based on vibration analysis, IEEE Instrum. Meas. Technol. Conf., pp. 572–577, 2001.
[18] Li, X. The analysis of vibration fault features and vibration mechanism caused by rotating machinery loosening," Adv. Mater. Res., vol. 518–523, pp. 3826–3829, 2012.
[19] Augeix, D. Analyse vibratoire des machines tournantes , Techniques de l'ingénieur Vibrations en milieu industriel, mesures, surveillance et contrôle, vol., TIB424DUO, Janvier 2001, bm5145.
[20] Chang, K-H. Motion Simulation and Mechanism Design with SolidWorks Motion 2013, SDC Publications, 2014, 132.
[21] Heng, J. (2002). Pratique de la Maintenance préventive» édition Dunod.

Machine Learning Model to Predict Slump, VEBE and Compaction Factor of M Sand and Shredded Pet Bottles Concrete

A.M Nadimalla*, S.A Masjuki, S.A Saad, M Ali

Department of Civil Engineering, International Islamic University Malaysia, 53100, Malaysia

altamashk1987@gmail.com

Abstract. A measure of how easy it is to transport, cast, compact and surface finish of fresh concrete without any segregation is directly proportional to workability of concrete. The consistency of fresh concrete indirectly measures its workability. VeBe test, slump test and compaction factor test are some of the methods that can be used to study the consistency of concrete. A quick determination of the concrete strength and fresh properties is an important element on large construction sites for mass concreting to save time and cost. It is important to develop a machine learning model to predict the strength and workability of concrete with the help of available data. This paper aims to develop the machine learning model with the help of experimental data such as slump, VeBe and compaction factor of concrete incorporated with shredded Polyethylene Terephthalate (PET) bottles, Manufactured Sand (M-sand) and River sand as fine aggregates replacement in concrete mixtures. The machine learning model is developed by using different machine learning techniques such as Multiple Linear Regression (MLR) and Decision Tree Regression (DTR). As per the results obtained the DTR model is performed well than MLR model.

1. Introduction

Workability is the property of concrete that dictates the effort needed for placement, compaction and finishing with minimal homogeneity loss. A quick determination of the concrete strength and fresh properties is an important element on large construction sites for mass concreting(Charhate, Subhedar, and Adsul 2018). This paper aims to ease the process of determination of fresh concrete properties while maintaining cost-effective material, its workability, life, etc. with the help of machine learning techniques with the help of available experimental data.

Machine learning is an AI type that allows a system to learn from data instead of explicit programming. Machine learning, however, is not an easy task. As the algorithms accumulate training data, precise models based on these data can be generated. The output provided by the machine learning algorithm is a machine learning model(Cihan 2019). Machine learning is used for prediction most often. The categorical variable values prediction is called classification, while the numerical variable value prediction is referred to as regression. Regression analyses the relationship between one or more independent and dependent variable(Cihan 2019). Chithra et al., 2016 develop the machine learning model to predict compressive strength and slump of high-performance concrete at curing ages of 1, 3, 7, 28, 56 and 90 days. The experimental data modelled using both Artificial Neural Networks and

Multiple Regression Analysis. Artificial neural networks have shown greater precision (Chithra et al. 2016).

Silva et al., 2020 generated a machine learning model with the help of literature data to predict the compressive strength of concrete using algorithms Artificial Neural Network (ANN), Random Forest, and Support Vector Machine (SVM). The results demonstrate that the Random Forest performed most effectively(Silva, Moita, and Arruda 2020). Dantas et al., 2013 developed a machine learning model using Artificial Neural Networks (ANNs) models with the help of data obtained from the literature. To predict the compressive strength of concretes containing Demolition Waste (CDW) at the age of 3, 7, 28 and 91 days. The results in the training and testing phases illustrate the potential use of ANN to predict compressive strength(Dantas, Batista Leite, and De Jesus Nagahama 2013). Khademi et al., 2017 developed models artificial neural network, multiple linear regression, and ANFIS models to predict the 28 days compressive strength of 173 different concrete specimens. The compressive strength was influenced by different concrete characteristics such as cement, gravel in kilograms, 3/4 sand, 3/8 sand, water-cement ratio and fineness modulus of sand. It is concluded that the ANFIS and ANN models could predict the 28 days compressive strength of concrete more accurately.

Furthermore, it is noticed that the ANN model could perform better than the ANFIS model in terms of R^2(Khademi et al. 2017). Cherickal et al., 2021 develop a machine learning models to predict the compressive strength of concrete by applying different regression techniques such as support vector regression, linear regression, random forest regression and decision tree regression on assumed data set. It was found that the accuracy of the random forest regression was more promising as per the outcome after applying several regression techniques(Cherickal et al. 2021). Bhanu P. Koyaa, 2021 develop machine learning models to predict six mechanical properties of concrete, namely Poisson's ratio, compressive strength, modulus of elasticity, modulus of rupture, coefficient of thermal expansion and splitting tensile strength are assessed by applying five different Machine Learning algorithms Decision Tree, Support Vector Machine, Linear Regression, Gradient Boosting and Random Forest models on the database of Wisconsin concrete mixes. These Machine Learning models were used to identify the most appropriate model that can persistently predict the mechanical properties of concrete. Coefficient of determination (R^2) and Root mean squared error (RMSE) results showed that the Support Vector Machine performed well than the other models applied on the datasets(Bhanu P. Koyaa, Sakshi Anejab 2021).

Many investigators have used the combination of different data for modelling the strength of concrete; only a few investigators have used the data to model the workability of concrete, such as slump. The main purpose of this study was to predict slump, VeBe, and compaction factor using Multiple Linear Regression and the Decision Tree Regression algorithms.

The Multiple Linear Regression and Decision Tree Regression models were trained to predict the slump, VeBe, and compaction factor. In addition, the mean absolute error (MAE), Mean Squared Error (MSE), and Root Mean Squared Error (RMSE) metrics were used to compare the models for prediction performance.

2. Materials and Methods

The testings for fresh properties (workability) of concrete were performed under laboratory conditions. Three concrete batches were prepared. The first batch was partially replaced with shredded PET bottle (SPB) Aggregate, the second batch partially replaced with M-sand and the third batch was partially replaced with M-sand with 1.5% incorporation of SPB as fine aggregate replacement. The proportion of SPB were 0.5%, 1%, 1.5% and 2%. The proportion of M-sand and M-sand with 1.5% incorporation SPB for both mixes were 25%, 50%, 75% and 100%. The mix design as per the Department of Environment (DOE) method. The fresh properties of concrete mixtures were determined by using the slump test, VeBe test and compaction factor test as per BS standards (Nadimalla et al. 2021). These three tests are used to determine the workability of concrete.

Figure 1. Concrete mixing

Figure 2. Mixed Concrete

3. Fresh Concrete Test (Workability tests)

The workability of each concrete sample in its fresh form in this project is tested using slump test, VeBe test, and compaction factor test in line with BS EN 12350-2:2009. These tests are necessary as it is a way to determine the consistency of each fresh concrete sample.

3.1. Slump Test

The slump test is conducted right after mixing. Each batch of concrete mix is tested by first filling the slump cone in three layers. After each layer is filled, the fresh concrete is compacted by tampering it 25 times with a tampering rod. The slump cone is then slowly removed to allow the concrete to recede. Finally, the distance by which the concrete fell is measured by using a scale.

3.2. VeBe Test

Just like slump test, VeBe test for each batch of concrete mix is tested after mixing. Fresh concrete is poured into a slump cone similar to slump test which is inside a metal cylindrical pot of the VeBe consistometer. The slump cone is removed and the swivel arm of the consistometer is adjusted so that the glass disk which is attached to it is rotated and placed on top of the fresh concrete. Next, the electrical vibrator is turned on and the time taken for the concrete to take up the shape of the cylindrical pot is measured. The time measurement is stopped when the glass disk is no longer transparent. Lastly, the electrical vibrator is turned off.

3.3. Compaction Factor Test

The compaction factor test is carried out after mixing just as slump test and VeBe test for all concrete mixes. Firstly, the trapdoors of the upper and lower hopper of the compaction factor apparatus are checked to make sure both are closed. Secondly, the upper hopper is filled with fresh concrete and levelled. Next, the trapdoor is opened to allow the concrete to fall into the second hopper. A tampering rod is used to tamper the concrete to fall through the hoppers in case it does not flow. After that, the trapdoor of the second hopper is released so that the concrete falls in to the cylinder of the compaction factor apparatus. Excess concrete of the cylinder is then removed. Subsequently, the cylinder containing the fresh concrete is weighted and subtracted by the known weight of the cylinder to get the weight of partially compacted concrete. As for the weight of fully compacted concrete, the concrete inside the cylinder is removed and filled back in three layers, with each layer tampered 25 times via a tampering rod for the purpose of compaction. Furthermore, it is weighted and subtracted by the known value of the cylinder to calculate the weight of fully compacted concrete. Finally, the compaction factor is found which is the ratio of partially compacted concrete to fully compacted concrete.

4. Machine Learning Techniques (Methodology)

In this research paper, MLR (Multiple Linear Regression) and DTR (Decision Tree Regression) models have been developed with the help of python application to predict concrete slump, VeBe and compaction factor. The experimental dataset used for this study consist of three input variables such as

River sand, SPB aggregates and M sand and three output variables, such as slump, VeBe and compaction factor of concrete as shown Table 2. The data obtained from experimental results used for the model training, 80% in the training phase and 20% in the testing phase. The MAE (Mean Absolute Error), MSE (Mean Squared Error), and RMSE (Root Mean Squared Error metrics used to measure the prediction accuracy, as shown in Table 1 and Figure 6. The evaluation of MAE, MSE, and RMSE metrics will conclude that which is the best suitable prediction model.

Figure 3. Compaction Factor Test

Figure 4. VeBe test

Figure 5. Slump Test

The machine learning model is developed to predict slump, VeBe and compactor factor with the help of experimental data using Multiple Linear Regression and Decision Tree Regression algorithms.
The most common form of linear regression analysis is Multiple Linear Regression. It is used to explain the relationship between one continuous dependent variable and two or more independent variables. The following is the equation for multiple regression.

$$y_s = -0.36682s + 0.76692p - 0.40010m + 217.79089 \quad (1)$$

Where y is the dependent variable and $x_{1....}x_n$ are the independent variables.
Decision tree used for classification and regression problems. A decision tree follows a set of if-else conditions to predict the data and classify it according to the conditions.

5. Results and Discussion

Table 1. Comparison of MLR & DTR model testing results for the slump, VeBe and compaction factor

Workability	Model	MAE	MSE	RMSE
Slump	MLR	3.7802	20.8538	4.5666
	DTR	1.3125	2.0312	1.4252
VeBE	MLR	2.6296	9.0925	3.0153
	DTR	0.3937	0.1828	0.4275
Compaction Factor	MLR	0.0328	0.0013	0.0364
	DTR	0.02625	0.0008	0.0285

IOP Conf. Series: Materials Science and Engineering 1244 (2022) 012023 doi:10.1088/1757-899X/1244/1/012023

Table 2. Experimental data to develop machine learning model.

Input Variables			Output Variables		
sand	PET	M sand	Compaction Factor	VeBe	Slump
531	0	0	0.93	6	31
528.345	2.655	0	0.9	6.94	30
525.69	5.31	0	0.88	6.66	30
523.035	7.965	0	0.87	7.2	29
520.38	10.62	0	0.85	7.2	27
398.25	0	132.75	0.86	12	13
265.5	0	265.5	0.83	15	7
132.75	0	398.25	0.81	20	6
0	0	531	0.76	26	5
390.285	7.965	132.75	0.91	6.39	28
257.535	7.965	265.5	0.9	7.32	26
124.785	7.965	398.25	0.88	7.69	26
531	0	0	0.95	5.7	32
528.345	2.655	0	0.92	6.64	31
525.69	5.31	0	0.9	6.36	31
523.035	7.965	0	0.89	6.9	30
520.38	10.62	0	0.87	6.9	28
398.25	0	132.75	0.88	11.7	14
265.5	0	265.5	0.85	14.7	8
132.75	0	398.25	0.83	19.7	7
0	0	531	0.78	25.7	6
390.285	7.965	132.75	0.93	6.09	29
257.535	7.965	265.5	0.92	7.02	27
124.785	7.965	398.25	0.9	7.39	27
531	0	0	0.91	6.3	30
528.345	2.655	0	0.88	7.24	29
525.69	5.31	0	0.86	6.96	29
523.035	7.965	0	0.85	7.5	28
520.38	10.62	0	0.83	7.5	26
398.25	0	132.75	0.84	12.3	12
265.5	0	265.5	0.81	15.3	6
132.75	0	398.25	0.79	20.3	5
0	0	531	0.74	26.3	4
390.285	7.965	132.75	0.89	6.69	27

257.535	7.965	265.5	0.88	7.62	25
124.785	7.965	398.25	0.86	7.99	25

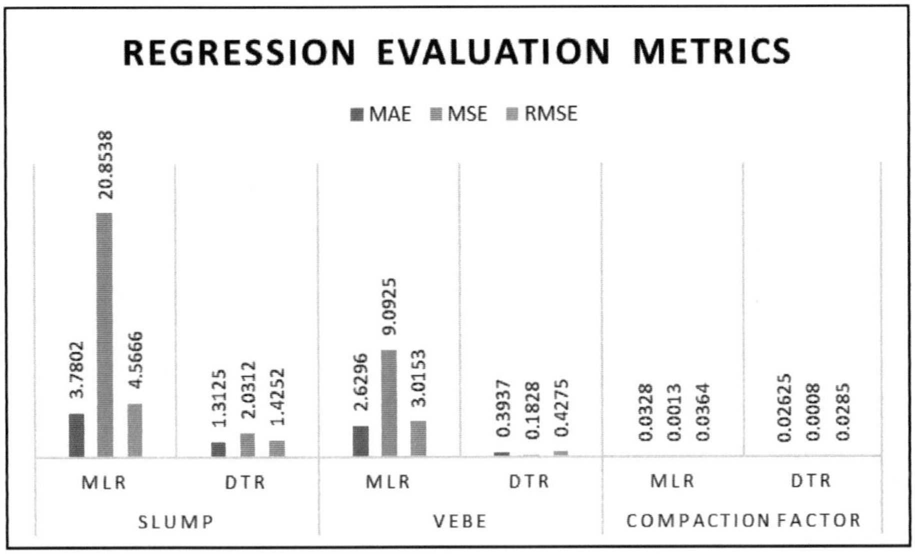

Figure 6. Comparison of MLR & DTR model testing results for the slump,VeBe and compaction factor

The results obtained using MLR and DTR models for the slump, VeBe and compaction factor are shown in Table 1 and Figure 6. It is observed that the error metrics in DTR are very low, indicating that DTR ability to predict the slump, VeBe and compaction factor values is better as compared to MLR. It is also evident that the MLR compaction factor model is better as compared MLR slump and VeBe model due to very low errors.

MLR model proposed three equations to predict slump, VeBe, and compaction factor are as follows.

$$y_s = -0.36682\text{s} + 0.76692\text{p} - 0.40010\text{m} + 217.79089 \quad (2)$$

$$y_v = 0.20546\text{s} - 0.43271\text{p} + 0.22724\text{m} - 99.85043 \quad (3)$$

$$y_c = -0.00031\text{s} + 0.00079\text{p} - 0.0004\text{m} + 1.05835 \quad (4)$$

It is noticed that the MLR model possesses more error than the DTR model due to difficulty in understanding the non-linear relationship. Hence, the DTR model was therefore developed to understand the nonlinear relationship.

6. Conclusion

A quick determination of the concrete strength and fresh properties is an important element on large construction sites for mass concreting; to save time and cost it is important to develop a machine learning model to predict the strength and workability of concrete with the help of available data.

The conclusions are as follows as per the above discussions:

- The DTR model is performed well than MLR model to predict slump, VeBe, and compaction factor because it possesses very low errors.
- The MLR model for compaction factor possesses very low error as compared to slump and VeBe model so it can be concluded that the MLR model is best suitable for compaction factor.

- The DTR and MLR model is proposed and also empirical formulae from the MLR model are proposed to predict the slump value, VeBe time and compaction factor for SPB, M-sand and M-sand with 1.5% incorporation of SPB concrete.

Acknowledgement
The authors would like to thank International Islamic University Malaysia for funding the research work under the Fundamental Research Grant Scheme for Research Acculturation of Early Career Researchers (FRGS-RACER) with project ID RACER/1/2019/TK06/UIAM//1. The authors also wish to extend profound gratitude to all personnel in the Civil Engineering Department of International Islamic University Malaysia kind assistance in this research.

References
[1] Bhanu P. Koyaa, Sakshi Anejab, Rishi Guptab &. Caterina Valeoa. 2021. "Comparative Analysis of Different Machine Learning Algorithms to Predict Mechanical Properties of Concrete." *Mechanics of Advanced Materials and Structures*.
[2] Charhate, Shrikant, Mansi Subhedar, and Nilam Adsul. 2018. "Prediction of Concrete Properties Using Multiple Linear Regression and Artificial Neural Network." *Journal of Soft Computing in Civil Engineering* 2(3):27–38.
[3] Cherickal, Phijo J., Vidhya V Kumar, Ranjini Mariam Philipose, S. Shailesh, and Tiji Thomas. 2021. "Techniques." 301–7.
[4] Chithra, S., S. R. R. Senthi. Kumar, K. Chinnaraju, and F. Alfin Ashmita. 2016. "A Comparative Study on the Compressive Strength Prediction Models for High Performance Concrete Containing Nano Silica and Copper Slag Using Regression Analysis and Artificial Neural Networks." *Construction and Building Materials* 114:528–35. doi: 10.1016/j.conbuildmat.2016.03.214.
[5] Cihan, M. Timur. 2019. "Prediction of Concrete Compressive Strength and Slump by Machine Learning Methods." *Advances in Civil Engineering* 2019. doi: 10.1155/2019/3069046.
[6] Dantas, Adriana Trocoli Abdon, Mônica Batista Leite, and Koji De Jesus Nagahama. 2013. "Prediction of Compressive Strength of Concrete Containing Construction and Demolition Waste Using Artificial Neural Networks." *Construction and Building Materials* 38:717–22. doi: 10.1016/j.conbuildmat.2012.09.026.
[7] Khademi, Faezehossadat, Mahmoud Akbari, Sayed Mohammadmehdi Jamal, and Mehdi Nikoo. 2017. "Multiple Linear Regression, Artificial Neural Network, and Fuzzy Logic Prediction of 28 Days Compressive Strength of Concrete." *Frontiers of Structural and Civil Engineering* 11(1):90–99. doi: 10.1007/s11709-016-0363-9.
[8] Nadimalla, A., S. A. Masjuki, S. A. Saad, M. Ali, N. M. Husain, W. N. A. W. Azahar, and N. Kasim. 2021. "Correlation Between Slump, {VeBe} and Compaction Factor of Concrete Containing Shredded {PET} Bottles, Manufactured Sand (M-Sand) and River Sand as Fine Aggregate." *{IOP} Conference Series: Materials Science and Engineering* 1051(1):12098. doi: 10.1088/1757-899x/1051/1/012098.
[9] Silva, Priscila F. S., Gray Farias Moita, and Vanderci Fernandes Arruda. 2020. "Machine Learning Techniques to Predict the Compressive Strength of Concrete." 1–14.

IOP Conf. Series: Materials Science and Engineering 1244 (2022) 012024 doi:10.1088/1757-899X/1244/1/012024

Monitoring The IIUM River Using Unmanned Aerial Vehicle And Image Classification

MOHAMAD NAZMI[1*], MOHAMED OKASHA[1,2*], AIZAT AASIM[1,2], MOUMEN IDRES[1]

[1] Department of Mechanical Engineering, International Islamic University Malaysia, P.O. Box 10, 50728 Kuala Lumpur, Malaysia

[2] College of Engineering, UAEU, UAE

* *Corresponding author:* nazmishah.work@gmail.com

Abstract. Prior research has shown viable methods towards identifying sources of pollution in rivers by utilizing Unmanned Aerial Vehicles (UAVs) combined with proper image classification techniques. This research attempts to develop and implement a novel approach to monitor the IIUM River whereby a Parrot Bebop 2 drone is utilized for data collection, while the Quantum Geographic Information System (QGIS) software is used for the supervised classification of the collected data. The image processing techniques of stitching or mosaicking, georeferencing and supervised classification are done using Adobe Photoshop, QGIS Georeferencing plugin, and QGIS Semi-Automatic Supervised Classification Toolbox, respectively. Results show that the classification process successfully recognized target objects, however, differing sun locations in datasets along with insufficient training data have led to some minor flaws. Despite these flaws, this research successfully achieved its objectives and will be vital for further investigations in the future.

1. INTRODUCTION

Malaysia's rapid growth in the agricultural, urban, and industrial sectors has led to significant impacts on the environment, particularly water pollution. In response, studies have been conducted to identify sources and measure their impacts to better identify challenges towards implementing a monitoring system that allows immediate preventive actions to be taken [1-7]. This study proposes the usage of unmanned aerial vehicles (UAVs) and image classification to help create a system for monitoring the IIUM river. Due to their ease of access and affordability, UAVs have been commonly used in multiple civil applications, including traffic monitoring, disaster management, wildlife monitoring and rescue missions [8]. Furthermore, conservation UAVs, also known as eco-drones, have been widely used for wildlife surveying, monitoring, and mapping land and marine ecosystems [9-11]. Image classification handled by artificial intelligence is also commonly used in monitoring. Its applications include satellite imagery, wire-photo standards conversion, medical imaging, videophone, character recognition, and photograph enhancement [12-14]. In this study, alternative approaches toward the image processing methods; stitching or mosaicking [15, 16], georeferencing [17], and supervised classification [18], will be used.

A common method for monitoring water pollution is to utilize satellite imagery. A review was made for oil spill remote sensing [19], exploring many existing ways for monitoring oil spills involving satellite imaging. Oil on water surfaces causes a polarizing effect on light, meaning it restricts the vibrations of the light wave. Thus, using polarized lenses can increase the contrast between oil and water and makes them easy to be observed. To increase this contrast, Brewster's angle was used. The review stated that, at that angle, the light is perfectly transmitted through the surface with no reflection which increases the contrast by up to 100%. Another common method for monitoring is by using hyperspectral imaging [20]. This method utilizes a spectrometer to capture images of a target object in multiple bands of wavelengths. Although the technique is very promising, it requires advanced processing and takes an extensive amount of time. Therefore, it is not so useful for real-time monitoring, especially on aerial platforms with limited processing power capability.

Satellite imagery for monitoring water pollution has several flaws that make it quite unreliable at times. The most noticeable flaw is the spatial resolutions of the images that a satellite can capture. In a study conducted to monitor the Chesapeake Bay [21], it is found that the multispectral satellite ESA Sentinel 2 has achieved up to 10m of spatial resolution - this is the distance that a pixel in the image represents in actual. This resolution is not sufficient when monitoring small water bodies such as the IIUM river which is only about 7 meters across at its widest section. Utilizing the same satellite in their approach which uses image segmentation to extract water bodies from satellite images, [22] found that some water bodies and rivers were not detected as water due to their tiny sizes. Determining the pollution in a river which is only few pixels across would be near impossible. Another flaw of satellite imaging is the availability. The mentioned satellite makes a pass over a region only once every 10 days. Hence, monitoring a rapidly changing pollution such as algal blooms in a river is a thing that satellite imagery cannot achieve. All these limitations can certainly be overcome with the use of UAVs as they have excellent spatial resolution, and they are always ready to use.

A study explored using an UAV to monitor water pollution in the southwest of China [23]. The UAV was equipped with a Canon 50D camera and an ACD multispectral camera. In this study, they utilized the three main classifications; supervised, unsupervised and object-oriented. They successfully calculated the total area of the water sediment pollution just by the images captured by the drone. In another research [24], where a Phantom 3-DJI drone was utilized in the urban lakes in Mexico, high-frequency spatial and temporal water sampling in two small lakes that experience cyanobacterial blooms were conducted. They measured the upwelling and downwelling irradiances over the water surface of the lakes by using a fast-scanning hyperspectral Spectroradiometer (GER-1500). It was found that the cyanobacteria distribution in the lakes are homogenous throughout the whole research process. In a similar study [25], a multirotor drone equipped with a Sony ILCE-6000 was used instead. Using the same detection method as in previously mentioned research, they successfully monitored the growth of the cyanobacteria. Additionally, they also utilized the UAV to create a time-lapse photography which helps in differentiating immobile aquatic and terrestrial macrophytes from the highly mobile cyanobacteria as their wavelengths are not too far apart.

This study aims to contribute to the IIUM River of Life project, and it provides a guideline for further research by institutions or individuals that are interested in this field of profession. The research objectives can be defined as follows:

1. To create a method that can be used to help monitor the water quality of the IIUM river and possibly find its source of pollution by using UAV and image processing.
2. To prove the usage of UAVs and image processing as an accessible alternative in tackling pollution problems.

This paper is organized as follow: First, Section 1 gives an introduction, a brief literature review and research objectives. Then, Section 2 discusses the image classification method along with data collection with the drone followed by image stitching, georeferencing, and lastly the classification itself. In addition, it includes a brief description of software and equipment. Section 3 presents the results of this research and discusses them in detail. Finally, Section 4 concludes the paper.

2. IMAGE CLASSIFICATION METHOD

In this study, Parrot Bebop 2 quadcopter drone is used as shown below in Figure 1. It has an RGB camera as the image sensor, a remote-control distance of up to 300 meters, 8 gigabytes of memory, and a 2700mAh battery which is capable of 25 minutes of flight time. The image sensor comes with a Fish-eye lens with 14 megapixels resolution [26].

Fig. 1 Parrot Bebop 2 drone [26].

The software is selected to create an easy way for monitoring rivers using UAVs and image classifications. The Quantum Geographic Information System (QGIS) software allows the use of the Semi-Automatic Classification (SCP) plugin [28]. It also has the Georeferencer plugin that allows the user to manually input the coordinates for easy georeferencing. The Adobe Photoshop software that has the photo merge function, allows the user to stitch images in few simple steps [29].

2.1. Image stitching/mosaicking

The drone is controlled using a smartphone through the Free Flight Pro application [27] with the Flight Plan feature for the autopilot and data collection. The image is set to be JPG Snapshot so that the images will be taken vertically from the drone to the ground. The drone will then automatically calibrate the images so the captured images by the drone would be exactly like the one seen from the smartphone during the control phase. In all flight tests, a set of blue plastic bags and paper plates are placed in different locations for each flight (except in the river itself as to not pollute it) to test the accuracy of the classification, and the reliability of the research. The pathway of the drone is set through the FlightPlan feature and by tapping consecutively where the drone should be as in Figure 2 below.

Fig. 2: Drone autopilot path and altitude set through the FlightPlan feature.

The altitudes are set by tapping on the circles, as shown in Fig. 1, where the paths are defined and adjusted with the slider bar on the right side of the screen. Additionally, the direction that the drone will be facing during the flight can be controlled. Then, on the top right of the screen, the settings of the drone during the flight are defined. The drone is set to take pictures at 7 second intervals. And the captured images during the flights need fist to be stitched together before projecting them onto a raster layer in QGIS [28]. This process is done using Adobe Photoshop software [29] with the feature of Photomerge in Automate section to create a panoramic image of all the images, taken during each flight. This process is autonomously completed by the software. Figure 3 shows a merged panorama image for one of the flight scenarios as captured by the drone.

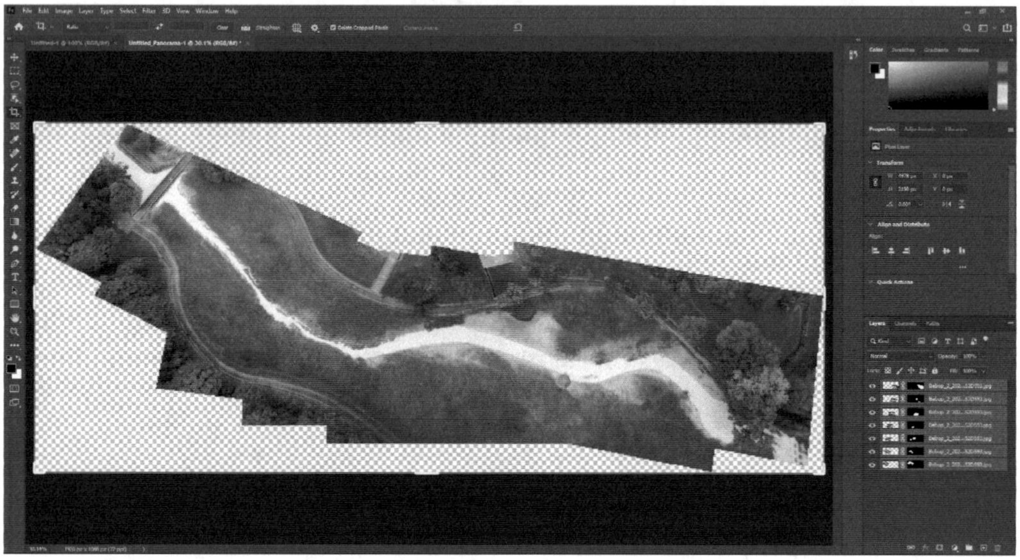

Fig. 3: The successfully stitched image for one of the flight Scenarios.

2.2. Georeferencing

The above stitched drone image now can be georeferenced. This is an essential to proceed further for the image classifications. Georeferencing is a process where the internal coordinate system of a digital map or aerial photo can be related to a ground system of geographic coordinates. A georeferenced digital map or image has been linked to a known Earth-coordinate system, so the users can determine where every point on the map or aerial photo is located on the real Earth's surface [30]. This process is done manually using the QGIS software with the Georeferenced plugin. After installing the Georeferencer and OSM place search plugins [31], the georeferencing process is done as by creating a new QGIS project. The OpenStreetMap layer from the premade layers is dragged and dropped on the left side of the workspace. Then, the OSM place search plugin is used to find IIUM area. Lastly, a point in the image is selected and then the corresponding location is selected in the From Map Canvas function. For this to work, at least 6 points are needed. Figure 4 shows the georeferenced drone image for one of the flights captured image.

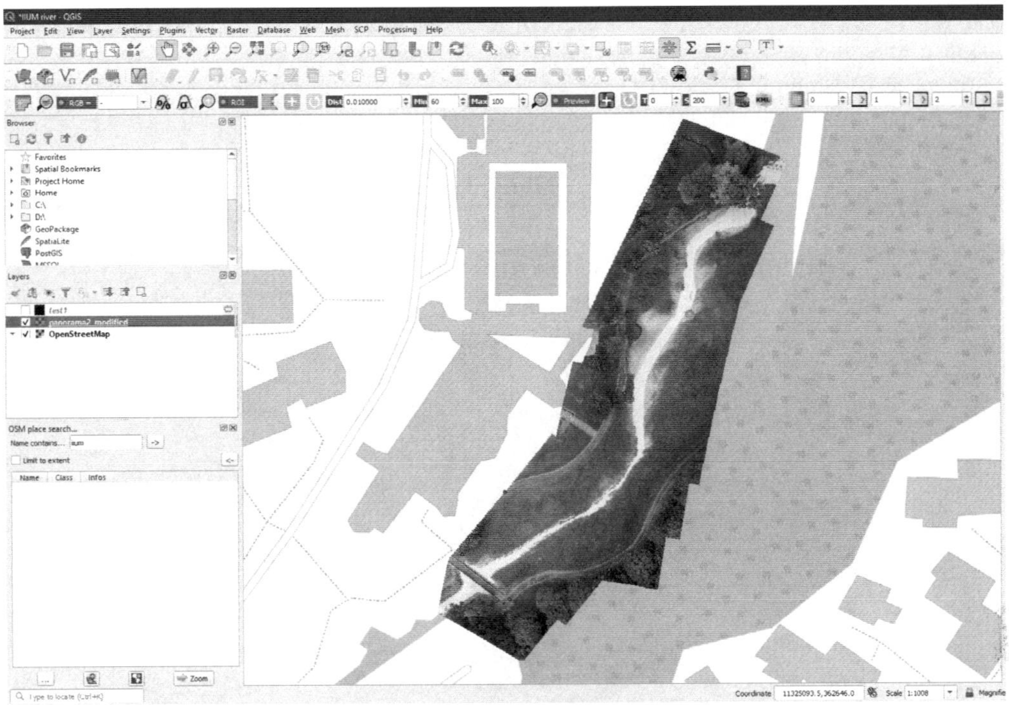

Fig. 4: Stitched and georeferenced drone image.

2.3. Classification

After the images have been completely processed according to the steps above, supervised classification can now be employed on them. First, the SCP (Supervised Classification Plugin) [31] needs to be installed. In the QGIS project, the images of interest are required to be dragged and dropped to the centre of the screen for classification. After the SCP plugin dock is enabled, the New Training Input function is used to create a new classification. After naming it, the Region of Interest (ROI) is used to select the area of an object. The named ROI will be included in the training algorithm where the software will use it as a training to recognize similar objects later. This step can be shown as below in Figure 5.

Fig. 5: Classifying a tree with the ROI function.

This process is repeated 4-5 times for all objects of the same class (in essence, another type of tree) and keep the class ID as the same number. They are then merged by highlighting all tree related objects and merging them in the training window. Then, this step is repeated for all other types of objects/classes that can be found in the flight images. After this is done, the workspace should look like as below in Figures 6 and 7.

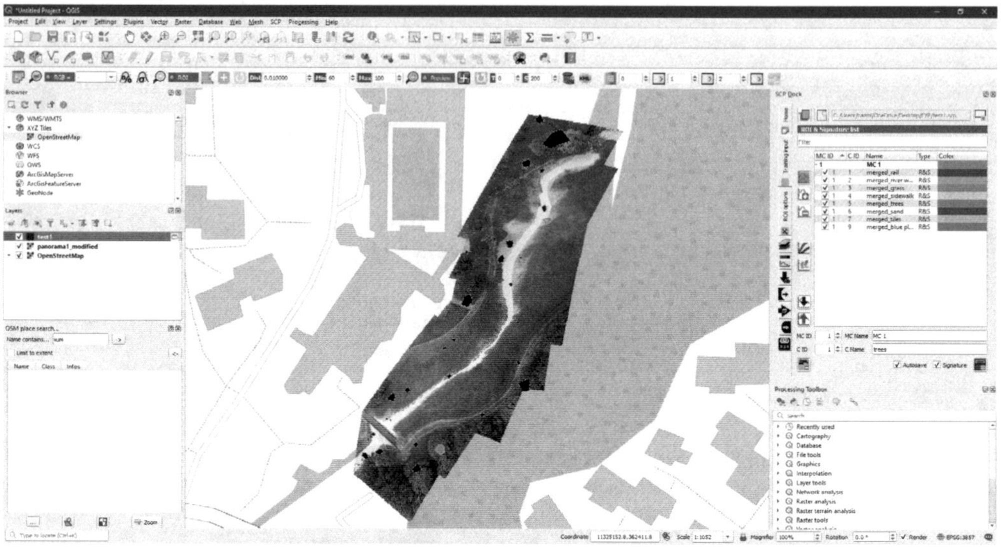

Fig. 6: Classification after merging all selected objects.

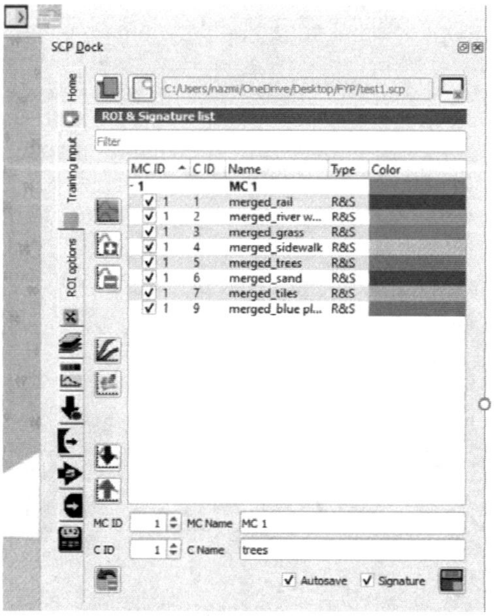

Fig. 7: Classes for the image classification training.

After all objects are classified, the trained algorithm can then be used to classify the other flight images. From here, the effectiveness of this research can be tested by looking at how accurate the algorithm can recognize all objects involved in all the used images.

3. RESULTS AND DISCUSSION

The comparison for the normal images with their classified counterparts is shown below through Figures 8-12.

3.1 Results

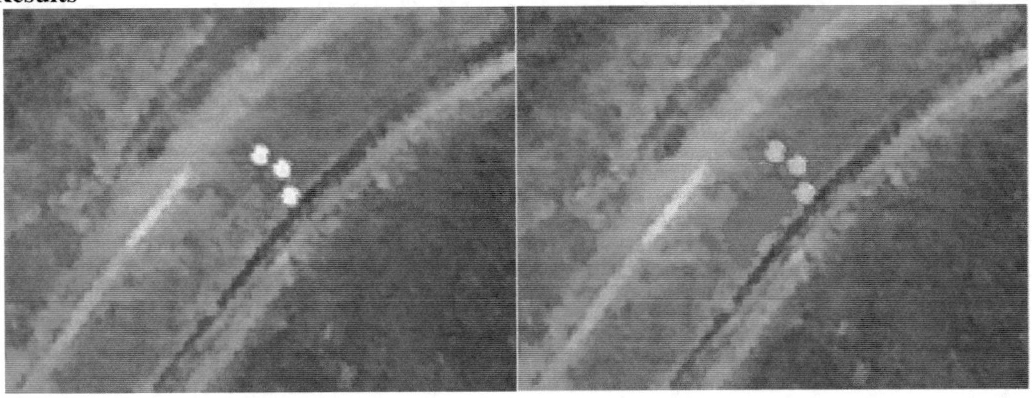

Fig. 8: Target objects in flight 1 recognized by the software.

In Figure 8, on the left, it shows the target objects, it includes 3 paper plates and 3 blue plastic bags placed on a pavement. On the right, it can be observed that the software successfully identified them by coloring them with the right class color as set earlier section.

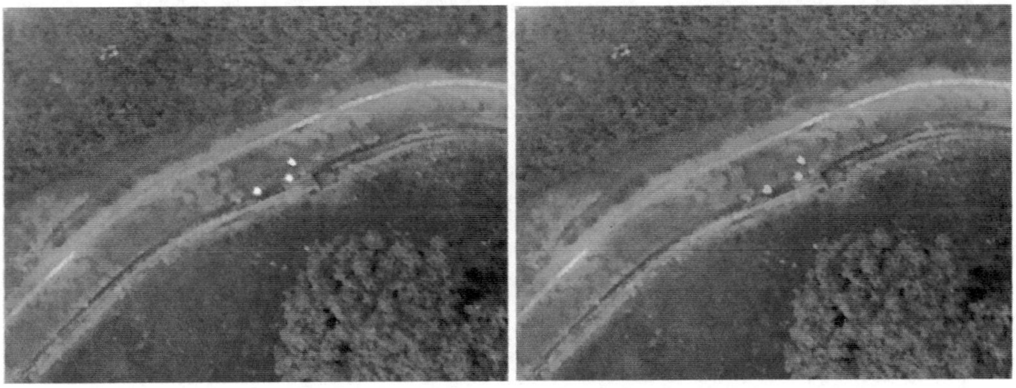

Fig. 9: Target objects in flight 2 recognized by the software.

In Figure 9, the target objects are moved slightly further from its previous location. Despite being moved, the software is still able to identify them successfully.

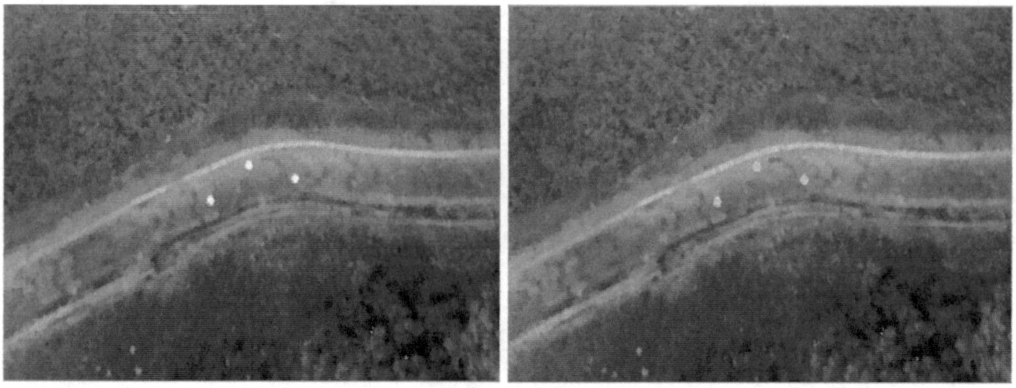

Fig. 10: Target objects in flight 3 recognized by the software.

In Figure 10, it is the same case as shown in Figure 8, but the target objects are relocated again and as show, they can be detected and classified.

Fig. 11: Target objects in flight 4 recognized by the software.

In Figure 11, the target objects are moved to a nearby bridge to further test the accuracy of the software. Once again, the software successfully identified them, but with some minor flaws where it is mistakenly some part of the bridge railing as the blue plastic bags.

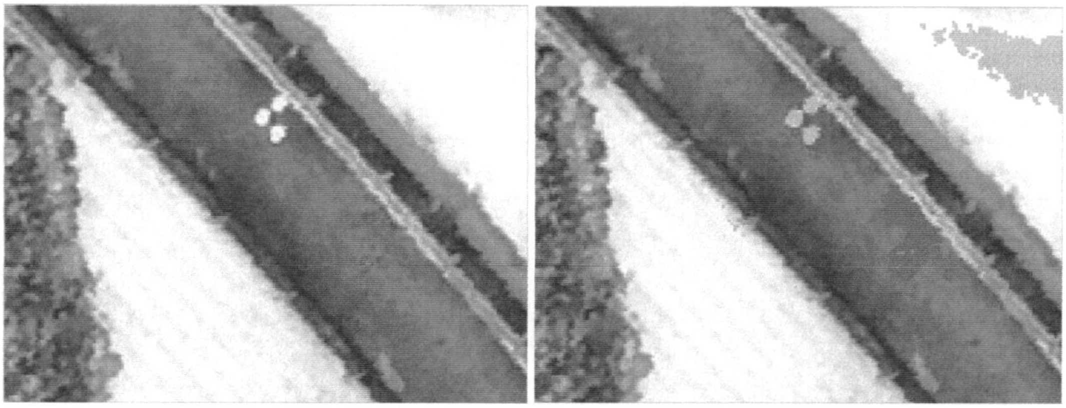

Fig. 12: Target objects in flight 5 recognized by the software.

In Figure 12, the target objects are moved slightly further on the bridge. The same result can be observed as the one in the flight before, but this time, it has a bigger and more apparent flaw where the software mistakes a large part of the river water surface as the paper plates.

3.2 Discussion

From the above scenarios, the software has successfully recognized the target objects in all 5 flight images. Despite being moved to different locations throughout the flights, the software is still able to identify them correctly. This shows that the training is effective for the program to classify objects, and this can be further enhanced by providing more data for training. There are several flaws that can be seen in the images, and this will be discussed below along with further details about how the classification software process works.

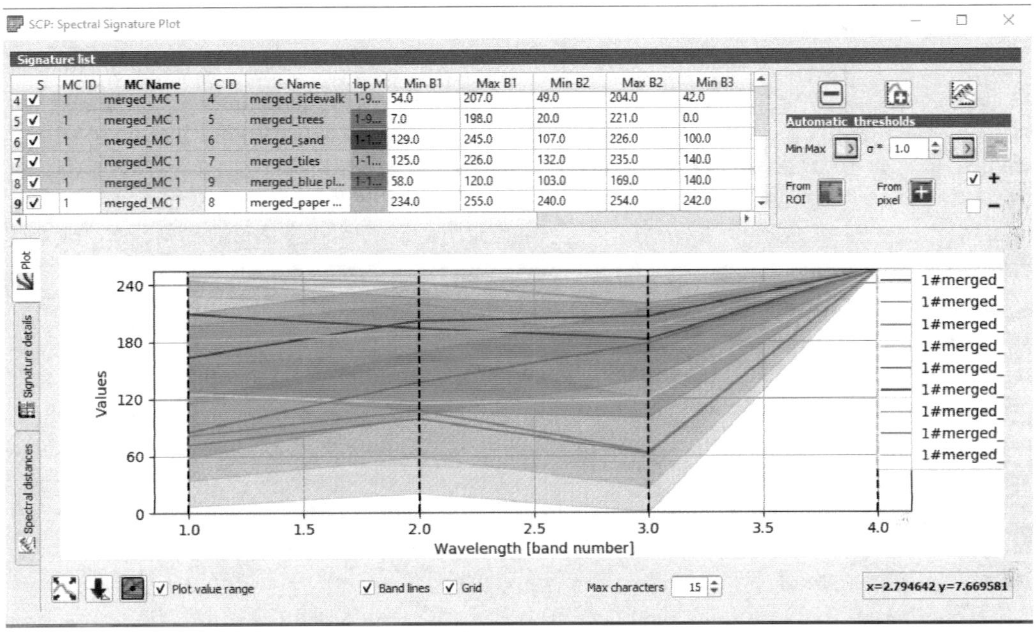

Fig. 13: Object classification based on their wavelengths.

The software utilizes the method of supervised classification. In this method, the user can choose sample pixels in an image that represents specific classes or objects and then uses those pixels as references and as training data for the classification of all the other pixels in the image. It is called supervised as it is based on the knowledge of the person using the program. As shown in Figure 13, the software analyses the wavelengths of groups of pixels in the images to classify them into the associated classes that are mostly corresponding to them during the training process.

In this paper, the Parrot Bebop 2 drone which has an RGB camera built in is used. This means the captured images only has 3 bands of wavelength data: red, green, and blue. The software will then classify the respective objects by the value of each used bands. If more wavelength bands are required, for example, using a multispectral imager, the software will be able to recognize more objects and reduce the flaws.

As mentioned in the literature review, in the example of layers of oil on a water surface, it can "shine" in the infrared spectrum. If two imagers (RGB and Multispectral) are used to analyse the images and then compare the effectiveness of the classification process, the imager that processes more band wavelengths will have more accuracy. Hence, using a drone with more bands of wavelengths in the images will enable the software to recognize even more objects with high accuracy. Furthermore, other factors that might affect the accuracy of classification is the drone's camera spatial resolution. In this study, the drone flies at an altitude of 50 meters when taking images of the river. If a drone that has a better spatial resolution is used instead, flying at the same altitude, more accurate results can be achieved. Additionally, more data for the training phase could reduce the reduce the errors and improve the accuracy. Additionally, the error could be because of insufficient data for the training phase of the classification. This research is conducted during the Covid-19 pandemic, which severely limits the time window for the data collection.

In the planning phase, the GPS errors of the drone during the flight needs to be considered. In this research, the drone autopilot is tested in an open field for around 15 minutes flight time. At the end of each flight, the drone is commanded to return to its take-off position. It is found that the drone landed about 2 meters away from its take-off position. The drone autopilot path for the data collection has considered this error to avoid collision with tall objects such as electrical posts and trees. This research is done in a small section of the IIUM river near Mahallah Hafsah in IIUM university. When involving the whole river, the battery capacity of the drone needs to be considered. Throughout the data collection phase, the weather was a bit windy which causes the drone to drain more battery power to stay stable in its supposed GPS location. This surveillance covered about 300 meters of river length and has used most of battery power due to the strong wind. Next planning of flights will need to consider the weather, sun location, and wind speed.

The georeferencing process was initially done by entering the latitude and longitude of certain locations in the image. However, these points were inaccurately recorded and unusable. The images were then manually georeferenced by eyeballing its respective points as mentioned in the methodology. This caused the images to be not very accurately georeferenced and some distortion was present. However, the georeferencing is still much better than by using only the latitude and longitude. This problem could be solved by georeferencing using the telemetry data of the drone flight instead.

Nevertheless, these results are accurate enough for the purpose of the surveillance task. Using this method, the software can help conduct the river surveillance remotely, without the need to be physically near the river, which can save energy and resources. Based on the surveillance tasks, if the amount of trash that is present in the vicinity of the river is significant, actions can be issued immediately. One of the best actions to take in this situation is to conduct immediate river cleaning procedure. Additionally, by using this method of surveillance, the sources of pollution in the river can be tracked.

4. CONCLUSION

To conclude, this research has successfully achieved its main objective to create a monitoring system for the rivers that could help to discover and reduce the pollution. It has successfully achieved its objectives by employing UAVs and image classifications as an affordable and accessible medium to be used by civilians for monitoring small water bodies compared to other conventional methods such as expensive satellite imagery. This research is open-ended in nature, its capability can be expanded by using the right equipment. In fact, drones equipped with infrared cameras have been widely used to monitor the normalized difference vegetation index (NDVI) in agriculture sector. This concept, when combined with the help of supervised classification, makes it a good alternative in monitoring water bodies where pollution is a big problem. Having mentioned the limitations and flaws of this system in the discussion, it is recommended to equip UAVs with higher resolution multispectral cameras along with using more objects and training data during the classification process to improve the performance and accuracy in future works.

REFERENCES

[1] R. Afroz, M. M. Masud, R. Akhtar, and J. B. Duasa, "Water pollution: Challenges and future direction for water resource management policies in malaysia," *Environment and Urbanization ASIA*, vol. 5, no. 1, pp. 63–81, Jan. 2014, doi: 10.1177/0975425314521544.

[2] L. Ramakreshnan, A. Rajandra, N. Aghamohammadi, C. S. Fong, and S. Nalatambi, "A preliminary insight into the environmental awareness of community in the vicinity of batik manufacturing units in Kelantan, Malaysia," *GeoJournal*, vol. 85, no. 6, pp. 1745–1753, Dec. 2020, doi: 10.1007/s10708-019-10046-w.

[3] M. S. Elias, S. Ibrahim, K. Samuding, S. A. Rahman, and A. Hashim, "The sources and ecological risk assessment of elemental pollution in sediment of Linggi estuary, Malaysia," *Marine Pollution Bulletin*, vol. 137, pp. 646–655, Dec. 2018, doi: 10.1016/j.marpolbul.2018.11.006.

[4] A. Razali, S. N. Syed Ismail, S. Awang, S. M. Praveena, and E. Zainal Abidin, "Land use change in highland area and its impact on river water quality: a review of case studies in Malaysia," *Ecological Processes*, vol. 7, no. 1. Springer Verlag, pp. 1–17, Dec. 01, 2018. doi: 10.1186/s13717-018-0126-8.

[5] K. M. Masum, A. Mansor, S. A. M. Sah, and H. S. Lim, "Effect of differential forest management on land-use change (LUC) in a tropical hill forest of Malaysia," *Journal of Environmental Management*, vol. 200, pp. 468–474, Sep. 2017, doi: 10.1016/j.jenvman.2017.06.009.

[6] M. Camara, N. R. Jamil, and A. F. Bin Abdullah, "Impact of land uses on water quality in Malaysia: a review," *Ecological Processes*, vol. 8, no. 1. Springer Verlag, pp. 1–10, Dec. 01, 2019. doi: 10.1186/s13717-019-0164-x.

[7] I. Lee, H. Hwang, J. Lee, N. Yu, J. Yun, and H. Kim, "Modeling approach to evaluation of environmental impacts on river water quality: A case study with Galing River, Kuantan, Pahang, Malaysia," *Ecological Modelling*, vol. 353, pp. 167–173, Jun. 2017, doi: 10.1016/j.ecolmodel.2017.01.021.

[8] O. Elijah *et al.*, "Application of UAV and Low Power Wide Area Communication Technology for Monitoring of River Water Quality," in *2018 2nd International Conference on Smart Sensors and Application, ICSSA 2018*, Nov. 2018, pp. 105–110. doi: 10.1109/ICSSA.2018.8535994.

[9] C. Harvey, "Drone Conservation: Studying and Saving Animals," Sep. 23, 2014. https://www.businessinsider.com/drone-conservation-studying-and-saving-animals-2014-9 (accessed Jun. 06, 2021).

[10] D. R. Green, J. J. Hagon, C. Gómez, and B. J. Gregory, "Using low-cost UAVs for environmental monitoring, mapping, and modelling: Examples from the coastal zone," in *Coastal Management: Global Challenges and Innovations*, Elsevier, 2018, pp. 465–501. doi: 10.1016/B978-0-12-810473-6.00022-4.

[11] A. Tripolitsiotis, N. Prokas, S. Kyritsis, A. Dollas, I. Papaefstathiou, and P. Partsinevelos, "Dronesourcing: a modular, expandable multi-sensor UAV platform for

combined, real-time environmental monitoring," *International Journal of Remote Sensing*, vol. 38, no. 8–10, pp. 2757–2770, May 2017, doi: 10.1080/01431161.2017.1287975.

[12] T. R. G. *et al.*, "A deep neural networks based model for uninterrupted marine environment monitoring," *Computer Communications*, vol. 157, pp. 64–75, May 2020, doi: 10.1016/j.comcom.2020.04.004.

[13] G. Cheng, X. Xie, J. Han, L. Guo, and G. S. Xia, "Remote Sensing Image Scene Classification Meets Deep Learning: Challenges, Methods, Benchmarks, and Opportunities," *IEEE Journal of Selected Topics in Applied Earth Observations and Remote Sensing*, vol. 13, pp. 3735–3756, 2020, doi: 10.1109/JSTARS.2020.3005403.

[14] C. Kyrkou and T. Theocharides, "EmergencyNet: Efficient Aerial Image Classification for Drone-Based Emergency Monitoring Using Atrous Convolutional Feature Fusion," *IEEE Journal of Selected Topics in Applied Earth Observations and Remote Sensing*, vol. 13, pp. 1687–1699, 2020, doi: 10.1109/JSTARS.2020.2969809.

[15] J. D. Shepherd, J. Schindler, and J. R. Dymond, "Automated mosaicking of sentinel-2 satellite imagery," *Remote Sensing*, vol. 12, no. 22, pp. 1–14, Nov. 2020, doi: 10.3390/rs12223680.

[16] X. Li *et al.*, "Generating high-quality and high-resolution seamless satellite imagery for large-scale urban regions," *Remote Sensing*, vol. 12, no. 1, p. 81, Jan. 2020, doi: 10.3390/RS12010081.

[17] A. Syetiawan, H. Gularso, G. I. Kusnadi, and G. N. Pramudita, "Precise topographic mapping using direct georeferencing in UAV," in *IOP Conference Series: Earth and Environmental Science*, Jul. 2020, vol. 500, no. 1, p. 12029. doi: 10.1088/1755-1315/500/1/012029.

[18] L. Frühe *et al.*, "Supervised machine learning is superior to indicator value inference in monitoring the environmental impacts of salmon aquaculture using eDNA metabarcodes," *Molecular Ecology*, vol. 00, pp. 1–19, 2020, doi: 10.1111/mec.15434.

[19] M. Fingas and C. E. Brown, "A review of oil spill remote sensing," *Sensors (Switzerland)*, vol. 18, no. 1. MDPI AG, p. 91, Jan. 01, 2018. doi: 10.3390/s18010091.

[20] M. B. Stuart, A. J. S. McGonigle, and J. R. Willmott, "Hyperspectral imaging in environmental monitoring: A review of recent developments and technological advances in compact field deployable systems," *Sensors (Switzerland)*, vol. 19, no. 14. MDPI AG, p. 3071, Jul. 02, 2019. doi: 10.3390/s19143071.

[21] D. Gong, C. Friedrichs, and R. Carnegie, "Drones: A New Tool for Water Monitoring," 2018.

[22] G. Kaplan and U. Avdan, "Object-based water body extraction model using Sentinel-2 satellite imagery," *European Journal of Remote Sensing*, vol. 50, no. 1, pp. 137–143, Jan. 2017, doi: 10.1080/22797254.2017.1297540.

[23] W. Zang, J. Lin, Y. Wang, and H. Tao, "Investigating small-scale water pollution with UAV Remote Sensing Technology," *World Automation Congress Proceedings*, pp. 2–5, 2012.

[24] R. Aguirre-Gómez, O. Salmerón-García, G. Gómez-Rodríguez, and A. Peralta-Higuera, "Use of unmanned aerial vehicles and remote sensors in urban lakes studies in Mexico," *International Journal of Remote Sensing*, vol. 38, no. 8–10, pp. 2771–2779, May 2017, doi: 10.1080/01431161.2016.1264031.

[25] P. Lyu *et al.*, “Autonomous cyanobacterial harmful algal blooms monitoring using multirotor UAS,” *International Journal of Remote Sensing*, vol. 38, no. 8–10, pp. 2818–2843, May 2017, doi: 10.1080/01431161.2016.1275058.
[26] “Parrot BeBop 2 Drone with Skycontroller (White) PF726103 B&H.” https://www.bhphotovideo.com/c/product/1206697-REG/parrot_pf726103_bebop_drone_2_with.html (accessed Jun. 08, 2021).
[27] “FreeFlight Pro - Apps on Google Play.” https://play.google.com/store/apps/details?id=com.parrot.freeflight3&hl=en&gl=US (accessed Jun. 08, 2021).
[28] “Welcome to the QGIS project!” https://www.qgis.org/en/site/ (accessed Jun. 08, 2021).
[29] “Photo, image & design editing software | Buy Adobe Photoshop.” https://www.adobe.com/sg/products/photoshop.html (accessed Jun. 08, 2021).
[30] “What does ‘georeferenced’ mean?” https://www.usgs.gov/faqs/what-does-georeferenced-mean?qt-news_science_products=0#qt-news_science_products (accessed Jun. 08, 2021).
[31] “QGIS Python Plugins Repository.” https://plugins.qgis.org/plugins/nominatim/ (accessed Jun. 08, 2021).

AUTHOR INDEX